Handbook of Engineering Management

The Engineering Management discipline remains complex and multidisciplinary, and has progressed and broadened in scope significantly over the last 10–20 years. Previously, the discipline has been fragmented and not aligned with the purposes of economic development, mega-project delivery, and technological progress. Digital engineering has revolutionized the field of engineering by introducing digital tools and technologies to the design, creation, operation, and maintenance of physical systems, products, and services. It has enabled more efficient, effective, and sustainable solutions, and has the potential to drive significant innovation and improve the way we design, build, and operate physical systems. This handbook will addresses new content of complexity by offering new engineering concepts such as simple, complicated, and complex, which have never been included in this discipline before and will generate interest from higher education, financial institutions, and technology companies.

Handbook of Engineering Management: The Digital Economy focuses on multidisciplinary integration and complex evolving systems. It discusses the incorporation of a system of systems along with engineering economic strategies for sustainable economic growth. This handbook highlights functional leadership as the main part of an engineering manager's competency and discusses how to form alliances strategically. In addition, it presents a comprehensive guide for the implementation of an environmental management system and shows how environmental and social impacts can be assessed in an organization applying digital tools. This handbook also brings together the three important areas of Engineering Management: Knowledge Management, the Digital Economy, and Digital Manufacturing.

In addition, this handbook provides a comprehensive guide to implementing an environmental management system and shows how environmental and social impacts in an organization can be assessed using digital tools. Based on the authors' practical experience, it describes various management approaches and explains how such a system can be used to prioritize actions and resources, increase efficiency, minimize costs, and lead to better, more informed decision making. It is essential to follow a systematic approach and to ask the right questions, whether the system is managed and implemented by humans, AI, or a combination of both. This handbook is laid out in a series of simple steps and dispels the jargon and myths surrounding this important management tool.

This handbook is an ideal read for engineering managers, project managers, industrial and systems engineers, supply chain engineers, professionals who want to advance their knowledge, and graduate students.

Handbook of Engineering Management
The Digital Economy

Lucy Lunevich

CRC Press
Taylor & Francis Group
Boca Raton   London   New York

CRC Press is an imprint of the
Taylor & Francis Group, an **informa** business

Designed cover image: © Shutterstock

First edition published 2024
by CRC Press
2385 NW Executive Center Drive, Suite 320, Boca Raton FL 33431

and by CRC Press
4 Park Square, Milton Park, Abingdon, Oxon, OX14 4RN

CRC Press is an imprint of Taylor & Francis Group, LLC

Library of Congress Cataloging-in-Publication Data
Names: Lunevich, Lucy, editor.
Title: Handbook of engineering management : the digital economy / edited by
Lucy Lunevich.
Description: First. | Boca Raton : CRC Press, 2024. |
Includes bibliographical references and index.
Identifiers: LCCN 2023028308 (print) | LCCN 2023028309 (ebook) |
ISBN 9781032448107 (hardback) | ISBN 9781032449975 (paperback) |
ISBN 9781003374879 (ebook)
Subjects: LCSH: Engineering–Management–Handbooks, manuals, etc.
Classification: LCC TA190. -H37 2024 (print) | LCC TA190 (ebook) |
DDC 620.0068–dc23/eng/20231018
LC record available at https://lccn.loc.gov/2023028308
LC ebook record available at https://lccn.loc.gov/2023028309

ISBN: 9781032448107 (hbk)
ISBN: 9781032449975 (pbk)
ISBN: 9781003374879 (ebk)

DOI: 10.1201/9781003374879

Typeset in Times
by codeMantra

Contents

Preface

The Engineering Management discipline remains complex, transdisciplinary and has advanced and broadened in scope significantly over the last 10–20 years. Yet, it does not meet the expectations of the society and businesses partly because the discipline itself has been fragmented, many subjects taught within this discipline have not been aligned with the purposes of sustainable economic development, mega projects delivery, and the technological progress society expects. As humanity, and in particular the Western economies, faces significant technological, economic, and social changes, moving from neo-liberal economic model to post-capitalism model economy, many social concepts will diminish, and new social concepts will emerge. One of them which requires a new transdisciplinary foundation is the Engineering Management discipline, and this discipline must be fundamentally changed in order to be valuable for industry, contributing and sustaining economic growth and economic complexity.

The traditional higher education system has been facing significant changes, rapidly adopting a blending model of learning and teaching, which requires a review of postgraduate programs, the composition of postgraduate courses, and the balance of practical experience needed for the future leaders like Engineering Managers. For instance, engineering managers must learn and understand the concept of complex evolving systems in order to be able to comprehend complex project situations he/she faces during working on mega projects, joint ventures, and strategic alliances.

Engineering Management profession is a life-learning discipline; it takes 20–30 years for Engineering Managers to reach full capacity and be effective as Engineering Manager. It does not have to be this way if the discipline is designed in different ways as suggested in this book. As a minimum, each Engineering Manager must be able to translate complexity into simplicity and distinguish between reality and concepts. This kind of academic education has never been tough before. It has to be changed if the engineering management discipline remains relevant and contributes to economic growth, social changes, and greater environmental awareness.

The major strengths of this book are:

1. Focus on greater transdisciplinary integration and cultural awareness
2. Focus on complex evolving systems
3. Focus on engineering economic strategy leading to sustainable economic growth
4. Focus on functional leadership and group leadership
5. Focus on the fundamental difference of doing environmental and social impact assessment, which leads to opportunities, not additional risks.
6. Focus on organizational development and knowledge management
7. Focus on the digital economy and digital manufacturing and challenges presented by future fuel and energy market.

The *Handbook of Engineering Management* is designed to assist design postgraduate programs, develop doctoral research programs, support Engineering Managers to deliver complex projects, train and develop staff, meet the challenges of the digital economy, and lead economic sustainable growth in the 21st century.

This book is organized into ten chapters. Chapter 1 covers the engineering management's past, present, and future and how this discipline has evolved over years. Chapter 2 offers a new topic of complex evolving systems, which has never been included in the education of Engineering Management before. Chapter 3 covers engineering economic strategy, based on the research and new development by Harvard University in this area, specifically Economic Atlas. Chapter 4 focuses on international engineering management, challenges and opportunities, new ways of project delivery, and communication strategy.

Chapter 5 covers the emerging paradigm of group leadership and functional leadership, which was written by Mr. Salicru, an international leadership coach and author of *Leadership Results* (John Wiley & Sons, 2017). Chapter 6 offers a new framework for decision-making with Big Data. Chapter 7 offers different perspectives on how to form strategic alliances and mitigate risks. Chapter 8 explains the concept of digital manufacturing and the challenges and opportunities it presents to Engineering Managers and organizational development. Chapter 9 discusses future fuel and technology, which presents new challenges for Engineering Managers.

Chapter 10 discusses international and European environmental management, systems, quality management, environmental and social impact assessments, and how engineering managers could optimize the process of environmental assessment, while winning the local community, and deliver real value for the society including indigenous groups.

We sincerely hope that this book will be of great value to you for many years to come in your professional and personal development.

Dr Lucy Lunevich
Melbourne, Australia, 2023

Editor and Contributing Author

Dr. Lucy Lunevich is a multidisciplinary researcher, best-selling author, well-known speaker and strategic advisor, senior lecturer, and program director of the Master of Engineering Management at RMIT University, Melbourne, Australia. Before joining RMIT University in 2017, she was a principal consultant and research manager at Shell Global, URS Corporation, and led multidisciplinary teams. Dr Lunevich earned her Doctorate in Science from Victoria University, Melbourne. She also holds a Master's in Resource and Environmental Planning from Massey University, NZ; a Master's and a Bachelor's in Sanitary Engineering and Public Health from Riga Technical University, Latvia; and and a graduate from an Advanced Study from Harvard Kenny School, USA. Lucy's research interest focuses on complex evolving systems including economics, social ecology, environmental systems, adaptive environmental management, climate change, and disaster management. She has been recognized as a scholar in critical digital pedagogy and published the book *Creativity in Teaching and Teaching for Creativity: Modern Practices in the Digital Era* (CRC Press, 2023). She has written for more than 36 publications, including 12 book chapters, 3 books, and 6 industry reports for the global resource sector.

Contributing Authors

Dr. Annelize Botes received her Doctorate in Mechanical Engineering from Nelson Mandela Metropolitan University in 2005 on the topic of *Laser Deformation of Dual Phase Steel Components*. During her career, she held various academic and research-related positions, and she is currently appointed as a research associate at Nelson Mandela University. Her research interest is in the field of laser material processing with a focus on the effect of processing parameters on the mechanical properties of alloys. She also acts as a technology advisor/consultant for manufacturing industries and has completed more than 300 industrial reports on topics varying from failure analysis, process development, and quality assurance.

Dr. Matthew C. Cook has spent the majority of his career working as an engineering manager and a technical authority in the aerospace and defense industries. He has strategically worked across the UK and Australia, where he has been responsible for the management and delivery of numerous major projects and leading large engineering teams. He received his Doctorate in Engineering from RMIT University in 2020. He continues to conduct research and actively publish in the areas of risk analysis, modeling of complex systems, amelioration of the design process, and decision-making under opacity. He is a chartered engineer with the British Engineering Council and a fellow of both the IMechE and RINA.

Dr. Oswald Eppers is a PhD chemist with over two decades of experience in environmental science and environmental consultancies conducted in Europe, South America, and Australia. During his career, he performed a variety of projects related to environmental impact assessments, contaminated site investigations and remediations, due diligence evaluations, hazardous waste characterization and management, as well as human health and ecological risk assessments. He was responsible in different organizations for the implementation of environmental management systems and advised the regional government of Arequipa in Peru during eight years in environmental policy and management. Currently, Dr. Eppers is Business Development Representative of K-UTEC Salt Technologies in South America, a company specialized in salt mining and processing.

Dr. John V. Farr is a professor emeritus of Engineering Management at the United States Military Academy (USMA) at West Point and was the founding director of the Center for Nation Reconstruction and Capacity Development upon his retirement in 2017. He currently teaches part time at the University of Central Florida and in the School of Business at Clarkson University and consults part time with Applied Research Associates conducting cost, decision, and risk analysis. From 2007 to 2010, he was a professor of Systems Engineering and Engineering Management and associate dean for Academics in the School of Systems and Enterprises at Stevens Institute of Technology (SIT). He was the founding director of the Department of Systems Engineering and Engineering Management at SIT from 2000 to 2007. Before coming to SIT in 2000, he was a professor of Engineering Management at the USMA at West Point, where he was the first permanent civilian professor in Engineering and Director of their Engineering Management Program. Prior to joining the faculty at West Point in 1992, he was the team leader of the Combat Engineering and Simulation Group and worked in nuclear weapons effects at the U.S. Army Engineer Waterways Experiment Station. He also worked in the design of offshore oil platforms throughout the Mideast and the United States and the design and maintenance of flood control structures in the Lower Mississippi Valley for the US Army Corps of Engineers. Dr. Farr is a former past president and fellow of the American Society for Engineering Management (ASEM) and a fellow of the American Society of Civil Engineers (ASCE). He is the former editor of the *Journal of Management in Engineering* and the founder of the *Engineering Management Practice Periodical*, and he has served as a reviewer for 18 refereed journals. He has authored or edited over 200 technical publications including 3 textbooks, 2 handbooks, 8 book chapters, and 98 refereed publications mainly on cost and decision analysis, infrastructure, engineering education and leadership, and systems engineering and thinking. Dr. Farr earned his undergraduate degree from Mississippi State University, and he earned his Master's and PhD in Civil Engineering from Purdue and the University of Michigan, respectively. Dr. Farr is also a member of Chi Epsilon and a founding member of Epsilon Mu Eta, Phi Kappa Phi honor societies, the International Council of Systems Engineering (INCOSE), ASCE, and ASEM. As recognition for his innovations and academic achievements, Dr. Farr has received numerous awards and recognitions such as the INCOSE Outstanding Service Award, 2017; Bernard R. Sarchet Award, ASEE, 2006; Franklin W. B. Woodbury Service Award, ASEM, 2005 and 2017; Henry Morton Distinguished Teaching Professor Award, SIT, 2006; Bernard R. Sarchet Award, ASEM, 2004; Merl Baker Award, ASEE, 2004; and numerous multiple government awards including the Meritorious and Superior Civilian Service Awards and the Commander's Award for Civil Service. Dr. Farr is also a registered civil engineer in Florida and Mississippi and a former certified project management professional (PMP). Dr. Farr has served on numerous defense national and academic advisory boards, including membership on the Army Education Advisory Committee (2019–2021) and Army Science Board (2002–2010), and as a member of both the Air Force

Studies Board (2006–2012) and the Board on Army Research and Development (2019–present) of the National Academies. He has served as a program evaluator in Canada. In the State of New York, he has served as an alternate member of the Board of Directors (2015–present), a two-term (2011–2016 and 2018–present) commissioner for the Engineering Accreditation Commission of ABET, and a Middle States evaluator. He has chaired or conducted 18 accreditation visits including numerous international evaluations. Dr. Farr has also served on numerous school, department, and program advisory boards including Norwich University, USMA, SIT, and Clarkson University. He has helped author two National Academies studies on systems engineering and engineering education. His research and consulting interests are in cost and data analysis and decision and risk analysis as applied to infrastructure, weapons of mass destruction, and other complex enterprises, especially in the arena of capacity development and complex systems. He has taught classes in modeling and simulation, decision analysis, engineering economics, cost estimation and management, technical leadership, and project management. He has served as a consultant to numerous companies and government agencies.

Dr. John Mo is Professor of Manufacturing Engineering and formerly Discipline Head of Manufacturing and Materials Engineering at RMIT University, Australia. Before joining RMIT in 2006, he was Senior Principal Research Scientist in CSIRO and led research teams including manufacturing and infrastructure systems. In his 11 years at CSIRO, his research team worked on many large-scale government- and industry-sponsored projects, including electricity market simulation, infrastructure protection, wireless communication, fault detection, and operations scheduling. He was the project leader promoting productivity improvement in the furnishing industry and consumer goods supply chain. Prof. Mo has written or contributed to more than 400 publications including 3 monographs, 150 journal articles, 220 refereed conference papers, 15 book chapters, and 12 public reports.

Sebastian Salicru is a registered psychologist, board-approved supervisor, and professional certified coach (PCC) working in private practice in Canberra, Australia. He holds a Bachelor of Applied Science (Psychology) and Postgraduate Diploma in Psychology, Curtin University; a Master of Science in Creativity and Change Leadership, State University of New York; and a Master of Management Research, University of Western Australia. He is also a professional certified coach (PCC) by the International Coach Federation (ICF); a fellow of the Institute of Coaching, McLean Hospital, Harvard Medical School; and a graduate from the 'Art & Practice of Leadership Development', Harvard Kennedy School. Sebastian's research interests include integrative psychotherapy, spirituality, executive coaching, and leadership development. As an author, Sebastian has published several papers in journals such as *Consulting Psychology Journal: Practice and Research*; the

Journal of Leadership, Accountability and Ethics; *OD Practitioner*; *Psychology*; *Open Journal of Depression*; *American Journal of Applied Psychology*; *Clinical Psychiatry*; and *World Journal of Psychiatry Mental Health Research*. He has also authored two book chapters and sole-authored a book – *Leadership Results: How to Create Adaptive Leaders and High-performing Organisations for an Uncertain World* (John Wiley & Sons, 2017).

Dr. Colin A. Scholes, CChem FRACI CEng MIChemE, is an associate professor in the Department of Chemical Engineering at the University of Melbourne. He is an expert in clean energy processing and membrane science, particularly in developing strategies to assist the transition to a low-carbon future as well as the generation and transportation of clean hydrogen carrier fuels. He also consults widely with the energy sector and authored over 100 publications on topics varying from separation technology, clean fuel generation, and carbon abatement strategies.

Dr. Milan Simic holds a PhD, Master's, and Bachelor's in Electronics Engineering, and a Graduate Diploma in Education. He is currently with RMIT University, School of Engineering. At the same time, he is Professor at MB University, Faculty of Business and Law, Belgrade, Serbia; Adjunct Professor at Kalinga Institute of Industrial Technology (KIIT), School of Computer Engineering, Bhubaneswar, Odisha, India; Honorary Editor at the *KES Journal*; Editor at the International Transactions on Evolutionary and Metaheuristic Algorithms; and Former Associate Director at Australia-India Research Centre for Automation Software Engineering, RMIT University. He has comprehensive experience in industry (Honeywell Information Systems), research institute, and academia from overseas and Australia. For his innovations and other contributions, he has received prestigious awards and recognitions. Dr. Simic is Member of a large number of engineering and science associations, like The Australasian Association for Engineering Education (AAEE) and International Knowledge Engineering Systems (KES). KES is a worldwide association involving about 5000 professionals, engineers, academics, students, and managers. As a KES silver member and *KES Journal*'s former General Editor, now Honorary, he has conducted strong international collaboration, while processing around 400 papers per year, with the support of more than 80 Associate Editors in the team and more than 800 Reviewers. Dr Simic has designed, accredited, and managed the first Mechatronics program of study at RMIT University. Among many other programs, he has also designed a Master's program of Engineering Management, face-to-face and online version, and successfully managed it till the end of 2018. This program is one of the most popular RMIT University programs and brings around $7 million per year. Dr. Simic has founded and managed RMIT CISCO Networking Academy. It is functioning as a profitable fee-for-service business attached to the University. CISCO networking labs are University facilities used for academic curriculum delivery. This is a great example of University and industry collaboration. CISCO Networking Academy is the largest educational institution in the world. In June 2015, Dr. Simic's research and development team won an international award in Germany for the development of fast 3D metal printing technology. The award is presented to new initiatives and start-up businesses across the world, each year by Robert Bosch Venture Capital (RBVC). Research Partner company is now producing 3D printers in Melbourne, Australia.

Technical Review Panel

Jeremy Joseph is an emirates professor; EurGeol, CGeol, MCIWM – hydrogeologist; environmental manager; and research scientist. He was Visiting Chair at three UK universities, including London. He was past Honorary Research Fellow at the Natural History Museum, London, and Research Scholar at the University of Melbourne, Australia. He has over 50 years of experience working in the private, public, and academic sectors, variously in the UK and Europe, Australia, and South East Asia. He has gained experience mainly in the water and waste industries, including consultancy to them, but with several years in the 1970s managing major database development as a full-time systems analyst in the public water sector. Work in the water industry included both water and wastewater treatment, as well as many supply aspects. In the waste industry, it included life-cycle analysis and recycling, as well as landfill and environmental management.

Dr. Allan Mclay, PhD, MEng, Grad Dip TT&L, MIML, is Honorable Senior Research Fellow, previously Program Director of the Executive Engineering Management program, RMIT University, Australia. He has served 50 years in the academic sector, initially as Senior Technical Officer/Engineer in physics research in the RAAF Academy, University of Melbourne and University College London, then as Manager Telematics/Computer education (RMIT) driving the introduction of satellite communications, the introduction of computers, and the development of both voice and video teleconferencing to support distance education in Australia. He has served as Senior Lecturer in the Faculty of Engineering, RMIT University, in the development and delivery of postgraduate programs including the Master of Engineering Management, the first of its kind in Australia. He has engaged in numerous leadership roles both within the University and elsewhere, including sitting on Boards of Management, writing Cabinet and Ministerial briefing papers (State Govt.), and advising government committees (State and Federal Govt.). He served as Acting Head of the School, Course Leader of the Graduate School of Engineering, Acting Head of the TAFE Off Campus Coordinating Authority (a TAFE College Director position), Representative of TAFE College Directors on Federal Govt. committees, and Active Member of RMIT University Appeals Committee. He managed the largest postgraduate coursework program in the School of Aerospace and served as Program Director of Mechanical and Manufacturing Engineering for 20 years.

Mr. William (Bill) Miller is Chief Consulting Engineer at GE Power, Zurich, Switzerland. He is a globally experienced engineer in safety, performance risk, and performance solutions, and he focused on real outcomes. He is experienced in the railway and power industries. He has more than 40 years of service in the power sector in private industry starting in Australia in junior engineering roles and progressing to international senior engineering roles in power plant areas such as systems design, research and development, new product introduction, field tuning, and distressed project finalization. He served for a short period in the rail industry to assist a troubled project to finalization. He is now serving as Chief Consulting Engineer for GE Power in Baden, Switzerland, for the past ten years. This is a global role concerned with the technical performance of a portfolio of projects.

Dr. Scott Wright, PhD, PE, PMP, is Associate Professor at the University of Colorado Boulder, Stanford University, USA. He has over 30 years of experience in government, private, and academic sectors leading, managing, and providing technical advice in the protection of human health and the environment; 21 years of active duty military experience in the Army Medical Department focused on managing projects and programs that delivered engineering solutions to prevent disease and non-battle injuries to military and civilian personnel in the United States, Germany, Australia, Korea, and Japan; three years of private sector experience as a project manager, technical team leader, and key client manager in Australia in manufacturing, food, pulp and paper, and defense industry sectors; and eight years as a university educator (tenured Associate Professor) involved in all aspects of course design, development, teaching in online and on-campus project management, business and sustainable engineering/renewable energy, and environmental sciences.

1 Future of Engineering Management in the Age of Data-Driven Decision-Making

John V. Farr and David Farr
United States Army

Lucy Lunevich
RMIT University

1.1 INTRODUCTION

For the last 50 years, the discipline of Engineering Management has been defined in terms of technology: manufacturing, industrial, information technology, automatization, biotechnology, nanotechnology, etc. Originally, the discipline was understood as the form of engineering, project management, cost-benefit analysis, opportunity realization, and operational management.

With rapid economic development across many countries, increasing economic complexity, and digitalization, the discipline of Engineering Management becomes intensively multidisciplinary.

It becomes the discipline for economic and social development instead of managing engineers and delivering complex projects. An engineering manager has a strategic role to play within an organization. This includes strategic recruitment, training staff to secure strategic opportunities, setting up alliances, joint ventures, developing teams for future opportunities, and making critical decisions in the complex evolving environment.

Engineering managers have two fundamental responsibilities: management and leadership. These two roles carried out by the engineering manager may conflict in some situations, posing both personal and organizational risks. The most difficult part of it is to know when to manage and when to lead. Highly developed interpersonal skills, self-awareness, emotional intelligence, critical thinking, self-reflection, and the ability to see business opportunities in the risks are the key to success for engineering managers.

Engineering Management should be able to manage and deliver projects with minimum costs, develop staff, and effectively allocate resources. It is expected that engineering managers can lead with confidence, manage multiple projects, and work across

DOI: 10.1201/9781003374879-1

1

different business environments (private and public). He/she should have a vision and understanding of different strategic approaches businesses can pursue in order to be successful endeavors. Engineering managers should understand the complexity of specific situation and must have the ability to translate this complexity into simplicity. Engineering managers must distinguish between different complex systems and system of systems, have cultural awareness and cultural intelligence, and have knowledge of various business practices, for instance, differences between Japanese and Chinese or Israeli and Saudi or Australian and United Kingdom business practices.

In the digital economy, the engineering manager involves in the process of creating knowledge and validates the knowledge from Big Data using tools such as artificial intelligence, robotics, and another algorithm. Knowledge becomes the potential asset of the company.

This book offers a new discipline typology by incorporating a chapter on engineering economic strategy, complex evolving systems, functional leadership, and group leadership. Many concepts have been introduced first time into the discipline of Engineering Management. It is anticipated that the new architecture of composition of various concepts and disciplines will assist engineering managers in developing complex projects and managing joint ventures, strategic alliances, and partnerships. Universities and higher education institutions can use this edition to review their postgraduate programs in Engineering Management.

This book intends to take the Engineering Management discipline to a new higher level so that engineering managers can meet the demands of the 21st-century digital economy. It covers a deficit in understanding, which not only is glaring but – given the current state of the world – has become patently unacceptable. Furthermore, this chapter is intended to assist higher education in designing new postgraduate programs, which cover a deficit in understanding complex project environments and complex evolving systems.

1.2   THE ENGINEERING MANAGEMENT DISCIPLINE

To understand the Engineering Management discipline, we must understand how the discipline relates to other disciplines. As shown in Figure 1.1, Engineering Management is often described as the bridge between the disciplines. Consistent with the definitions provided in the previous section, Engineering Management has traditionally been described as the "bridge" (Kotnour and Farr, 2005) between the traditional disciplines of science/engineering and management and business development.

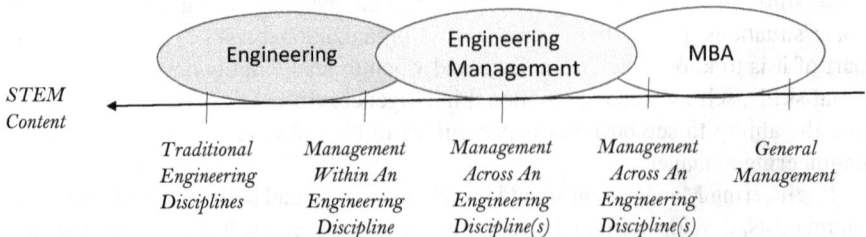

FIGURE 1.1   Management across disciplines.

Engineering Management, as the name implies, is a multidisciplinary technical profession. An engineering manager often coordinates and/or works on multidiscipline teams working to solving a problem. They also must work across multiple fields such as marketing, accounting, and operations, as well as coaching and mentoring project managers. Their world has become significantly more complicated because of complexity and technology and higher risks to manage. The vast amount of information, often available in real time, and the challenges of a dispersed and diverse workforce require additional skill sets than needed in past years to be able to differentiate between the immediate, urgent, and the important for progressing a project.

1.2.1 THE DIGITAL ECONOMY

The digital economy according to Lane (1999) can best be described as "the convergence of computing and communication technologies in the Internet and the resulting flow of information and technology that is stimulating all of electronic commerce and vast organizational change". Table 1.1 contains a list of 12 elements or characteristics of the digital economy. Even though this list is almost 30 years old, these characteristics still hold true today.

1.2.2 DATA IS AN ASSET THAT DRIVES INNOVATION

Big data and the digital economy are forcing almost every business toward a data-driven model and management, for example, big data helps industrial companies to deliver products that are less prone to failure and more economically produced by being able to analyze many different real and extrapolate to simulated failure modes. In this way, the engineering margin can be reduced where it does not matter and added where it does. Innovation typically means bringing something new to the market and is categorized as disruptive (i.e., a new technology that provides more efficient and accessible alternatives), incremental (i.e., continuous improvement of existing products), architectural (i.e., new products using existing components/systems together), or radical (i.e., new products/services based upon breakthroughs in science and technology) (see Dieffenbacher, 2022). Innovation in any of the four forms can be applied to processes, products, services, and management. Whether being used for reporting, better decisions, market recognition, personalized experiences, etc., big data is being used in all aspects of innovation. Ayers (2019) states that there are four essentials to innovating with big data:

1. Talent comes first – you need to hire technical training for collecting and analyzing data, but the ability to translate those technical skills into tangible, real-world outcomes.
2. Act on new data now – all businesses must be able to pivot quickly.
3. Understand your environment – big data can be used to serve and keep customers.
4. Optimize customer service – every business needs new products and customers.

TABLE 1.1

Twelve Elements of the Digital Economy

Twelve Elements of the Digital Economy	Characteristics
Knowledge	Smart everything, lack of privacy
Digitization	Information, communication, etc., have been reduced to electrons
Virtualization	Physical things can now be virtual – books, education, entertainment, jobs, shopping, etc.
Molecularization	Corporations are being disaggregated and transformed into smaller entities
Integrated/interworking	Network economy that can overcome scale and size
Disintermediation	Middlemen at all levels are being eliminated – producers to consumers, within organizations, packagers of information, etc.
Convergence	Dominant economic sector created by the convergence of computing, communications, and content
Innovation	Cycle times have decreased, you must quickly obsolete your own products
Prosumption	The gap between consumers and producers blurs
Immediacy	Byte based economy requires immediacy
Globalization	Global economy is driving and being driven by the digital economy
Discordance	Unprecedented social issues are causing upheaval and conflict

Modified from Tapscott, D., *The Digital Economy Anniversary Edition: Rethinking Promise and Peril in the Age of Networked Intelligence*, McGraw-Hill, New York, 1995. With permission.

These essentials have fueled innovation by increasing productivity and a better understanding of market conditions and consumer habits. Big data might not be able to create a culture of innovation, but it can help create innovative new products, services, and processes. Nevertheless, sometimes it is the responsibility of the engineering manager to create a culture of innovation in order to improve productivity.

1.2.3 ROLE OF ENGINEERING MANAGER

What is the role of the engineering manager in the digital economy – provides the product or service in a more efficient and sustainable and profitable way so consumers will buy it whether it is a consumer good or an industrial project. Big data is a tool to get there. The properly informed engineering manager is in a reasonable position to navigate this environment. In 1990, only IBM (CNN Money, 2022) was in the top 20 that did not manufacture traditional products. In 30 short years in the US business models, services, products, organizational structures, and required job skills have dramatically changed. The question that exists is what skills do engineering managers need to provide value added in the digital economy globally while being affected

TABLE 1.2
Fourteen Habits of Lifelong Learners

• They read daily	• They take various courses	• They actively seek opportunities to grow	• They take care of their bodies
• They have diverse passions	• They love making progress	• Challenge themselves with specific goals	• Embrace change
• Believe it's never too late to start something	• Their attitude to getting better is contagious	• Leave their comfort zone	• Never settle down
• Choose the right career	• Aren't afraid of failure		

Modified from Nowik, O., "14 Powerful Habits of People Dedicated to Lifelong Learning," *Lifehack*, 1 August 2022, accessed 17 January 2023 at https://www.lifehack.org/articles/communication/12-signs-you-are-lifelong-learner.html. With permission.

by the 12 characteristics listed in Table 1.1. Also, who should be providing those skills – educators, trainers, on-the-job mentors, etc.? The engineering manager is also responsible for ongoing training, coaching, mentoring, communication of staff and management, building relationships across the organization and stakeholders and, frequently, community. Furthermore, an engineering manager is responsible for knowledge creation (process quality) and validation (process efficiency) so that the organization remains sustainable and productive in the future.

1.2.4 PROFESSIONAL AND PERSONAL DEVELOPMENT

Our formal education has provided us with basic science and engineering skills and occasionally some management skills to be successful. Lifelong learning is how we improve every day. Most successful people improve both professionally and personally and strive to "get better" every day. Table 1.2 lists 14 habits of lifelong learning. All too often we are busy with life and focus on professional careers. In the digital age, we must leave our comfort zones, continue to take various classes to keep up with rapidly changing technology, and most importantly embrace changes. Most engineering managers have roots in traditional engineering. Embracing lifelong learning is key to a successful and impactful career.

1.3 ENGINEERING MANAGEMENT ENTERPRISE

1.3.1 DEFINITION OF ENGINEERING MANAGEMENT

Many definitions of engineering manager exist. The WWW presents hundreds of definitions – many of them by universities' marketing degrees for their specific engineering manager curriculum. Farr (2011) proposed that a suitable definition might be:

In today's global business environment, engineer managers integrate hardware, software, people, processes and interfaces to produce economically viable and innovative products and services while ensuring that all pieces of the enterprise are working together.

Even this definition is probably too narrow given the emergence of the digital economy. In some respects, engineering manager needs to have both a skill set and a discipline, such as statistics and mathematics versus a Master of Business Administration (MBA), which implies a degree title. Anyone involved in the digital economy must work at the interface of technology and management. To say that this is under the purview of any one discipline is probably pointless. Engineering management skills are ubiquitous in the digital economy. The digital economy demands from engineering managers more knowledge than one could learn at universities. It requires learning at a university environment and on jobs in a challenging project environment. A more applicable term for engineering manager might be something along the lines of:

Engineering Management can best be defined as both a multi-discipline and a set of skills needed to solve complex problems involving technology, people, processes, finance, and/or interfaces to rapidly produce economically viable and innovative products and services.

1.3.2 ROLE OF ENGINEERING ORGANIZATION

As previously discussed, the roles of the technical organization and the Engineering Manager have dramatically changed and continue to evolve. The digital economy pushed technical and engineering organizations to a new higher level of productivity. The 21st-century technical organization must be concerned with:

1. Maintaining an agile, high-quality, and profitable business base of products or services in a fluctuating economy;
2. Hiring, managing, and retaining a highly qualified and trained staff of engineers, scientists, technicians, and others in a rapidly changing and more complex technological environment;
3. Be adept in all aspects of the digital economy to include an evolving economic system (i.e., crypto, blockchains, and non-fungible tokens [NFTs]) while understanding enablers (i.e., WWW, computers/communications, digital marketing, and social media); and
4. Demonstrating a high level of capability maturity while ensuring that the requirements for all stakeholders are satisfied.
5. Utilize effectively productive capital, human resources, and data and create high-quality asset.

Engineering education has been slow to evolve to the new realities of a digital economy. Even traditional engineering is being driven by digitization, computer-aided design and manufacturing, dispersed teams, etc. Engineers often enter the job market not as traditional engineers but as project managers, technical sales, and systems engineers involved with conceiving, defining, architecting, designing, integrating, marketing, and testing complex and multifunctional technology-centric systems (Abel, 2005). This includes all the traditional engineering disciplines. Combined with the fact that the modern engineering enterprise is now characterized by geographically dispersed and multicultural organizations, engineering manager is more relevant than ever. Because of the blurring of boundaries between technical and management roles, engineers must continue to redefine their roles to remain relevant in the modern economy. Like most technical professions, engineering manager has evolved dramatically because of the interdisciplinary and multidisciplinary nature and complexity of modern systems.

The engineering manager profession traditionally mirrors both trends in business and education. Early business engineering focused on the civil and mechanical engineering disciplines. Early formal education for engineering manager focused on manufacturing that dominated the discipline through the 1990s. Rapid advances in information technology in the 1980s and organizational changes in all engineering practices led to a decline in the specialist engineer and a rise in the generalist engineer. This was often at the expense of traditional engineering content that was replaced with more technology-centric topics. Often, productivity and breadth were the focus versus technical depth often to the detriment of organizations that "forget" how to do the traditional engineering and hollow themselves out by outsourcing much of the value-added activity. Table 1.3 shows how the history of the industrial revolution in the 20th and 21st centuries. There are lots of these types of categorizations. When coupled with Table 1.4, the evolution of engineering education follows these changes in technology.

The fourth industrial revolution is already underway and evolving at an exponential rate and transforming entire systems of businesses, especially manufacturing and entertainment, governance, and knowledge sharing. It is characterized by the blurring of lines between the physical, digital, and biological spheres (World Economic Forum, 2016). A metaverse where users interface with a virtual-reality space and can

TABLE 1.3
History and Driving Characteristics of the Industrial Revolutions

First Industrial Revolution (1784)	Second Industrial Revolution (1870)	Third Industrial Revolution (1969)	Fourth Industrial Revolution (?)
Standardization	Machine age, mass production	Electronics	Imagination age
Steam, water	Electricity	Information technology	Metaverse
Mechanical	Mass production	Automated production	Fusion of technologies
Production equipment	Division of labor		

Modified from World Economic Forum, "The Fourth Industrial Revolution: What It Means, How To Respond," 14 January 2016 accessed 16 January 2023 at https://www.weforum.org/agenda/2016/01/the-fourth-industrial-revolution-what-it-means-and-how-to-respond/. With permission.

TABLE 1.4

Elements of Formal Engineering Manager Education Needed for the Four Industrial Revolutions

First Industrial Revolution (1784)	Second Industrial Revolution (1870)	Third Industrial Revolution (1969)	Fourth Industrial Revolution (?)
		Technical	
Work measurements	Systems engineering Operations research	Big data/data science	Additive manufacturing
Domain-specific content (i.e., civil, construction, mechanical, computers, etc.)	Quantitative methods	Software	Digital manufacturing
Statistics	Statistics	Knowledge management	Digitization
	Quality control Simulation	System of systems	AI-enhanced decision-making
		Management	
Engineering economics	Management	Project management	Strategic management
	Engineering economics and finance	Engineering economics and finance	Internal with multicultural issues
	Organizational behavior	Creativity	External with diverse stakeholders
	Team skills	Communication and social media	Dispersed and virtual workforce
	Emotional intelligence	Quality management	
	Cultural intelligence	Risk management	

interact with a computer-generated environment and other users is used in medicine, training, gaming, and many other areas in common.

What Tables 1.3 and 1.4 show is how the technical content for formal engineering manager education and training has changed. Engineering Managers must be prepared to work in an environment where the definition of products and services are software and communications driven. Long gone are the days of working in isolation where the engineers drive the requirements. The engineer's role is to fulfill the requirements.

1.4 TRADITIONAL ENGINEERING MANAGEMENT SKILLS

Figure 1.2 contains a systemigram of the skills needed by most entry or early engineers. What has changed are the domains that these are practice. The technologies needed to develop and support products and services for the digital economy now dominant the job market.

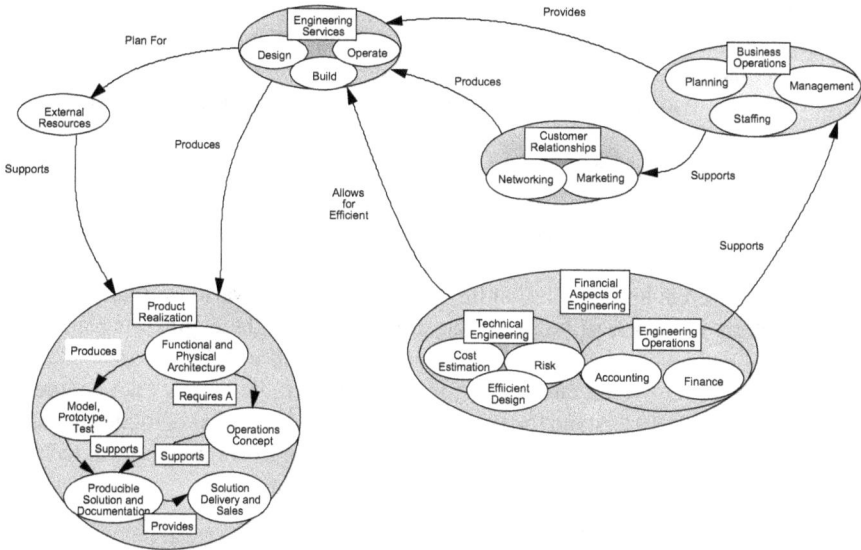

FIGURE 1.2 Engineering skills for the 21st century. (From Farr, J. V. and Faber, I. J., *Engineering Economics of Life Cycle Cost Analysis*, CRC Press, Boca Raton, FL, November 2018. With permission.)

As conveyed with the systemigram in Figure 1.2, the world that a modern engineer must operate in is complex and interrelated. In practice, engineering has become more interdisciplinary as reflected by the emergence of new engineering interdisciplinary disciplines (i.e., systems, software, biomedical, robotics, and mechatronics). Journeymen engineers still pursue two career paths. One is focused mainly on technical issues in which they eventually assume positions of technical management. However, with technology and complexity driving most products and services, these positions have more of a multidisciplinary focus. Others grow into the non-design financial aspects of the business as they progressed into the ranks of management. However, given the number of small design firms, interdisciplinary nature of engineering, flattening of organizations because of technology, complexity, and computer-based design requiring engineers to be involved throughout the project life cycle, engineers must often work as an engineering manager earlier in their career.

1.5   COMPETENCIES AND TOOLS FOR ENGINEERING MANAGERS

The digital economy will have a profound effect on the technical competences of all workers in the technology workforce. However, management skills will also evolve with many evolving as fast as technology and are still unexplored. For example, Malhotra et al. (2007) listed six skills needed to manage virtual teams, which include:

1. establish and maintain trust through the use of communication technology;
2. ensure that distributed diversity is understood and appreciated;
3. manage virtual work–life cycle (meetings);

4. monitor team progress using technology;
5. enhance the visibility of virtual members within the team and outside in the organization;
6. enable individual members of the virtual team to benefit from the team.

Digital literacy is also critical for all managers in the modern economy. This is limited to not just organizational but also individual skills. Neumeyer and Liu (2021) propose that there are three dimensions of digital literacy to include cognitive, social, and technical at four different levels: basic usage, application, development, and transformation. Different levels of digital literacy are needed for various organizations.

Technology literacy will be the greatest challenge for the engineering manager. Simply put, technology is becoming more complex, and unless you are immersed you are obsolete. However, engineering mangers need to be literates not competent users. They need to understand the uses, limitations, and potential of modern analytical techniques driven by big data. They must be smart buyers or procurers of technology both externally from within our own teams/organizations.

1.6   SUMMARY AND CONCLUSIONS

The complexity and skill set needed to work at the interface of technology and management have dramatically changed in the last 30 years. We have to manage in a different way, and the skills needed for most jobs have changed. The same basic elements of being a good leader have not changed; however, the environment, enablers, and skills needed are very different. Unfortunately, engineering manager education has been slow to react. What needs to change? The authors believe that:

1. Engineering Management classes must dramatically change to address the power of social media, disbursed and diverse workforce, managing the virtual worker, technology tools, etc.
2. Some of the traditional industrial engineering content that dominates Engineering Management education probably needs to be enhanced with systems engineering, software systems, etc.
3. We need to understand and embrace technology enablers for creativity, organizational management, communications, productivity, etc.

Society is in an age of rapid technology changes, i.e., the digital economy. Like any product/profession, we must adapt or become irrelevant. Engineering managers must help drive this transformation. Like any modern technical profession, their education is just the start of a lifelong learning experience.

REFERENCES

Abel, K., *An Analysis of Stevens Engineering Management Graduates, 1990-2004*, Stevens Institute of Technology, Hoboken, NJ, 2005.

Ayers, R., "How to Innovate with Big Data: 4 Essentials," *Innovation Management*, 11 April 2019 accessed 18 January 2023 at https://innovationmanagement.se/2019/04/11/how-to-innovate-with-big-data-4-essentials/.

CNN Money, 2022 accessed 1 November 2022 at https://money.cnn.com/magazines/fortune/fortune500_archive/full/1990/.

Dieffenbacher, S. F., "Types of Innovation in Business – How to Choose Yours?" *Digital Leadership*, 7 June 2022 accessed 17 January 2023 at https://digitalleadership.com/blog/types-of-innovation/.

Farr, J. V., *Systems Life Cycle Costing: Economic Analysis, Estimation, and Management*, CRC Press, Boca Raton, FL, 2011.

Farr, J. V. and Faber, I. J., *Engineering Economics of Life Cycle Cost Analysis*, CRC Press, Boca Raton, FL, November 2018.

Kotnour, T. and Farr, J. V., "Engineering Management: Past, Present and Future," *Engineering Management Journal*, 17(1), 15–26, 2005.

Lane, N., "Advancing the Digital Economy into the 21st Century," *Information Systems Frontiers*, 1(3); ProQuest, 317, October 1999.

Malhotra, A., Majchrzak, A., and Rosen, B., "Leading Virtual Teams," *Academy of Management Perspective*, 21(1), 60–70, 2007.

Neumeyer, X. and Liu, M., "Managerial Competencies and Development in the Digital Age," *IEEE Engineering Management Review*, 49(3), 49–55, 2021.

Nowik, O., "14 Powerful Habits of People Dedicated to Lifelong Learning," *Lifehack*, 1 August 2022, accessed 17 January 2023 at https://www.lifehack.org/articles/communication/12-signs-you-are-lifelong-learner.html.

Tapscott, D., *The Digital Economy Anniversary Edition: Rethinking Promise and Peril in the Age of Networked Intelligence*, McGraw-Hill, New York, 1995.

World Economic Forum, "The Fourth Industrial Revolution: What It Means, How to Respond," 14 January 2016 accessed 16 January 2023 at https://www.weforum.org/agenda/2016/01/the-fourth-industrial-revolution-what-it-means-and-how-to-respond/.

2 Engineering Economic Strategy and Problems of Economic Complexity

Lucy Lunevich
RMIT University

2.1 INTRODUCTION

Engineering economic strategy (EES) crosses many disciplinary boundaries. EES is a transdisciplinary body of knowledge that facilitates economic growth and higher productivity and improves human conditions via effective engineering solutions, product developments, and connection to services. It consists of a minimum of three core interrelated disciplines: (1) Engineering Enterprise, (2) Economic Development, and (3) Strategic Focus.

EES deals with the effective implementation of a set of strategic steps by an engineering or technology company in specific market conditions and specific geographic places over a time period. A good example of an effective EES is the $70 billion US acquisition of British Gas (BP) by Shell Royal Dutch (Shell Global) in 2015. It took nearly 100 years to accomplish this acquisition. An effective engineering strategy allowed Shell Global to increase its competitive advantages by securing the one in 100 years' strategic opportunity through the purchase of high-quality strategic assets and resources, improving business sustainability and leading to the company's continuous growth for many years to come.

1. The Engineering Enterprise discipline embraces system thinking and social science, meaning an organisation. It considers an organisation as (1) a system in a broader sense, (2) a social system, and (3) a system functioning according to clearly defined operational and communication rules (Janssen, 2016). Furthermore, an organisation is a complex, evolving system, communicating with and receiving the external energy of the universe.
2. Economic Development is relevant to time, space, and geographic locations. It depends on the technological, social, political, and moral states of society. It comprises several disciplines, such as microeconomics, macroeconomics, trade, managerial economics, distribution, and systems (models, simple, complex, or complicated).

DOI: 10.1201/9781003374879-2

TABLE 2.1

Engineering Economic Strategy

Engineering Enterprise	Economic Development	Strategic Focus
Organisational Technologies	Systems (models, social, simple, complex, complicated, complex evolving systems)	Who benefits Risks
Organisational Structures	Generate, distribute	Productivity
Quality data, quality management system	Managerial economics	Inequality
Systems of systems	Trade	Export, import
Processes, procedures	Capital	
System thinking	Value	$, Jobs, ideas
Vision, mission	Microeconomics (organisation, industry)	Markets
	Macroeconomics (organisation, industry, country)	
Sustainable business strategy	Economic complexity	Productive capital

3. Strategic Focus is one of three disciplines because an engineering or technology company has a limited time to implement technology or deliver the project. A strategic opportunity has a limited time for accomplishment. Therefore, it is important to build businesses around effective strategies.

This chapter discusses various aspects of EES from various perspectives, allowing readers to unfold the complexity of this transdisciplinary subject.

Table 2.1 summarises some sub-disciplines of EES.

2.2   ECONOMIC DEVELOPMENT

2.2.1   Economic Complexity

Why do some countries produce many complex products and provide complex services, such as X-ray machines and medical devices, while other countries sell only natural resources or agricultural products? Why do some countries or regions have highly diversified industries, products, and services and others do not? This is a problem of economic complexity, which is partly captured by the Atlas of Economic Complexity, developed by the Harvard Growth Lab (Hausmann et al., 2013; Hidalgo, 2021).

Economic complexity is important because it translates into productivity. Increasing productivity leads to increasing wages. The second problem of economic development is connectivity to infrastructures and services, which affect the wider population. This happens across the world. However, both problems present significant opportunities for engineering and technology companies. If more people can be connected to networks (internet, power, water, electric grid, affordable transport) and provided with higher-density work infrastructure (economic zones, for instance), then overall productivity will increase substantially, and more people will have access to good jobs.

Engineering and technology companies usually have multiple opportunities to join regional or international economic development. However, they often lack a strategic approach or vision due to their focus on project delivery (budget, time, and scope). Short-term thinking is the biggest problem for engineering companies and across businesses. However, big engineering and technology companies cannot afford to miss strategic opportunities and should scan for other strategic opportunities while working on projects that are not strategic in nature. Therefore, businesses should encourage their people to make business deals "every day;" constantly scan for strategic opportunities to improve existing and develop new products; and move to new markets through collaborations, joint ventures, and alliances which will deliver profits for many years to come.

Economic complexity is a measure of the collective knowledge in society as expressed in the products it makes and the services it provides. The economic complexity of a country, region, city, or organisation can be calculated based on the diversity of products and services it produces. Figure 2.1 summarises the concept of economic complexity.

The Economic Complexity Rankings and the Economic Complexity Index (ECI) are two indicators used to describe economic complexity.

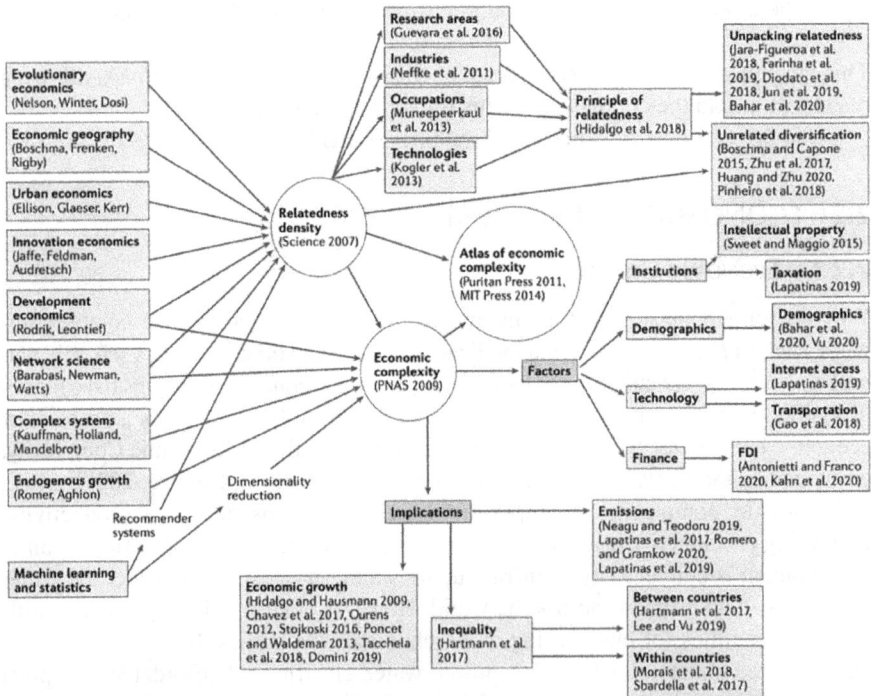

FIGURE 2.1 Summary of literature on economic complexity and relatedness. (Adopted from Hidalgo, C., *Nat. Rev. Phys.*, 3, 92–113, 2021. https://www.nature.com/articles/s42254-020-00275-1. With permission.)

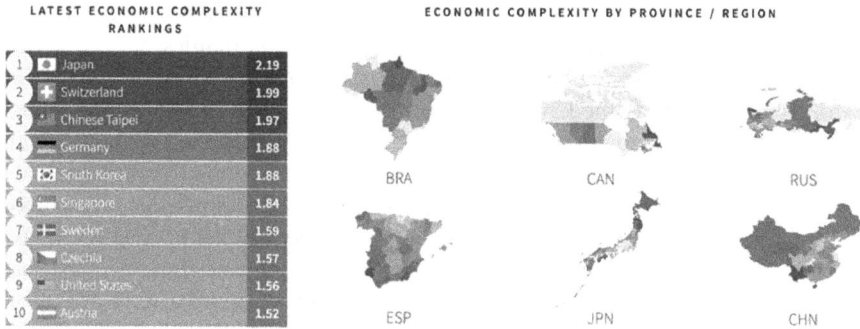

FIGURE 2.2   The Economic Complexity Ranking in 2022 (OEC, *The Best Place to Explore Trade Data*, 2022. Retrieved from https://oec.world/en: https://oec.world/en. With permission.)

The Economic Complexity Rankings have been offered by several institutions (Harvard University, OECD) and have emerged as a scientific discipline from various trading data.

The Economic Complexity Ranking shows the countries and regions that are expected to grow in the future. The Economic Complexity Index (ECI), which is a powerful dimensionality reduction technique, is used to predict and explain future economic growth, income inequality, and the impact of greenhouse gas emissions on economic growth. Figure 2.2 explores the latest economic complexity rankings for countries, products, states, and provinces, according to the OEC (2022)

The key to economic complexity, which translates to higher productivity, is tacit-knowledge.

2.2.2   THE TACIT-KNOWLEDGE ECONOMY

Studies show that all rich countries are "rich because they exploit technological progress" (Hausmann, 2013; Hidalgo, 2021). These countries moved the bulk of their labour force out of agriculture and into the cities, where knowledge can be shared more easily (Hausmann, 2013), (Hidalgo, 2021). Their families have fewer children and "educate them more intensively, thereby facilitating further technological progress" (Hausmann, 2013). As a result, technological progress leads to the development of more complex economies. Complex economies have better institutions, more educated workers, and more competitive environments, so these approaches are not completely at odds with each other or ours. In fact, institutions, education, competitiveness and economic complexity emphasise different aspects of the same intricate reality. The Economic Complexity Ranking indirectly captures information about the quality of governance in a country; It also tries to capture the total amount of productive knowledge that is embedded in a society as a whole, and that is related to the diversity of knowledge that a society holds.

Thus, productivity progress is a complex, steady process with the ripple effect of technological development on social and legal practices, which translates into business culture change, professional development, and, eventually, business models.

Many innovative ideas and initiatives get lost during the implementation phase simply because organisations are not aligned with strategic opportunities. Therefore, it is important for the managers of engineering companies to comprehend these changes.

As mentioned above, the key to economic complexity that translates into higher productivity is tacit knowledge. In business, it is called strategic capability. Tacit knowledge is implicit knowledge stored in the brains of people. Translating tacit knowledge into explicit knowledge is really hard, as tacit knowledge is acquired internally through doing. The bearer of tacit knowledge may not be able to explain why they do things in a particular way. This is how we train musicians, doctors, academics, engineers, and scientists. Consider how long it takes to learn to be a medical doctor – 20 years! Consider tennis players like Federer and Nadal; If they began training at five years old, by 17 years old, they have been training for 12 years already just to attempt to play at junior-level competitions. Studies show that an individual requires 13,000 hours of learning and practice to achieve professional expertise in a specific field. It is why good companies train people constantly, designing and carefully considering training programmes for all staff. This is also the reason why some companies enter alliances and joint ventures, as this allows organisations and staff to gain knowledge and skills in different business environments.

Tacit knowledge is vast and growing, so only a minuscule fraction of it is made explicit and known to others. As a result, most products require teams of people with different pieces of knowledge, similar to a symphonic orchestra (Figure 2.3).

It is good to think about your team as a symphonic orchestra. What makes an orchestra so unique? The most important thing is not the skills, as all musicians have great skills, but the ability to listen to each other. Listening to each other brings harmony, fosters productivity in the team, and improves business culture.

Recent research at Harvard University's Center for International Development (CID) suggests that tacit knowledge flows through amazingly slow and narrow channels (Hausmann et al., 2013; Hidalgo, 2021). Hausmann stated that it is easier to move brains than move tacit knowledge into brains (Hausmann, 2013). When new

FIGURE 2.3   A symphonic orchestra (adopted from Leading Economic Growth, Hausmann R., Harvard University, 2021).

industries are launched in Germany and Swedish cities, it is mostly because entrepreneurs and firms from cities move in, bringing with them skilled workers with relevant industry experience.

The point is that urbanisation, schooling, and internet access are insufficient to effectively transmit the tacit knowledge required to be productive or increase productivity. Knowledge resides in brains, and emerging and developing countries should focus on attracting skilled employees instead of erecting barriers to skilled migration (Hausmann, 2013; Hayes, 2011). In this way, firms can gain knowledge and skills that are not easily available in the local market. According to Hausmann, companies should tap into their diasporas, attract foreign direct investment in new areas, and acquire foreign firms, if possible, from Germany, France, and Italy, where the higher education systems remain the best in the world. Knowledge and skills move when people do (Hausmann, 2013; Hidalgo, 2021).

2.2.3 DIGITAL PROGRESS

Digital progress has become possible because of substantial discoveries in neuroscience. The human brain is a complex system that allows individuals to function in multi-meaning and conflicting environments. Brains can adjust to significant rates of changes. This is important for the digital economy when the ability to verify the truth (a big data case) is disappearing. The world is becoming no longer measurable by the human, but instead by universal algorithms, such as DeepMind, AlphaGo, and AlpaZero. These universal algorithms can compete with humans when it comes to decision-making. Contrarily, Professor Savelyev pointed out that the human brain function is a set of complex biochemical reactions unique to each individual. According to him, it is not possible to reproduce the functions of the brain by just code and signals. This means civilisation is still quite far from real artificial intelligence, but depends on advances in brain research. Jeff Hawkins proposed an alternative paradigm of how the human brain works. In his view, the brain is not a Turing machine that manipulates symbols according to a table of rules, which is the model on which computers and artificial intelligence are based. Instead, the brain is a giant hierarchical memory that constantly records what it perceives and predicts what will come next (Hausmann, 2013). The brain makes predictions by finding similarities between patterns in recent sensory inputs and previous experiences stored in its vast memory (Hausmann, 2013). It matches current fragmentary sounds in a sea of noise with a known song or the face of a person in disguise with that of your child. The idea is similar to the auto-complete function in the Google search box, constantly guessing what you will enter next based on what you have already typed in (Hausmann, 2013; Hayes, 2011). ChatGPT is a more advanced tool, but it has a similar approach, based on predicting what combinations are next. To see the hierarchy in this mechanism, consider that you can predict the word by perceiving just a few letters; by looking at a few words, you can predict what the sentence, or even the paragraph, means (Hausmann, 2014). The hierarchy allows one to understand the meaning, whether the input enters the brain through reading or listening. The brain is, thus, an inductive machine that predicts the future based on finding similarities, at many different levels, between the present and the past. The process involves adding and combining new and existing capabilities to support more diverse and complex activities (Hausmann, 2013, 2014).

2.2.4 ECONOMIC COMPLEXITY INDEX

The Atlas of Economic Complexity was developed by the Harvard Economic Growth Research Lab and provides information on the level of economic complexity of countries and cities (https://atlas.cid.harvard.edu/) (Hausmann et al., 2013), (Hausmann R., 2014). This section is just a short summary of it.

According to Hausmann, goods can be made with machineries, raw materials, and labour, but products can be made with knowledge. Countries or industries with more diverse and deep knowledge can make higher-complexity products, which only few places in the world might be able to. The true value of products can be measured by the knowledge embedded in the product. Hausmann pointed out that when we think about products in terms of the knowledge embedded into products, markets take on a different meaning (Hausmann et al., 2013). Markets and organisations allow the knowledge that is held by a few to reach many, and make us collectively wiser. The job of an engineering manager is to learn about economic complexity and have the right strategy to navigate markets. There is no effective EES without understanding economic complexity and the kind of knowledge embedded into products.

However, the amount of knowledge embedded in a society does not depend on how much knowledge each individual holds. It depends on the diversity of knowledge across individuals and their ability to combine this knowledge and make use of it through a complex web of interaction. Hausmann explained that people collectively use large volumes of knowledge in the modern society (Hausmann et al., 2013). In fact, this trend is gradually increasing.

There is a difference between explicit and tacit knowledge. Explicit knowledge can be transferred easily by reading a text or listening to a conversation. Hausmann pointed out that the problem is that critical parts of knowledge are tacit and are, therefore, hard to embed in people. Tacit knowledge resides in the brains of people. Unarticulated tacit knowledge is what constrains the process of growth and development in countries and many businesses. Economic progress is collective know-how or "a collective phenomenon" which leads to economic progress. (Hidalgo, 2021). As individuals, we are not more capable than our ancestors, but as societies, we have developed the ability to make many things!

Modern societies can amass large amounts of productive knowledge because they distribute bits and pieces of knowledge among their many members. However, this knowledge has to be put back together through organisations and markets to be used. Our most prosperous modern societies are wiser, not because their citizens are individually brilliant, but because these societies hold a diversity of know-how and because they can recombine it to create a larger variety of smarter and better products (Hidalgo, 2021).

With a rapidly progressing digital economy, tacit knowledge becomes a competitive advantage for countries and organisations because embedding tacit knowledge is a long and costly process. This is why people are trained for specific occupations and why organisations become good at specific functions or products. In business language, embedded knowledge is called business capabilities. Some of these capabilities have been modularised at the level of individuals, while others have been grouped into organisational and even into networks of organisations of industries. It is why organisational design (vision, mission, organisational structure, and knowledge management system) is critical for effective knowledge utilisation.

The complexity of an economy is related to the multiplicity of useful knowledge embedded in it. For an advanced society to exist and sustain itself, people who are knowledgeable about design, marketing, finance, technology, human resources, Big Data, operations, and trade law must be able to interact and combine their knowledge to make products and services. These same products cannot be made in societies that are missing parts of this capability set. Economic complexity, therefore, is expressed in the composition of a country's product output and reflects the structures that emerge to hold and combine knowledge (Hausmann et al., 2013; Hausmann, 2014). Hausmann stressed that knowledge can only be accumulated, transferred, and presented if it is embedded in networks of individuals and organisations that put this knowledge to productive use (Hausmann et al., 2013; Hausmann, 2014).

For a society or an organisation to increase economic complexity, it must be able to hold and use a large amount of productive knowledge. Therefore, embedded knowledge can be measured by the complexity of produced products. For example, Figure 2.4 shows an extract from the Atlas, including the five top and bottom products by complexity.

For instance, X-ray machines are very complex products produced only by the United States and Germany. Cotton fabrics are produced by many countries. Some countries benefit from resource sectors like oil and gas and have low embedded knowledge. In contrast, some countries possess advanced technologies because of a significant level of embedded knowledge that is expressed in productive diversity. Therefore, the ubiquity of a product reveals information about the volume of knowledge required for its production.

The Atlas of Economic Complexity, developed by Hausmann et al. (2013), is an attempt to visualise the Economic Complexity of various countries by differentiating simple and complex products and export and import products. Figure 2.5 allows us to understand the approximate measuring of economic complexity.

2.3 STRATEGIC FOCUS

The goal of any business is to deliver superior sustainable performance while providing benefits to its shareholders and community and building a sustainable business in the long term (10–20 years). This is a big statement; How is it possible to achieve this? "Performance" means a return on investment (Janssen, 2016). "Sustainable" means profits over the long term rather than just meeting the next quarterly earnings targets. Delivering a sustainable stream of income is much harder. "Superior" means better than competitors (Janssen, 2016). Firms who always strive to win in any market condition and firms who are less likely to be blindsided by competition can achieve these goals. Winners also gain better access to resources: finance, projects, better people, and lucrative markets. The strategy formulation and implementation phases

TABLE 2.2.1: TOP 5 PRODUCTS BY COMPLEXITY

Product Code (SITC4)	Product Name	Product Community		Product Complexity Index
7284	Machines & appliances for specialized particular industries	Machinery		2.27
8744	Instrument & appliances for physical or chemical analysis	Chemicals & Health		2.21
7742	Appliances based on the use of X-rays or radiation	Chemicals & Health		2.16
3345	Lubricating petrol oils & other heavy petrol oils	Chemicals & Health		2.10
7367	Other machine tools for working metal or metal carbide	Machinery		2.05

TABLE 2.2.2: BOTTOM 5 PRODUCTS BY COMPLEXITY

Product Code (SITC4)	Product Name	Product Community		Product Complexity Index
3330	Crude oil	Oil		-3.00
2876	Tin ores & concentrates	Mining		-2.63
2631	Cotton, not carded or combed	Cotton, Rice, Soy & Others		-2.63
3345	Cocoa beans	Tropical Agriculture		-2.61
7367	Sesame seeds	Cotton, Rice, Soy & Others		-2.59

TECHNICAL BOX 2.2: WHO MAKES WHAT?

When associating countries to products it is important to take into account the size of the export volume of countries and that of the world trade of products. This is because, even for the same product, we expect the volume of exports of a large country like China, to be larger than the volume of exports of a small country like Uruguay. By the same token, we expect the export volume of products that represent a large fraction of world trade, such as cars or footwear, to represent a larger share of a country's exports than products that account for a small fraction of world trade, like cotton seed oil or potato flour.

To make countries and products comparable we use Balassa's definition of Revealed Comparative Advantage or RCA. Balassa's definition says that a country has Revealed Comparative Advantage in a product if it exports more than its "fair" share, that is, a share that is equal to the share of total world trade that the product represents. For example, in 2008, with exports of $42 billion, soybeans represented 0.35% of world trade. Of this total, Brazil exported nearly $11 billion, and since Brazil's total exports for that year were $140 billion, soybeans accounted for 7.8% of Brazil's exports. This represents around 21 times Brazil's "fair share" of soybean exports (7.8% divided by 0.35%), so we can say that Brazil has revealed comparative advantage in soybeans.

Formally, if X_{cp} represents the exports of country c in product p, we can express the Revealed Comparative Advantage that country c has in product p as:

$$RCA_{cp} = \frac{X_{cp}}{\sum_c X_{cp}} \bigg/ \frac{\sum_p X_{cp}}{\sum_{c,p} X_{cp}} \qquad (1)$$

We use this measure to construct a matrix that connects each country to the products that it makes. The entries in the matrix are 1 if country c exports product p with Revealed Comparative Advantage larger than 1, and o otherwise. Formally we define this as the M_{cp} matrix, where

$$M_{cp} = \begin{cases} 1 & if\ RCA_{cp} \geq 1; \\ 0 & otherwise. \end{cases} \qquad (2)$$

M_{cp} is the matrix summarizing which country makes what, and is used to construct the product space and our measures of economic complexity for countries and products. In our research we have played around with cutoff values other than 1 to construct the M_{cp} matrix and found that our results are robust to these changes.

Going forward, we smooth changes in export volumes induced by the price fluctuation of commodities by using a modified definition of RCA in which the denominator is averaged over the previous three years.

FIGURE 2.4 Extract from the Atlas. (Adopted from the Atlas, Hausmann, R., et al., *The Atlas of Economic Complexity: Mapping Paths to Prosperity* (C. M. Press, Producer), 2013, Oct 23. Retrieved from: https://atlas.cid.harvard.edu/. With permission.)

require the right people; The right leadership to ensure that the right choices are made, the right assets are deployed and the right actions are taken at the right time! But how does one know if it is a right time to deal? This is the strategy that prompts this answer and the people who drive it.

2.3.1 STRATEGY IS A SET OF ACTIONS

Strategy is a set of actions towards the goal a firm wants to accomplish, considering that the goal is aligned with the company's vision and mission. According to the definition of Strategic IQ, firms must constantly be steering purposefully in a winning

If we define M_{cp}, as a matrix that is 1 if country c produces product p, and 0 otherwise, we can measure diversity and ubiquity simply by summing over the rows or columns of that matrix. Formally, we define:

$$Diversity = k_{c,0} = \sum_p M_{cp} \tag{1}$$

$$Ubiquity = k_{p,0} = \sum_c M_{cp} \tag{2}$$

To generate a more accurate measure of the number of capabilities available in a country, or required by a product, we need to correct the information that diversity and ubiquity carry by using each one to correct the other. For countries, this requires us to calculate the average ubiquity of the products that it exports, the average diversity of the countries that make those products and so forth. For products, this requires us to calculate the average diversity of the countries that make them and the average ubiquity of the other products that these countries make. This can be expressed by the recursion:

$$k_{c,N} = \frac{1}{k_{c,0}} \sum_p M_{cp} \cdot k_{p,N-1} \tag{3}$$

$$k_{p,N} = \frac{1}{k_{p,0}} \sum_c M_{cp} \cdot k_{c,N-1} \tag{4}$$

We then insert (4) into (3) to obtain

$$k_{c,N} = \frac{1}{k_{c,0}} \sum_p M_{cp} \frac{1}{k_{p,0}} \sum_{c'} M_{c'p} \cdot k_{c',N-2} \tag{5}$$

$$k_{c,N} = \sum_{c'} k_{c',N-2} \sum \frac{M_{cp} M_{c'p}}{k_{c,0} k_{p,0}} \tag{6}$$

and rewrite this as :

$$k_{c,N} = \sum_{c'} \widetilde{M}_{cc'} k_{c',N-2} \tag{7}$$

where

$$\widetilde{M}_{cc'} = \sum_p \frac{M_{cp} M_{c'p}}{k_{c,0} k_{p,0}} \tag{8}$$

We note (7) is satisfied when $k_{c,N} = k_{c,N-2} = 1$. This is the eigenvector of $\widetilde{M}_{cc'}$, which is associated with the largest eigenvalue. Since this eigenvector is a vector of ones, it is not informative. We look, instead, for the eigenvector associated with the second largest eigenvalue. This is the eigenvector that captures the largest amount of variance in the system and is our measure of economic complexity. Hence, we define the Economic Complexity Index (ECI) as:

$$ECI = \frac{\vec{K} - <\vec{K}>}{stdev(\vec{K})} \tag{9}$$

where < > represents an average, stdev stands for the standard deviation and

$$\tag{10}$$
$\vec{K}$ = Eigenvector of $\widetilde{M}_{cc'}$ associated with second largest eigenvalue.

Analogously, we define a Product Complexity Index (PCI). Because of the symmetry of the problem, this can be done simply by exchanging the index of countries (c) with that for products (p) in the definitions above. Hence, we define PCI as:

$$PCI = \frac{\vec{Q} - <\vec{Q}>}{stdev(\vec{Q})} \tag{11}$$

where

$$\tag{12}$$
$\vec{Q}$ = Eigenvector of $\widetilde{M}_{pp'}$ associated with second largest eigenvalue.

FIGURE 2.5 Measure of economic complexity. (Adopted from the Atlas, Hausmann, R., et al., *The Atlas of Economic Complexity: Mapping Paths to Prosperity* (C. M. Press, Producer), 2013, Oct 23. Retrieved from: https://atlas.cid.harvard.edu/. With permission.).

direction (Nells, 2012; Janssen, 2016). Those with moderate IQs keep up with the pack, but the smartest firms do not simply react to change. They drive it, shaping the competitive environment to their advantage (Nells, 2012). Firms that fail to change their strategies in a timely fashion put themselves in great danger. The longer the delay, the bigger the strategic problems become, and the harder they are to fix (Nells, 2012). The more a firm invests in tactical responses that do not address the underlying strategic problems, the more resources are diverted from much-needed strategic change, and the more the firm is distracted from the strategic issues it should be addressing (Nells, 2012).

It is easy to get caught in this trap. Sales and profits can continue to grow for many years before strategic weakness shows. Firms become complacent and defer expensive and painful changes until later. However, once financial results collapse, shareholders have little patience with investing heavily in long-term problems, rather than wanting a quick fix. It becomes very difficult to make the necessary changes, and the firm struggles, squeezed between impatient investors and an increasingly hostile competitive environment until it finally falls (Nells, 2012). Nells divided this problem into various levels of strategic intelligence.

2.3.1.1   Low Strategic Intelligence

Developing and implementing a strategy is a non-trivial task, so it is not surprising that firms are reluctant to change once they have discovered one that works. However, the problem is more serious than this for some; these companies have never had a strategy and do not know what one is. Most medium and small businesses across all countries and industries face this problem. Some of these businesses consistently deliver profitable growth without knowing why. Nells (2012) pointed out that they are "the strategically blind, blissfully ignorant" and sit at the bottom of the Strategic IQ ladder. This means that these companies operate only in non-competitive environments. When competitive pressure rises, the need for strategy becomes more apparent, but "blissfully ignorant" do not know how to deal with it. When profits suffer, business culture deteriorates, and it is too hard to invest time and money in figuring out what to do. When these firms arrive at the time to build alliances, joint ventures, and learn new business models, it is too late to think about a strategy.

Strategically incompetent firms are one rung up on the IQ ladder from those who are in strategic denial because they admit they have a strategic problem, but they do not have the competence to solve it. Some firms are lost in the dark; everyone can "feel" the problem, but they do not know what it is. Others find themselves squabbling because there are a wide range of strongly held views on the issue and no real agreement on how to proceed. About 50%–60% of companies are probably in this category.

2.3.1.2   Elements of a Strategic Business Model

To commit to building strategic competence, firms must recognise that strategy is important and understand what it involves. The hardest part is that it requires both parallel building business capability and developing strategy. It cannot be one or the other; it is an entwined interactive process of social learning and strategic testing (Nells, 2012).

The choice of where to compete is key because some businesses and business segments are more attractive than others. Firms must pick the right battlefields. The elements of a strategic business model are summarised in Table 2.2.

A firm's ability to build an advantage will depend on the assets it has at its disposal and how it organises these assets. It may choose to invest in some activities and let third parties perform others.

The strategic business model documents the causal logic of the strategy, explaining the linkages between the drivers of advantage and the level of advantage expected. In addition to the logic, firms require strategic metrics showing the size of their advantage relative to competitors, the rate at which this is changing to see who is changing faster, the goals the firm has set itself, and milestones along the way. Metrics help to test the veracity of the model and to ensure that everything is on track.

TABLE 2.2
Elements of a Strategic Business Model

External scope	Competitive advantage
(the battlefields we choose to fight in)	(the advantage(s) we seek)
Customers/channel	Lower cost
Products/services	Better
Geographies	Faster
Vertical scope	Smarter
Internal scope	
Activities	
Assets, architecture	
Functional strategies	
System causal logic	Strategic scorecard
If we do X then Y happens	Relative measures of success
Links between activities	Rate of change
Virtuous circles	Goals and milestones on the way
Crucial assumptions driving choices	
If they turn out to be wrong, then the strategy needs to change	
We know how we would change it	

The strategic business model should identify the crucial assumptions on which the strategy is based (Nells, 2012).

2.3.1.3   Moderate Strategic Intelligence

The first step on the road to moderate Strategic IQ is for a firm to make a real commitment to developing skills in strategy formulation and implementation as one of its core assets (Nells, 2012) (Janssen, 2016). The process of strategy formation requires a different set of skills and involves different kinds of people. The goal is a high level of competence. Hiring consultants to develop a strategy does not suffice (Nells, 2012). While this may provide a quick and very necessary fix, it does not prepare the firm for the next time it needs to make a change or put it on the path towards high Strategic IQ, where it must constantly strive for better strategy (Janssen, 2016).

Developing high strategic competence is a long and challenging journey for any firm. While individual journeys differ, firms pass through various stages of strategic enlightenment as they add strategic knowledge and skills. The early stages typically focus on building expertise in strategy formulation, while the latter are more concerned with strategy execution (Nells, 2012; Janssen, 2016). Table 2.3 summarises three levels of strategic journey.

TABLE 2.3

Climbing the Strategic IQ Ladder

High IQ	Distributed intelligence Synched thinking-acting Mindset of change	Many firms that reach the top end of the moderate band still view strategy as a one-shot deal. Their goal in developing a strategy is to spend a short time thinking intensely about what they should do, make the changes as quickly as possible, and then switch off their strategic minds and focus on execution. Strategy for them is a commitment to a particular course and any deviation is a distraction, an admission of their failure to develop a good strategy in the first place.
Medium IQ	Debating when to change Competent to change Clear model of success	The first few steps include a focus on strategy formulation. They must learn to conduct a rigorous external strategic review to identify the range of opportunities offered by the competitive environment and an internal strategic review to test the firm's ability to exploit these opportunities. This helps to build strategic awareness. The next level is to learn how to synthesise all this information into viable strategic options. A range of options is helpful, but it means the firm must now make clear choices and inform people.
Low IQ	Incompetent In denial Strategically-blind	Strategically incompetent firms are one rung up on the IQ ladder from those in strategic denial. They admit they have a strategic problem but do not have the competence to solve it. Some firms are lost in the dark. Everyone can "feel" the problem, but they do not know what it is.

2.3.2   High Strategic Intelligence

High IQ firms are never satisfied with their current model. Everyone in the organisation seeks for strategic improvement (Nells, 2012). They are driven by lofty and inspiring goals to deliver higher performance, always seeking to improve their current model while setting aside time and resources to test radical new approaches. They generally have many strategic options and superior decision-making processes in choosing options. They seek to align their organisations continuously by focusing on measures related to strategic success and rewarding those who deliver it. These companies operate on the edge, always prepared to change and learn from experience, and seek new opportunities for growth. Shell Royal Dutch created the Joint Venture with Petroleum China (PetroChina) in 2010 to deliver Liquid Natural Gas (LNG), a mega project in Queensland, Australia. This allowed Shell Royal Dutch access to cheap finance through PetroChina for the mega project, the ability to secure the best proven innovative exploration technologies, and the chance to develop a new business model. In 2015, Shell Royal Dutch completed the 1 in a 100-year strategic acquisition of British Gas (BG) for as low as $70 billion US dollars. Why were all these possible? The company created a strategic opportunity for itself, showing that high strategic intelligence and a continuous search for new strategic opportunities bring rewards. Firms with this changing mindset see strategy as a dynamic, ongoing process rather than a one-shot deal. In this case, the distinction between strategy formulation and implementation disappeared, and everything was addressed strategically. The firm was always looking to increase advantage, thinking and acting at the same time. The processes that drive better strategy are developed as cherished assets and embedded in the firm's architecture (Nells, 2012). The firm developed a fast, efficient, and effective process for strategic change and worked tirelessly to improve it (Nells, 2012). It allocated resources to the change process and was financially disciplined in its approach, always expecting payback from its efforts.

2.3.3 CORPORATE ENTREPRENEURSHIP

The concept of corporate entrepreneurship is valuable because it encourages a creative attitude within firms. Corporate entrepreneurship refers to radical change in an organisation's business, driven principally by the organisation's own capabilities. Bringing together the words 'entrepreneurship' and 'corporate' underlines the potential for significant change or novelty not only by external entrepreneurship but also by reliance on internal capabilities within the corporate organisation. Corporate entrepreneurship can enhance a business strategy that recognises opportunities, mobilises resources, and creates value in response to rapid changes in social, political, and economic conditions (Feldman, 2014). Feldman (2014) pointed out why investments in certain places yield jobs, growth, and prosperity, while similar investments in seemingly identical places fail to produce the desired results. Michael Porter's Competitive Advantage of Nations (Porter, 1990) focuses on five well-known forces that restrict strategy to concern about competition only (Feldman, 2014). Entrepreneurial strategy is frequently a collaboration between two competitors, leading to new business models. In this way, companies can advance the known patterns of competitive strategy (Feldman, 2014).

The fortunes of companies, industries, and regions are deeply intertwined. Places benefit when industries and firms grow, and places suffer when firms and industries decline. One prevailing explanation relies on the dynamics of the industry life cycle (Hausmann et al., 2013; Lunevich, 2022; Feldman, 2014). Notably absent are considerations of the actions of entrepreneurs as agents of change and the role that entrepreneurs, or more broadly, firm strategy, might play in regional economies and the vibrancy of a place. The key to developing a successful entrepreneurial strategy is to build relationships with regional authorities (Feldman, 2014; Lerner, 2009; Hausmann et al., 2013).

In the book *Art of Winning an Unfair Game*, Lewis (2004) challenges the conventional wisdom that a baseball team would be unable to compete by attempting to buy the best players. Winning teams like the New York Yankees spent three times as much as other teams on payroll (Lewis, 2004; Schumpeter, 1934). Talented players identified and cultivated by the As were subsequently recruited to more profitable teams. A new strategy was needed in order to win! A team made up of players who were undervalued by the market could win games with its constrained budget (Lewis, 2004). Baseball insiders thought this strategy would not work, but it did. The team went on to playoffs, and in the process, changed the way that baseball recruiting is done (Lewis, 2004).

In seeking opportunities, it might be useful to step outside of the known territory and learn from other market spaces, i.e., football games. The football teams who attend European or World games try many strategies, as they have a relatively short time to select their best players and combine teams and coaches. Paying attention to their media briefings, language used, and hidden signals during media presentations teaches a lot about their strategy.

There has been great debate about how to best allocate resources to achieve economic progress over the next 20 years. The theory of economic development argues that economic development positions the economy on a higher-quality growth trajectory and is achieved through innovation and entrepreneurship (Feldman, 2014; Lewis, 2004).

With so much at stake, there is a need to strive for consensus about the role of government in the modern economy and how to best move society forward. Building successful regional economies is a complex and long-term endeavour (Feldman, 2014). Governments around the world are engaged in providing technology-based economic development incentives to stimulate innovation and entrepreneurship. When government investments yield high rates of return, the allegation of preferential treatment or picking winners is raised (Feldman, 2014; Janssen, 2016). Alternatively, when government investments are in the poorest places and the short-term rates of return are low, the allegation is that money has been wasted. Therefore, the main purpose of entrepreneurial strategy is to assist governments (regional and federal) in creating high returns on investment projects (Hausmann et al., 2013). If a company has an entrepreneurial strategy, it could go a long way in securing cheaper finance and better market opportunities, while delivering something valuable to the community and economy (Lunevich, 2022).

2.4 ENGINEERING ENTERPRISE

2.4.1 ORGANISATION AS SYSTEM

Founding and subsequent engineering managers face the problems of building organisational capabilities and changing organisational structures for effective project delivery, aligning teams for strategic objectives, or other reasons. The problem is not that changes are by definition bad. The problem occurs when managers apply changes without first studying the heart of the organisation.

Organisations have been regarded as systems for decades for good reason. However, there has been less emphasis in academic studies about organisations being social systems and that there are social relationships between staff.

This chapter goes further by taking an organisation as a complex, evolving system. Some social systems only exist and function because they are made up of people and the way people work together. Without cooperation, nothing will come about. Thus, the success and productivity of an organisation are defined by the level of cooperation between people. It appears that the cooperation can be researched and described in detail. According to Janssen (2016), cooperation takes shape via a universal pattern referred to as a "transaction" in enterprise engineering. It is how two people reach an agreement, achieving a certain result (Janssen, 2016; Joseph, 2020). Before the final result is achieved, many more agreements need to be made, each having a partial result. All those agreements, the results, and how agreements and results are linked form the construction of an organisation (Joseph, 2020). What conclusions can be drawn from these? Organisation structures, strategies, and processes can be well-defined and clear. However, if there are no agreements between people, the organisation will be less productive or even dysfunctional. An organisation is a social system because it is made of imperfect humans. Organisations or teams can be dysfunctional as a result of miscommunication, cultural differences, and level of professional and personal experiences.

Enterprise engineering is not a method for defining the best company strategy, but works on the premise that a strategy has already been adopted. The next step would be to align the organisation's capability with this strategy. This requires (1) an alignment of organisational structure and (2) the identification of strategic expertise needed for the strategy implementation. Enterprise engineering is a paradigm change that allows us to change the references framework and look at the organisation differently.

The concepts of "system" and the "systemic" way of thinking, referred to as "system thinking" are key to understand any paradigm. It strongly emphasises the relations between the elements. An organisation is a system because it is formed by elements that are somehow related to each other (Janssen, 2016).

A human organism consists of subsystems that complement each other. If an organisation is healthy, people do not fight each other, similar to the way different healthy body parts do not fight each other. These systems are complex biochemical processes that allow the human body to regenerate, renew, reproduce new generations, treat diseases, and heal body and mind. An organisation is a system too, and consists of subsystems connected by activities and agreements between people.

2.4.2 LEARNING ORGANISATION

According to Senge's book *The Fifth Discipline – The Art and Practice of the Learning Organisation*, organisations do not learn because they have "learning disabilities" (Senge, 2006). In other words, they keep making the same mistakes, not because the people working in those organisations are incompetent, but because they are placed in structures where the same mistakes are repeated over and over (Janssen, 2016). According to Senge, the problems experienced today are caused by "solutions" of the past. This has to do with the dominant and incorrect idea that there is a linear-direct-relation between cause and effect that are close to each other in terms of time and space (Janssen, 2016; Senge, 2006).

The learning organisation focuses on people, people's learning, and people learning together, and the organisation's role is to nurture and support people's learning journey. The organisation learns when people inside the organisation learn. The freedom of individual and organisational learning is to maximise the level of knowledge shared and created in an organisation. A learning organisation is an organisation that has a bigger picture of how to maximise people's learning ability and potential toward a shared vision of both individual and organisational growth. In analysing Senge's rhetorical vision of the learning organisation, Jackson (2000), using a fantasy theme analysis, a method of rhetorical criticism underpinned by the symbolic convergence theory, identified four major fantasy themes, i.e., living in an unsustainable world, getting control but not controlling, new work for managers, and working it out within the micro world. He stated that the dramatic qualities of Senge's socially rooted vision, which embraces community and altruism, and its ability to inspire followers have helped Senge's learning organisation to stand out from other competing conceptions.

Senge's learning organisation is so inspirational that it possesses the power to nurture creative practitioners to make it true. He attempted to build organisations that serve humans, not enslave them (Amidon, 2005). Therefore, a learning organisation, in Senge's view, is an organisation that can mobilise and integrate the power of systems learning from the individual level to the team level and then the organisational level, with an appreciation of the wider context of which the organisation is a part. Senge's model of learning organisation seems to include the definition and the five disciplines (personal mastery, mental models, team learning, shared vision, and systems thinking).

2.4.2.1 Personal Mastery

Senge (1990:141) defines personal mastery as:

> the discipline of continually clarifying and deepening our personal vision, of focusing our energies, of developing patience, and of seeing reality objectively...Personal mastery goes beyond competence and skills, though it is grounded in competence and skills. It goes beyond spiritual unfolding or opening, though it requires spiritual growth. It means approaching one's life as a creative work, living a life from a creative, as opposed to reactive, viewpoint.

Senge considers personal mastery to be the spiritual foundation of the learning organisation.

2.4.2.2 Mental Models

According to Senge (2006), mental models are "deeply ingrained assumptions, generalisations, or even pictures or images that influence how we understand the world and how we take action." In other words, they are constructed by individuals based on their personal life experiences, perceptions, and understandings of the world (Senge, 2006).

Mental models are powerful in influencing human behaviour (Senge, 2006). However, it seems to be one of the most abstract disciplines in Senge's learning organisation philosophy. Senge (2006) stated that reflective practice is the most crucial factor contributing to mental models (Senge, 2006). (Bui, 2010) conceptually proposed three antecedents of mental models, including organisational culture, leadership, and organisational commitment. Later, Bui and Baruch empirically tested these three antecedents of mental models with data from the international context of higher education (Bui, 2010). The findings supported their conceptual framework.

2.4.2.3 Team Learning

Team learning is a "process of aligning and developing the capacity of a team to create the results its members truly desire" (Senge 2006:236). It is regarded as a fundamental unit of learning organisations (Senge, 2006). There is a large body of literature identifying the antecedents of team learning. For example, in their conceptual paper, Bui and Baruch proposed four antecedents of team learning, i.e., goal setting, team commitment, leadership, and development and training (Bui, 2010). In an empirical study, Bui (2010) proposed and tested a list of antecedents, including goal setting, team commitment, leadership, organisational culture, development and training, and individual learning. The empirical results show that development and training is not likely to be an antecedent of team learning (Senge, 2006).

2.4.2.4 Shared Vision

Shared vision is a vision to which people throughout an organisation are truly committed (Senge 2006). Shared vision is important for bringing people together and fostering a commitment to a shared future, improving environmental performance and promoting innovation (Senge, 2006). It is "vital for learning organisations because it provides the focus and energy for learning" (Senge 2006:192). However, there is a misunderstanding in the literature on a leader's visioning ability, in which influencing strategies come from the top down. Bui and Baruch argue that shared vision must be approached from top-down, bottom-up, and horizontally across the organisation (Bui, 2010).

2.4.3 Stages of the Development of an Organisation

2.4.3.1 Stance 1: Doing Things Well

The world of "good management" is, perhaps, dated but is still with us. The emphasis is on doing things well – well enough to satisfy the dominant purpose, determined by the dominant stakeholders – senior managers, owners, politicians, and trustees. A commercial enterprise's purpose will most likely be to make a profit. Government agencies seek to implement government policy to satisfy the minister. Not-for-profits want to achieve specific aims. Employees at all levels may have a different picture of the organisation's actual or desirable purpose, but in Stance 1, they have little say.

Power is distributed hierarchically: command and control leadership. Information flows downward or not at all; the hierarchical assumption is that knowledge is a function of "seniority." The "lower" you are, the less you know. Formal individual learning is also "downward," passing on from those who know to those who do not have the "requisite" knowledge and skills. This is done through various forms of expository teaching – manuals, instruction, checklists, and practice with feedback.

2.4.3.2 Stance 2: Doing Things Better

In a Stance 2 orientation, the organisation continues to focus primarily on itself, but now the competition is the prevailing ethos. The fundamental shift of mind is towards doing things better – better than before and better than competitors. Similarly, internally, individuals and teams are set to compete with each other for prizes, rewards, and survival. This is a quest for individual supremacy – of the individual person, the individual team, or the individual organisation.

Stance 2 continues to be based on a hierarchy, with stipulated "leaders" who still determine purpose at the top: "winning" in the eyes of key stakeholders, primarily the owners and customers, and perhaps suppliers. This means increasing market share, greater profits, raising stock prices, and doing well on numerous variables in league tables of performance. There is considerable effort to brief people, now seen as "human resources" rather than merely "hands," about the "mission and vision." This aims to gain employees' enthusiastic commitment, "winning their hearts and minds," perhaps with a dose of fear. Upward questions are allowed for clarification rather than to challenge or contribute to strategy or policy. Efficiency, productivity, and performance are measured and highly valued, and play a significant part in the reward system, which focuses on various forms of performance-related pay, including individual targets, management by objectives, and bonuses. Allegedly rational formulae and algorithms are used to create a "fair" system of linking effort, contribution, and achievement to "fair" rewards, very often based on some form of competition. In practice, these cause many negative "unintended consequences," including demotivation, cheating, and keeping information, skills, and ideas secret.

2.4.3.3   Stance 3: Doing Things Together Better

In practice, key decisions will need to be made about whom to involve in tackling a particular problem. We need to learn to work together through relational practice, perhaps overcoming a history of mutual suspicion or confrontation; or we may be largely unaware of each other, never having recognised our mutual interconnections. Either case calls for the development of empathy, the ability to appreciate – though not necessarily agree with – what and why others think, feel, and want to make happen; understanding why people hold those positions; why they make sense, seem reasonable, legitimate, and important to them. Achieving this requires collective creativity through dialogue, not debate, with a shift away from dwelling in the past to engaging with the emerging future.

2.4.3.4   Stance 4: Doing Things that Matter – To the World I

In Stance 3, the emphasis is on working collaboratively with stakeholders to tackle organisational problems in pursuing organisational priorities such as profit and cost-effectiveness while doing no environmental, ecological, or social harm – even doing a bit of good.

SOME REFLECTIONS FOR YOU:

1. Why individual assumptions about how people should behave or act within organisation play an important role in strategy implementation?
2. What is the difference between organisational performance and productivity if we compare stances 1 and 4?
3. What is your personal view about the meaning of learning organisation?
4. Will the concept of learning organisation change over time?

2.4.4   KNOWLEDGE MANAGEMENT SYSTEMS

An organisation is a complex, evolving system. Information processing capability is a central concept in designing and building organisations. It is central to knowledge acquisition and communication among decision-makers. According to Galbraith and others, the role of the organisational structure is to increase the organisation's information processing capacity to deal with internal complexity and environmental uncertainty (Galbraith, 1974) (Tushman, 1978) (Joseph, 2020). Joseph and Gaba (Joseph, 2020) conclude that existing research is divided into two directions: aggregation and constraint. The aggregation view reflects how different types of structures enable individuals to interact to make collective decisions. The constraint view reflects how the context established by the organisational structure enables or constrains individual decision-making (Galbraith, 1974; Joseph, 2020). In this context, Galbraith (1974) identified four organisational design strategies for better information processing. Two aim to reduce the information necessary for management, while the other two increase an organisation's ability to process information (Table 2.4).

TABLE 2.4

Organisational Information Processing Strategies

Information Processing Strategy

Creation of redundant resources	Creation of self-contained tasks	Investment in information systems	Creation of knowledge management system
Reduce the need for information processing		Increase the capacity to process information	

Adopted from Galbraith, J.R., *Interface*, 4, 28–36, 1974. With permission.

Galbraith (1974) points out that the creation of slack resources is a regular task in solving job-scheduling problems when completion dates can be extended until the number of exceptions that occur is within the existing information-processing capability of the organisation (Janssen, 2016). However, from the three popular managerial techniques, namely, Theory of Constraints (TOC), just in time (JIT), and lean manufacturing (LM), these resources are losses. All managerial approaches aim to reduce uncertainty, use available resources efficiently, and reduce the need for extra resources. In the Theory of Constraints (TOC), such excess resources are considered buffers (Galbraith, 1974). TOC justifies that a buffer is needed only before the least productive node of the production chain since it determines the throughput of the entire line. The Creation of Redundant Resources strategy contrasts the desire of management, arises from a lack of information, and leads to the inefficiency of organisations in general (Galbraith, 1974).

The second strategy to reduce the amount of information processed (Table 2.4) is the Creation of Self-Contained Tasks. It is the decomposition of the system into loosely-coupled modules grouped around similar products or services (Galbraith, 1974). Such a module should have all the necessary resources to cover the entire value chain. After that, it can be considered a "black box" that hides internal information flows. Some believe that this approach shifts the basis of the authority structure from one based on input, resources, skill, or occupational categories to one based on output or geographical unit. By using this approach, origination and teams can be designed as flexible manufacturing cells, agile project teams, and temporary units. It could become necessary to combine several teams for a more complex task, as information processing may require more effort than in the case of non-autonomous groups.

Galbraith argues that the organisation can invest in a mechanism that allows it to process information acquired during task performance without overloading the hierarchical communication channels (Galbraith, 1974). This tool is called a Vertical Information System. (Galbraith, 1974) suggested that the effect of such systems is achieved by formalising a decision-making language that simplifies information processing in the authority hierarchy. An example of such a language is the accounting system. More often than not, providing more information overloads the decision-makers.

"Classical" enterprise resource management systems offer an optimised model of processes, which reduces the complexity of choosing an operating model at a strategic level. Secondly, these systems prescribe certain actions to workers that are rigidly integrated into the software, thus reducing the uncertainty at the operational level.

Thirdly, such systems provide a wide range of reports of uncertainty at the middle and higher levels. New IT, often referred to as technology enabling digital transformation, opens up new ways to reduce information overload. According to many, the Investment in Information Systems strategy can improve information capabilities without causing information overload (Janssen, 2016).

Creating a knowledge management systems strategy moves the decision-making down the levels to where the information exists but does so without reorganising into self-contained groups. This is achieved through lateral relationships. The concept of knowledge management (KM) is the ability to access available information. In the first stage (1960–1980), there was the concept of knowledge as a tool that impacts the performance of organisations. In the 1990s, knowledge was viewed as a process. The third generation of research (the 2000s) linked knowledge management to the success of organisations in general. In 2010, KM role is identified more as a social process than a management system that should be designed.

In a broad sense, the modern KM system is the technology and managerial methods that support the development of social capital, motivate corporate culture, and stimulate information exchange (Galbraith, 1974). Technologically, knowledge management systems can be based on both traditional communication systems and social networks. The new paradigm of social networks corresponds exactly to the model of social capital, which is defined through structural (horizontal relationship at the work level), cognitive (shared codes and language), and relational components (trust, norms, and obligations). In summary, the main purpose of KMS is not to provide all the necessary knowledge to a specific employee but to quickly find someone who has the competencies required within or outside the organisation. It deals with who knows what rather than who knows everything. Table 2.5 presents all of the organisation's design strategies in terms of information processing, their benefits, and their limitations.

Hayes (2011) highlighted that key IT associated with information processing and knowledge management could be classified into three main groups: (1) integration systems that provide storage and retrieval (document management, data mining, directories, expert systems, workflow systems, (2) interactive systems that support the interaction of people, the distribution, creation, and use of knowledge (emails, forums, social networks, blogs, and other web 2.0 systems), and (3) platforms (groupware, intranet, and enterprise 2.0) that offer general principles for building infrastructure (Galbraith, 1974; Janssen, 2016; Hayes, 2011).

Davenport (2005) and Wiig (2004) proposed a classification of organisational technologies that support the activities of various classes of employees. The system considers two dimensions – the complexity of the work performed (from performing routine procedures to expert activity) and the level of independence from other employees (from an individual activity to large group interaction) (Davenport, 2005). Wiig proposed a more detailed classification of work complexity – from routines to actions in a completely unpredictable situation. Based on the integration of the approaches of these researchers, it is possible to construct a classification of information systems used to support various types of activities related to information systems that automate the performance of routine procedures and require the employees only to know their duties. The general process, the purpose of data, and their further use may not be known to them (Zelenkov, 2022).

TABLE 2.5

The Organisation's Design Strategies in Terms of Information Processing, Their Benefits, and Their Limitations

Strategy	Benefits	Limitations
Creation of redundant resources	This strategy does not produce any benefits. According to TOC, the creation of redundant resources (buffers) is justified only in front of the least productive nodes of the job chain to guarantee their stable load.	This strategy arises from a lack of information and leads to the inefficiency of organisations in general.
Creation of self-contained tasks	The moving of the decision-making down in the level to where tasks proceed and information exists. The organisation consists of a set of "black boxes" that hide internal complexity. There is no information exchange between "black boxes" and, therefore, no need for coordination and synchronisation.	It is complicated to implement such as system in practice fully. Small autonomous teams can solve only small problems. If it becomes necessary to combine several teams for a more complex task, the information processing may require more effort than in the case of non-autonomous groups.
Investment in information systems	Simplifying information processing in the authority hierarchy by the formalisation of a decision-making language. It can be realised without IT.	May lead to information overload. IT-based applications can reduce this overload due to process and rules standardisation of algorithms that must make a decision.
Creation of knowledge management system	The moving of the decision-making down in the level to where tasks are processed and the information exists. Establishing a context that supports information and knowledge exchange between workers and groups.	It requires significant changes in a corporate culture. This can become an insurmountable barrier for many organisations.

Adopted from Galbraith, J.R., *Interface*, 4, 28–36, 1974. With permission.

An engineering manager is responsible for knowledge creation (processes and quality management systems) within the organisation. It is important for the engineering manager to understand the current knowledge management system and pay attention to its improvement, as the processes will change over time for various reasons, including strategic, business, re-organisation, and other reasons (Zelenkov, 2018).

As shown in Figure 2.6, business process management systems (BPMS) support small and medium group collaboration within rigidly-defined models. At the same time, collaboration systems (e-mails, messengers, forums, and social networks) do not impose any restrictions on the processes. Personal knowledge management systems include tools that allow an employee to save his existing digital objects and the connections between them – from merely storing documents in a file system (Zelenkov, 2018, 2022). According to Zelenkov, the effectiveness of personal information management is determined by the motivation of the employee and his ability

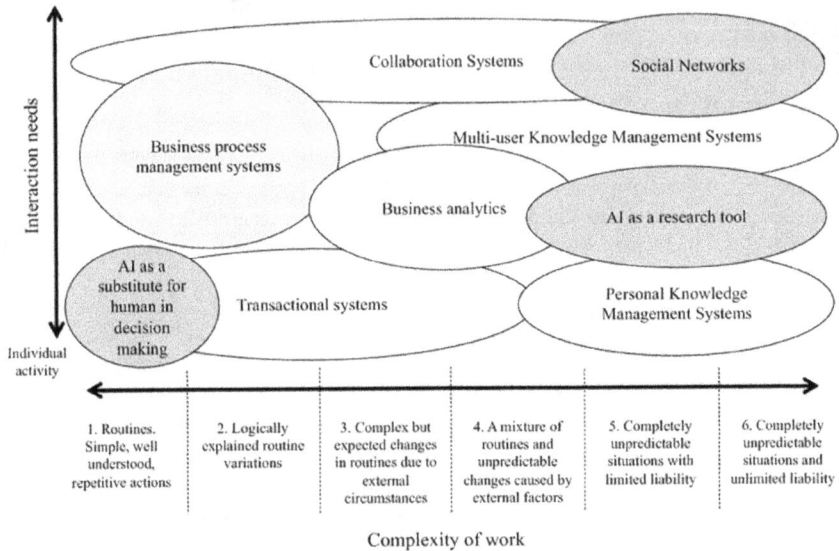

FIGURE 2.6 Classification of information and complexity of work. (Adopted from Zelenkov, Y., Explaining the IT Value through the Information Support of Decision-Making. In E. Zaramenskikh, A. Fedorova, *Digitalisation of Society, Economy and Management* (pp. 29–48). Springer, London, 2022. With permission.)

to manage information. (Zelenkov, 2022; Wiig, 2004). Multi-user knowledge management systems should provide tools for working with metadata (data about data), advanced search tools, and the ability to analyse the relationships between elements of the system (Zelenkov, 2015).

Each type of information system, shown in Figure 2.6, has its own purpose both in terms of the complexity of the supported processes and in terms of the employees involved in them (Zelenkov, 2022). It is expected that EM will be able to review the existing flow of information and KM system and adapt it as it suits a specific project because this system could support or prevent productivity within the team. Correct access to information by the right team and people is key to productivity and successful project delivery. Obviously, one information system cannot satisfy all an organisation's needs. Therefore, the systems must comply with the organisation's information design strategy (Wiig, 2004).

2.5 LEADING ECONOMIC GROWTH

2.5.1 REFUSAL TO ACCEPT LIMITS

In the book *Pathology of the Capitalist Spirit: An Essay on Greed, Hope and Loss*, David Levine (2013) states that "capitalism is not about self-interest but self-doubt. It depends on the choices we make, our talents and interests, and the opportunities available." These are exactly the roles and responsibilities of engineering managers. When opportunities are available, and choices are well made to match our talents and interests (company talents and strategy), the results are lives with a reasonable degree of gratification and a sense of place. For an organisation, this means a sustainable business model that fits well into the industry ecosystem, and contributes to and benefits from it.

However, if we cannot settle into a way of life, if the alternatives we do not choose remain as compelling or more so than those we do, if we do not know what interests us, if we are unable to assess our talents accurately, then we never gain satisfaction from what we do and what we own" (Levine, 2013). For a company, this means a strategic failure; Bankruptcy.

Capitalism was originally understood as the form of economic organisation that created a new world of wealth and alleviated poverty, or at least created the possibility that there might be such a world. Growth and freedom are considered essential preconditions for capitalism (Levine, 2013). Levine pointed out that capitalism is essentially an institutional legal mechanism for enabling a few people to gain wealth at the expense of the many. Central to this moral-political divide has been how we understand, assess, and cope with a primitive form of desire; The desire for more and more (Levine, 2013). The refusal to accept limits is closely linked to the matter of deprivation. In this system, the appearance of success in gaining satisfaction hides a reality of loss (Levine, 2013; Feldman, 2014). Loss is the reality of the system. This is not because so many lose but because loss becomes the primary end.

Capitalism is a system organised around a desire that defies satisfaction. In other words, capitalism is a loser's game (Levine, 2013; Feldman, 2014). Greed tends to foster activity that imposes a loss on others, as greedy desire can never be satisfied. Levine added that the attack on desire shifts from deriving satisfaction from the object to seeking pleasure in the deprivation of others (Levine, 2013). Redefining loss as gain has been the strategy built into social systems to ensure that they endure and thrive on the deprivation they are designed to impose (Levine, 2013).

2.5.2 Productive Capital

As Marx tells us, it is one thing to own wealth, and something else to own capital (Levine, 2013). Capital is the part of our wealth that has the capacity to produce more wealth for us or our company. According to Levin and others, owning capital means owning wealth, but the power to create wealth is a self-moving entity. The quality of wealth is linked to the matter of agency or subjectivity. Marx describes capital as a process in which value is the subject and its independently acting agent (Levine, 2013; Tushman, 1978). According to Marx, the most important thing is not that labour produces value, but the progressive loss of the significance of labour and the elimination of the human element as the vital factor in work that produces the special world man creates for himself.

So, what is capital? The idea of capital is the idea of a good that does not get old, does not lose its ability to satisfy needs, is not limited in that ability, and does not erode over time (Feldman, 2014). Therefore, capital, in the form of productive capacity, is the only enduring good with the potential to hold and increase its value. Furthermore, capital has to do with how wealth endures across times, and subsumes individuals into a sequence beyond their particular desires and finite lives (Levine, 2013). As Keynes (Feldman, 2014) pointed out, the value of capital depends on our expectations about an unknown future. Some insist that we must produce something that has a worth measured in value. Levine (2013) put it this way:

when we fall under the illusion that we can separate the creation of value from the creation of things, the entire system of production and need satisfaction is put at risk, which happens when the fantasy develops that financial assets are not limited by the productivity of the real assets they represent.

In other words, difference drains objects of their moral standing and prevents their owners from gaining moral standing by owning them (Levine, 2013; Feldman, 2014).

The definition of "productive capital" according to OECD is "Productive capital stock is the stock of a particular, homogenous, asset expressed in 'efficiency' units. The importance of the productive stock derives from the fact that it offers a practical tool to estimate capital services" (2001).

2.5.3 PRODUCTIVITY MEASURES

According to OECD's guide (2001), productivity is commonly defined as the ratio of a volume measure of output to a volume measure of input use. There is no single measure of productivity. Productivity is also defined by industry and region business culture, and there is no common measure for different types of industry, countries, and regions.

Types of productivity measurement include technology, efficiency, and real cost savings.

  i. Technology – A frequently stated objective when measuring productivity growth is to trace technical change (OECD, 2001). Technology has been described as "the currently known ways of converting resources into outputs desired by the economy" and appears either in its disembodied form (such as new blueprints, scientific results, new organisational techniques) or embodied in new products (advances in the design and quality of new vintages of capital goods and intermediate inputs) (OECD, 2001).
  ii. Efficiency – The quest to identify changes in efficiency is conceptually different from identifying technical change. Full efficiency in an engineering sense means that a production process has achieved the maximum amount of output that is physically achievable with current technology, given a fixed number of inputs (OECD, 2001). Technical efficiency gains are, thus, a movement towards "best practice," or the elimination of technical and organisational inefficiencies (OECD, 2001). However, not every form of technical efficiency makes economic sense, and this is captured by the notion of allocative efficiency, which implies profit-maximising behaviour on the side of the firm (OECD, 2001).
  iii. Real cost savings. Real cost savings is a pragmatic way to describe the essence of measured productivity change. Although it is conceptually possible to isolate different types of efficiency changes, technical changes, and economies of scale, this remains a difficult task in practice (OECD, 2001). Productivity is typically measured residually, capturing not only the above-mentioned factors but also changes in capacity utilisation, learning-by-doing, and measurement errors of all kinds (OECD, 2001).

In this sense, productivity measurement could be seen as a quest to identify real cost savings in production and re-investing it into other business
needs in practice.

Other productivity measures include product and service complexity, level of investment in research and development, and speed of commercialisation of innovations
(both products and services).

2.5.4   PRODUCTION FUNCTION

To discuss the different approaches towards productivity measures, it is useful to
refer to the production function. A production function relates the maximum quantity of gross output (Q) that can be produced by all inputs including primary inputs
(X), i.e., labour and capital, and intermediate ones (M). The function also contains
a parameter $A(t)$ that captures disembodied technological shifts. Disembodied technical change can be the result of research and development that leads to improved
production processes, or technical change can be the consequence of learning-by-
doing or imitation. It is called "disembodied" because it is not physically tied to any
specific factor of production. Rather, it affects inputs proportionally. This form of
technical change is also called "Hicks-neutral" and is "output augmenting" when
it raises the maximum output that can be produced with a given level of primary
and intermediate inputs, without changing the relationship between different inputs.
Under this assumption, the production function can be represented as:

$$Q = H(A, X, M) = A(t) \cdot F(X, M).$$

2.5.5   CREATIVE DESTRUCTION

This section discusses the process of creative destruction relevant to engineering
managers, who frequently need to make decisions in a highly uncertain environment.

The creative destruction concept is a universal concept and applies to economic,
political, biological, and geopolitical systems. It leads to the selection of more intelligent matter over less intelligent matter. Moreover, it eliminates less intelligent matter
(biological species, for instance) to create higher intelligent matter (Lunevich, 2022).
According to Levine (Levine, 2013) the notion of creative destruction captures the
two-sided quality of the urge underlying the capitalist process well. On one side, it is
an urge to make a new world that is better than the old. It approximates more closely to
the ideal expressive of the hope that limits may be overcome, and mortality set aside.
On the other side, it is an urge to destroy the world as it is because that world falls so
far short of the ideal. Indeed, it represents not only an obstacle to it but its negation. In
this process, destruction is the hidden truth of creativity (Lunevich, 2022). As Levine
puts it, *"what we create to replace what we have to destroyed cannot provide the
infinite satisfaction we seek but instead just the latest form of that given world whose
reality represents the denial of the wished-for gratification"* (2013).

Creative destruction expresses itself in a real process of the contradiction being
embedded in the desire for the infinite, which is the contradiction embedded in the

urgent need to make real what cannot be real – the perfect world in which all of the good is ours and is never-ending (Levine, 2013; Hayes, 2011). The creativity that associates with economic development is not only inseparable from destruction but also from the destruction of all ways that might actually exist in the world (Lunevich, 2022). Marx highlighted that creative destruction arises ultimately from the organisation of economic institutions to serve greed (Levine, 2013). In contrast, Schumpeter stated that creative destruction owes its origin to what he terms the entrepreneurial spirit: the urge and ability to act with confidence beyond the range of familiar beacons and to overcome resistance. Schumpeter stated, as cited in Levine (2013), that the entrepreneurial function "does not essentially consist in either inventing anything or otherwise creating the conditions which the enterprise exploits. It consists in getting things done" (Schumpeter, 1934). According to him, the entrepreneurial function is an expression of will, specifically the will to get things done, but evidently not just anything (Lerner, 2009; Feldman, 2014). Rather, the entrepreneurial function lies in clearly creating the new and destroying the old (Levine, 2013). This emphasis on will is clear when Schumpeter considers the erosion of the capitalist spirit, which occurs when "innovation itself is reduced to a routine" and, therefore "personality and will count for less."

Will is the force that exists only where there is resistance to be overcome. The entrepreneurial spirit expresses itself as the act of overcoming the resistance of interests vested in the status quo. Interests vested in the status quo are those attached to already existing ways of life. So, overcoming interests is another way of talking about overcoming the inertia of attachment to how things have been done and how life has been shaped in the past. According to Schumpeter, this resistance has diminished over time. Indeed, it has well-high vanished, and the entrepreneurial spirit is no longer needed for innovation and change and as a result (Levine, 2013; Schumpeter, 1934).

Within the cycle of creative destruction, that process becomes an end of itself and not the means to achieve satisfaction (Lunevich, 2022). In order words, the real end is to create dissatisfaction, to make it clear to all that what we have, indeed what can be had, is never good enough. The capitalist always experiences the state of the world as a constraint to be overcome rather than as a reality into which he can settle and live his life (Lunevich, 2022).

2.5.6 THE HIDDEN TRUTH OF CREATIVITY AND INNOVATIONS

Destruction is the hidden truth of creativity. What we create to replace what we have destroyed cannot provide infinite satisfaction. Creative destruction expresses in a real process the contradiction embedded in the desire for the infinite, which is the contraction embedded in the urgent need for what cannot be real. Therefore, the creativity we associate with capitalism is not inseparable from destruction; According to Levin, it is inseparable from the destruction of all ways of life that might actually exist in the world (Levine, 2013; Schumpeter, 1934).

The creation of a new world takes place through innovation, implementation, and work. In other words, it operates on the temporary plane of the real. With the growing sophistication of financial markets, the tendency towards speculative movements is enhanced by the development of new and more complex financial instruments (Janssen, 2016). This process tends to further disconnect the creation of claims over wealth from the capacity of capital investment in real assets to produce wealth. The difficulty in sustaining this disconnect over time eventually leads to a downward adjustment in asset values, the magnitude of which is no more limited by the real potential to produce wealth than the original upward adjustment, sometimes referred to as a speculative boom. (Schumpeter, 1934; Levine, 2013; Janssen, 2016). This is because fantasy's attack on reality is an expression of the human creative potential. It is an expression of that peculiarly human capacity to create reality as it might be negating reality as it is.

The link of capitalism to fantasy means that capitalism represents the release of man's creative potential. The release of creative potential always involves an attack on what is and, therefore, on what we experience as real. Creativity shares this attack on the real, and therefore this destructive potential, with the speculative process. Levine stated that *"capital needs to be understood, not only as the source of the destruction that creates a new world, but also of the destruction that does not"* (2013).

In the cycle of creative destruction, there is first the creation of something new and better than what we had before, along some important dimension. But this new and better object, product, or service soon becomes old and inferior. When this happens, the satisfaction afforded by the once-new object is revealed for what it was all along; an inferior sort of satisfaction. While we are in the habit of thinking that this process is all about the creation of an object capable of a higher order of satisfaction, what the process does is drain our experience of the satisfaction these objects originally promised to provide after owning and using them. In fact, it is a process not of gaining but of losing the satisfying object.

2.6   CREATIVITY AND MORAL DEVELOPMENT

2.6.1   Knowledge Economy

A knowledge economy is one in which growth depends on the quantity, quality, and accessibility of the information available rather than the means of production. A knowledge economy is an economy in which the production of goods and services is based primarily upon knowledge-intensive activities. For instance, an engineering manager needs to decide on a choice of knowledge management system: should this system be high information intensity or low, and why so. In the knowledge economy, a large portion of economic growth and employment is a result of knowledge-intensive activities – how much information should an employer process, and what is the risk for the organisation and the manager?

Some characteristics of a knowledge economy are growth in high technology investment and industries. There is growth in knowledge-intensive service sectors such as education, communications, and information. Knowledge is a non-finite resource. Capital gets used up but knowledge is not limited and can be shared without losing it. A knowledge economy demands creative people because many problems cannot be solved in ordinary ways.

2.6.2   Creativity and Problem-Solving

Digital economy is demanding creative people for solving complex problems. Therefore, creativity is the ability to develop and express ourselves and our ideas in new ways (Lunevich & Wadaani, 2023).

- Being creative means solving a problem in a new way.
- It means changing your perspectives and others'.
- Being creative means taking risks and ignoring doubt, and facing fears.
- It means breaking with routine and doing something different for the sake of doing something different.

Creativity is going beyond the usual – stepping outside of the box. It can be defined in many ways, such as how a person explores ideas or uses different ways to solve issues – and how one experiences life. There are many forms of creativity, including:

- Deliberate and cognitive creativity requires a high degree of knowledge and lots of time.
- Deliberate and emotional creativity requires quiet time.
- Spontaneous and cognitive creativity requires stopping work on the problem and getting away.
- Spontaneous and emotional creativity probably cannot be designed for.

Examples of creative thinking skills include problem-solving, writing, communication, and open-mindedness. All these skills are in demand by the knowledge economy. In addition:

i. Creativity is the intellectual ability to make creations, inventions, and discoveries that bring novel relations, entities, and/or unexpected solutions into existence (Wang, 2009).
ii. Creativity is the gifted ability of humans in thinking, inference, problem-solving, and product development.
iii. Creativity is the act of turning new and imaginative ideas into reality.
iv. Creativity is characterised by the ability to perceive the world in new ways, to find hidden patterns, to make connections between seemingly unrelated phenomena, and to generate solutions.
v. Creativity skills can be learned, not from sitting in a lecture, but by learning and applying creative thinking processes.
vi. Creativity is a skill that can be developed and a process that can be managed.
vii. Creativity begins with a foundation of knowledge, learning a discipline, and mastering a way of thinking.

In addition, creative people are curious. They ask questions all the time. Creative people like challenges. They do not run away from challenges; they tackle them head on. Creative people are not afraid to experiment. They have higher standards and know how to accept and give constructive criticism.

Therefore, intelligence matters. It demonstrates an ability to gather knowledge and use it effectively. Creativity is the ability to go beyond the intelligence frame and capitalise on seemingly random connections of concepts. In conclusion, expert creatives do not need to be more intelligent than the average person. Creativity can accelerate a company's profits and growth beyond that of its less-innovative competitors (e.g., Google, Amazon, and Apple). The added benefit is that the creativity and the resulting innovation are unique to the creator – the individual or company that came up with the idea. Businesses should identify creative people within the organisation and create policies to support them so that their creativity can be translated into company profit and strategic growth.

2.6.3 Moral Development

Creativity is required for humans to reach full moral development (Lunevich, 2022). On the contrary, lack of moral development leads to inequity in society (Plato, I. c. 428–348 BCE). According to (Kohlberg, 1981) only 10%–15% of people are capable of the kind of abstract thinking necessary for stages 5 and 6 of post-conventional morality. Therefore, having an environment that supports business learning and enhances creativity is required to develop abstract thinking capability and make sense of innovations (Lunevich, 2022).

REFERENCES

Amidon, D. M. (2005). Knowledge Zones Fueling Innovation Worldwide. *Research Technology Management*, 48(1), 6–8.

Bui, H. B. (2010). Creating Learning Organisations: A System Perspective. *Emerald Insight: The Learning Organization*, 17(3), 23–34.

Davenport, T. H. (2005). *Thinking for a Living: How to Get Better Performances and Results from Knowledge Workers* (pp. 3–23). Boston, MA: Harvard Business School Press.

Feldman, P. M. (2014). The Character of Innovative Places: Enterpreneurial Strategy, Economic Development, and Prosperity. *Small Business Economics*, 43, 9–20.

Galbraith, J. R. (1974). Orgnisational Design: An Information Processing View. *Interface*, 4, 28–36.

Hausmann, R. (2013, Oct 30). The Tacit-Knowledge Economy. *Project Syndicate*, p. 1.

Hausmann, R. (2014, Jan 29). A Brain's View of Economics. *Project Syndicate*, p. 12.

Hausmann, R., et al. (2013, Oct 23). *The Atlas of Economic Complexity: Mapping Paths to Prosperity*. (C. M. Press, Producer). Retrieved from: https://atlas.cid.harvard.edu/

Hayes, N. (2011). Information Technology and the Possibility of Knowledge Sharing. In L. M. M. Easterby-Smith, *Handbook of Organisational Learning and Knowledge Management* (pp. 83–95). London: WILEY

Hidalgo, C. (2021). Economic Complexity Theory and Applications. *Nature Reviews Physics*, 3(2), 92–113.

Janssen, T. (2016). *Enterprice Engineering: Sustained Imporvement of Organisations*. London: Springer.

Joseph, J. G. (2020). Organisation Structure, Information Processing, and Decision-Making: A Retrospective and Road Map for Research. *Academy of Management Annals*, 14, 267–302.

Kohlberg. L. (1981). *The Philosophy of Moral Development: Moral Stages and the Idea of Justice*. San Francisco: Harper & Row.

Lerner, J. (2009). *Boulevard of Broken Freams: Why Public Efforts to Boost Enterpreneurship and Venture Capital Have Failed and What To Do About It*. New York, Princeton, NJ: Princenton University Press.

Levine, D. (2013). *Pathology of the Capitalist Spirit: An Essay on Greed, Hope and Loss*. New York: Palgrave.

Lewis, M. (2004). *The Art of Winning an Unfair Game*. New York: W. W. Norton & Company.

Lunevich, L. (2022). Critical digital pedagogy: Alternative ways of being and educating, connected knowledge and connective learning. *Creative Education, 13*(6), 1884–1896. https://doi.org/10.4236/ce.2022.136118

Lunevich, L. & Wajaani, M. (2023). *Creativity in Teaching and Teaching for Creativity: Modern Practices in the Digital Era*. New York: CRC Press.

Nells, J. (2012). Smart Strategy. In J. Nells, *Stragic IQ: Creating Smarter Corporation* (pp. 3–33). New York: John Wiley & Sons.

OEC. (2022). *The Best Place to Explore Trade Data*. Retrieved from https://oec.world/en

OECD. (2001). *Productivity Manual: A Guide to the Measurement of Industry-Level and Aggregate Productivity Growth*. Paris: OECD.

Porter, M. (1990). *The Competitive Advantage of Nations*. Boston, MA: Simon and Schuster.

Schumpeter, J. A. (1934). *The Theory of Economic Development: An Inquiry into Profits, Capital, Credit, Intrest, and the Business Cycle* (pp. 30–91). London: Transaction Publishers.

Senge, P. (2006). *Fifth Discipline: The Art and Practice of the Learning Organisation*. New York: Doubleday.

Tushman, M. N. (1978). Information Processing as an Integrating Concept in Organsiational Design. *The Academy of Management Review*, 3, 613–624.

Wang, L. (2009). *Advances in Transport Phenomena 2009*. Springer. https://www.springer.com/series/8203.

Wiig, K. (2004). *People-Focused Knowledge Management: How Effective Deision-Making Leads to Corporate Success*. London: Elsevier.

Zelenkov, Y. (2015). Critical Regular Components of IT Strategy. Decision Making Model and Efficiency Measurement. *Journal of Management Analytics*, 2, 95–110.

Zelenkov, Y. (2018). Agility of Enterprise Information Systems: A Conceptual Model, Design Principles and Quantitative Measurement. *Business Information*, 2, 30–44.

Zelenkov, Y. (2022). Explaining the IT Value through the Information Support of Decision-Making. In E. Zaramenskikh, A. Fedorova, *Digitalisation of Society, Economy and Management* (Vol. 53, pp. 29–48). London: Springer.

3 Engineering Management – Cultural Intelligence

Milan Simic and Vuk Vojisavljevic
RMIT University

3.1 INTRODUCTION

International engineering management is dealing with the implications of economy and policy issues related to business strategy, organisational structure, manufacturing, materials management, marketing, research and development, human relations, and financial management that arise in a multinational engineering and technological organisation. Economic, political, and cultural environments and sustainability are the key points of consideration. Thanks to the rapid development of modern transport systems, and even more, quicker information communication systems' development and deployment, the world is becoming a global village, more interconnected and more interdependent than ever. Following that comes the globalisation of the world economy. It is a consequence of the continually changing nature of international trade.

There are currently 235 countries in the world as given in the "Countries in the world by population (2023)" on the Worldometer site (Worldometer, 2023). They are all different in many characteristics. Some of the very important are population, geographical/geopolitical parameters, like natural resources, climate, but also culture, moral, and ethics views. All countries produce, consume, and trade various goods and services, at different levels of economic complexity. The world is experiencing a strong trend of globalisation, regionalisation, polarisation, and digitalisation simultaneously. It is a move to a more integrated and interdependent world economy that should be managed by well-qualified and experienced international engineering managers, if companies and regions want to prosper.

3.2 FACTORS OF PRODUCTIONS

Production in a country is specific, and it depends on various parameters or factors. Factors of production are defined traditionally as land, labour, capital, and entrepreneurship. Land, as a factor of production, has a broad meaning that includes all environmental resources found in the terrestrial location. It includes water, plants, wood, and other vegetation, but also oil, gold, lithium, and other minerals. In addition to those natural resources, land is also used for many other purposes. On the same land,

DOI: 10.1201/9781003374879-3

manufacturing plants and commercial buildings, as well as residential housing, are built. Agriculture is a traditional usage of land. For example, Australian agriculture accounts for 55% of the land use, where additional land use for water extraction is not included (Australian Government, Department of Agriculture, Fisheries and Forestry, 2023).

There are two types of natural resources. First, there are renewable resources that could be refilled, such as water, vegetation, wind energy, hydro, and solar energy. Water is circling in the world, while all energy is coming from the sun. Non-renewable resources are resources that can be exhausted in supply. There are oil, coal, and natural gases, which all store energy from the sun accumulated through centuries. Some estimates are that they might be depleted by the next century, if used in the same pace as today, but that will not happen soon (Todorovic and Simic, 2019a).

Labour, as a factor of production, refers to the human resources, i.e., their skills and knowledge that they use to produce products or perform services. Labour goes through training and education to be able to perform and achieve the best possible productivity and efficiency.

Capital, as a factor of production, refers to the money that is used to purchase items used to produce goods and perform services. That also includes computers, and other information and communication technology (ICT) systems (Reddy et al., 2020), other equipment, properties, and production and commercial buildings. This is often referred as physical capital. At the same time, financial capital, as money, is needed, as contents on bank accounts, stocks, and bonds. Capital productivity tells us how efficiently a company uses physical capital in providing goods and services.

Entrepreneurship is seen as a combination of the other three factors, through the introduction of innovative ideas for the creation of new values and production, and the ability to overcome old practices.

Another way to look at the factors of production is to see division as traditional (tangible) and intangible factors. Land, resources, labour, and available capital are seen as traditional factors, while information, collaboration, and entrepreneurship are seen as intangible.

All factors of production are required to create goods or deliver services, which are measured by a country's gross domestic product (GDP). GDP is the total market value of all final goods and services produced within a country in one year. Gross domestic product measures the production in the country that occurs within a nation's boundaries no matter who owns the factors of production. That could be domestic or foreign residents. GDP is given by the equation (3.1):

$$GDP = C + G + I + (Ex - In) \tag{3.1}$$

where,

C refers to the private consumption in the national economy,
G represents the sum of government spendings,
I is the sum of all country's business spendings on capital investments,
Ex is the total export and
In is the total import.

TABLE 3.1

World GDP Ranking of Top 15 Countries in 2023 Based on IMD Data

Rank	Country	GDP in Billions of US$ (thousand)
1	United States	26.19
2	People's Republic of China	19.24
3	Japan	4.37
4	Germany	4.12
5	India	3.82
6	United Kingdom	3.48
7	France	2.81
8	Canada	2.33
9	Russian Federation	2.14
10	Brazil	2.06
11	Iran	2.04
12	Italy	1.99
13	Republic of Korea	1.79
14	Australia	1.79
15	Mexico	1.48

Shandwick, W. (2022), "Reputation Accounts for 63 Percent of a Company's Market Value," *Weber Shandwick*. https://www.prnewswire.com/news-releases/reputation-accounts-for-63-percent-of-a-companys-market-value-300986105.html (accessed 28 January 2023). With permission.

Gross domestic product is the measure of the size of an economy. According to the International Monetary Fund list for 2023 (Shandwick, 2022), the ranking for the strongest 15 world economies is shown in Table 3.1.

GDP plus any income earned and brought into the country, I_{in}, by residents from overseas investments, minus income, I_{out}, earned within the domestic economy by overseas residents is known as Gross National Product (GNP), as shown by equation (3.2):

$$GNP = GDP + I_{in} - I_{out} \qquad (3.2)$$

National income is the total amount of money earned in a country. Gross National Income (GNI) measures the value of the incomes of residents, no matter where the income is earned, in the domestic market or in foreign markets. GNI is the sum of a nation's GDP plus net income received from overseas. GNI is the sum of value added by all producers who are residents, plus any product taxes (minus subsidies) not included in the output, plus income received from abroad such as employee compensation and property income.

GNI per capita is the gross national income, in US$, divided by the midyear population. GNI per capita is a common measure of economic development. It is seen by the citizens as the general standard of living in a country.

Economic growth of a country refers to the increase in productive capacity and national output, measured by the rate of increase in GDP. Every country in the world has an inflation rate, which is basically the rate of diminishing national currency. Having that in mind, real GDP is defined, as an inflation-adjusted measure of GDP, in US dollar values, after considering changes in value owing to price changes.

There are few sources of economic growth. In the longer term, sources of economic growth include the availability of more resources and factors that lead to higher productivity. They include the application of new technologies, increase in labour skills and knowledge, innovative products, expanding markets, and scale of economies. Companies aim to achieve the lowest possible average costs of production. Per capita GDP, or income per capita, is a measure of national well-being. There are other measures as well, and they include the following:

- Net Economic Welfare (NEW)
- Net Social Welfare (NSW)
- Human Development Index (HDI)
- Gross National Happiness (GNH)

Net Economic Welfare (NEW) is a more precise measure of welfare than the gross national product. It is adding value to positive, nonmarketable activities, such as relaxation and leisure time, and subtracting negative factors, like degradation of the environment through greenhouse gases pollution. They contribute to respiratory diseases while, for example, power production plants contribute to higher GDP and GNP.

Net Social Welfare (NSW) is the increase in the welfare of a society resulting from particular courses of actions. There are actions or ways how communities are organised, like access to public schooling, education, medical services, or social justice that cannot easily be quantified.

Human Development Index (HDI) has parameters like long and healthy life, with life expectancy index, knowledge, or years of education, measured with education index and a decent standard of leaving, expressed through GNI per capita, i.e., GNI index.

Gross National Happiness (GNH) as a measure of collective happiness in a nation was introduced in 1972 by Bhutan's fourth Dragon King, Jigme Singye Wangchuck. The four pillars of GNH were defined as:

1. Sustainable and reasonable socio-economic development,
2. Environmental sustainability,
3. Preservation and promotion of culture,
4. Good governance and equity before the law.

3.3 ENGINEERING ETHICS

Ethics is a set of moral principles or values that guide and shape our behaviour. Decisions must be made, as shown in Figure 3.1, having in mind consequences, or sometimes not, depending on the chosen business ethics. The importance of ethics

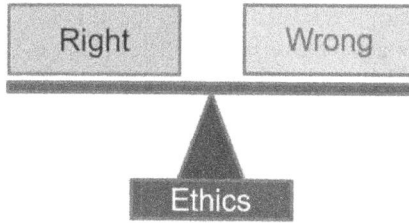

FIGURE 3.1 The role of ethics in management.

in business can be seen through the fact that the reputation accounts for 63% of the company's market value (Shandwick, 2022). Many consumers will avoid doing business with a company that they do not trust and will buy from a company they trust. Most of the employees planning to find a new job are motivated by the loss of trust in their employer.

In history, there are many high-profile examples of ethical failures. One of them is Enron ethical collapse. Top managers at Enron abused their power and manipulated information. They were involved in the unethical treatment of internal and external parties putting their own interests above all other employees and the public (Johnson, 2003).

In order to prevent ethical misconducts, governments are taking actions. In the United States, The Foreign Corrupt Practices Act (FCPA) defines the prohibition of the payments of bribes to foreign officials (U.S. Securities and Exchange Commission, 2011). There is also the Organisation for Economic Co-operation and Development (OECD) Convention on Combating Bribery of Foreign Public Officials in International Business Transactions (OECD). Apart from government bodies, there are other institutions that also take extreme care of business ethics in their domains of action. There are ethics research institutes and professional codes of practice for each profession, like medical practitioners, engineers, or others. Australian Computer Society (ACS) Code of Ethics was created in 2017. There are six core ethical values given in the ACS's code of ethics. This Code of Ethics applies to all ACS members regardless of their role or specific area of expertise in the ICT industry.

1. Priorities of the public interest – Interest of the public is above personal, business, or sectional interests.
2. Enhancement of quality of life – "You will strive to enhance quality of life of those affected by your work."
3. Honesty – "You will be honest in your representation of skills, knowledge, services and products."
4. Competence – Work competently and diligently.
5. Professional development – "…enhance your own professional development, and that of colleagues and staff."
6. Professionalism – "You will enhance the integrity of the Society and the respect of its members for each other."

Many systems of ethics are related to religion but not all of them. Religion is a system of shared beliefs and rituals, i.e., a vision of the world, life, and behaviour. Some of the major religions and ethical systems are Christianity, Islam, Hinduism, Buddhism, and Confucianism.

Different attitudes to business ethics are presented here.

Duty-based ethics, referred as *absolutism*, tell us to perform duties using the following algorithm:

Do the right thing,
Do it because it is the right thing to be done,
Do not do the wrong things,
Avoid them because they are wrong.
It is often assumed that the things are right, or wrong. Employees have a duty to act according to those rules, regardless of the future good, or bad consequences from their actions.

Consequence-based ethics, or *utilitarianism*, is the opposite of absolutism. With this attitude, there is always thinking about the consequences of the actions that could be taken. Sometimes, todays' good and right choice might prove to be bad in the future. The future cannot be seen but could be better predicted using modern computer-based tools available, like fuzzy logic (Todorovic and Simic, 2019b) and artificial intelligence. Multi-Attribute Decision Making (MADM) is already used in business to select the best choices (Todorovic and Simic, 2019c).

A country that has 10% of the world's reserves of lithium, and the largest reserves in Europe, had to make a hard decision. The question was about lithium mining, having in mind environmental issues, and degradation, or to stay with agriculture, i.e., keep green solution, for the life and business in that region. Respecting public opinion and having in mind consequence-based ethics and consequences to the environment, in 2021, government decided not to proceed with the lithium project known as Jadar – Rio Tinto (Rio Tinto, 2022).

Friedmanism teaches us that the duty of business is to maximise profits within the law and increase returns to shareholders. This ethics approach is established by American economist Milton Friedman (Shultz et al., 2020).

The opposite trend to Friedman (1970) and Adam Smith (Liu, 2022) was introduced by the Roundtable business group based in New York. More than 200 CEO of global businesses signed the declaration about well-being of their employees, society, and community on 19 August 2019.

Cultural relativism declares that all cultures are worthy in their own right and are of equal value.

Consequence-based ethics and the Australian Computer Society (ACS) Code of Ethics were applied when a technology acceptance model (TAM) was created. It is used in the management of the transition to autonomous vehicles (AV) (Aldakkhelallah et al., 2022). The key stages of the model are shown in Figure 3.2.

It appears that the AV technology is ready from the engineering point of view, but the other three stages, especially Ethics are more complex, and solutions depend on the country, i.e., community norms, traditions, cultural, social, and religions

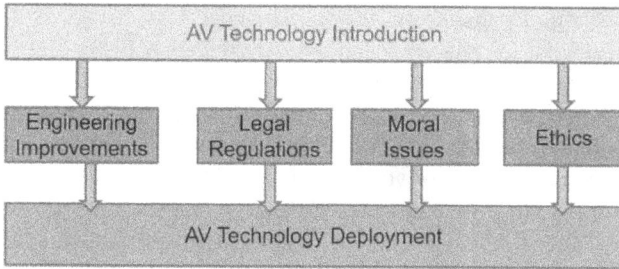

FIGURE 3.2   Key stages in autonomous vehicles technology acceptance model.

acceptance. Communities around the world have different views on many ethical questions, and all of those should be embedded in the artificial intelligence of autonomous vehicles. Most likely that the AVs will have dedicated software, i.e., AI for different countries. That will be similar to left-hand and right-hand driving. There are two mechanical design solutions developed to accommodate road traffic rules in each particular country. It is interesting to notice that with the transition to AV technology, car designers will have no problem with mechanical design. The design of the cars in left-hand and right-hand driving countries could be the same. The only difference will be in the AI control system. It will just have to follow traffic regulations, ethics norms, and principles of the particular country and that is all in the software development domain.

Many companies have developed codes of ethics based on a system of moral values. Generally, they cover right or wrong conduct, by directors, shareholders, management, and staff. That may extend to include partners, contractors, and suppliers. The culture of a society influences workplace practices. This is especially important for multinational and international organisations. There are a number of valuable studies of how cultures differ and their impact on businesses.

3.3.1   Engineering Manager in a Virtual Environment

International Business Management or International Engineering Management should be performed by experienced management teams. The main problem related with human resources in multinational companies is selecting the right people for sending overseas and additionally repatriating them back, into workforce, when they finish the task and return to their home country. There are certain adaptability criteria that apply to staff and managers that plan to relocate overseas. Adaptability, or the ability to adapt to changes, is one of the most important characteristics. There could be many problems, on all levels of work and life, remotely from the base, or home location. It is a common understanding that managers need at least one year to accommodate. Sometimes, they call that, a *listening phase*. The hypothesis is that men are accommodating faster than women, on average. Another hypothesis is that a man over 40 years of age accommodates faster than a younger man. Multicultural life experience, frequent travels, knowledge of foreign languages, and good problem-solving skills are excellent predispositions to become a good international manager.

Important criteria for managers' selection are education, the age of the candidate, and experience. Like for any other job, proper education is the first key selection criterion, but there is experience, as well. Overseas work often demands managers to make on-the-spot decision, i.e., in real time. A high level of experience is needed in the process that needs and assumes independence in making real-time decisions. Finally, there is an advantage of younger age candidates and their desire to explore new environments. However, most experienced managers, not the youngest ones, show a higher level of stability. Like any other criteria, education, age, and experience are under scrutiny. There are many large, successful, international companies, running well-known Internet businesses, established and managed by younger people. Some of them did not even manage to obtain degree qualifications, but they are successfully managing their multimillion-dollar companies. Of course, regardless of not having a proper educational level, they invest a lot of money into research in many different domains. This is extremely positive for their own countries and for all other countries where their multinational companies have subsidiaries.

Other important parameters for international engineering managers are family status and health status. International managers must have good physical and emotional health. Having a healthy family life and full family support is extremely important. Relocation is a big change in life and could affect the family and professional performance of an engineering manager. Those issues can have a significant impact on the working efficiency of expatriates.

Motivation and leadership are closely related to the potential commitment to the new job in the new country. For example, motivation factors include desire for adventure, desire to increase chances for promotion, desire to improve financial status, need for challenges, and success. There are few more hypotheses, given as follows:

- Married are more willing to accept job overseas,
- Married without children are also more ready to go,
- Prior international experience is very important and helpful,
- Career and attitudes of spouses are very important.

All the hypotheses mentioned here are good research questions for the research in *Global Management* that could be conducted through worldwide survey and statistical analysis of the collected primary data.

Thanks to the Internet and the large number of associated applications, businesses could be managed remotely. Physical locations of offices are not that relevant anymore and international business is running 24/7. Communication is one of the most important aspects of remote, virtual management. When working virtually, around the clock, which means asynchronously, interactions between the leader and team members are not frequent as in an ordinary office environment. The use of online project management tools is extremely important in this scenario. There is now new culture of using email and online tools, asynchronous working, and dedication to the duties assigned. Attending and chairing remote/virtual meetings require new sets of skills and knowledge, apart from the need to have reliable communication and audio/video facilities. Managers should try not to run back-to-back meetings. Presentations and meetings should not run for more than one to two hours without a break. Instead of happy hours on Fridays, as in face-to-face environment, managers should organise virtual team meeting building activities. There is a large variety of possible activities.

Large companies, such as Ericsson, Microsoft, Apple, IBM, Amazon, and many others, have offices around the globe, and there are always staff members and managers who can take over the business during their local working hours. Some of the most popular virtual business environments are Microsoft Teams, Cisco Webex, Cisco Jabber, Zoom, Google Hangouts, Jira, and others.

3.4 CULTURAL INTELLIGENCE

3.4.1 INTRODUCTION

In the 21st century, the world economy has become increasingly interconnected, with a continuously growing need for an even higher level of collaboration to achieve better resource management and overcome problems with sometimes very different business environments. The globalisation of the world economy can be observed by witnessing a significant increase in the international trade of products, capital, and numerous services. Thus, the global dimension of the related process and the complexity of the relationships between subsystems of the global economy demand support in improved project management.

The company's success in the global economy is closely related to the capacity of companies' leadership and the ability of business leaders to perform optimal actions, develop strategic plans, and lead others who might have different cultural values and beliefs from their own.

Nevertheless, the role of modern business leaders demands effective work in cross-border situations, with sometimes very different economic, political, and cultural practices. Therefore, business leaders are still expected to be successful and productive in domestic contexts.

3.4.2 LEADERSHIP AND CULTURAL INTELLIGENCE

Leadership is an integral part of modern management, and consequently, there is a strong need for the new kind of business leaders. Apart from the ability to solve complex business model problems, effective leaders must have the ability to solve complex social and behavioural problems. The new leaders must be capable of working effectively out of their comfort zone and in new environments, usually with partners from a wide spectrum of diverse cultural backgrounds (Shin et al., 2007; Rockstuhl et al., 2011).

The strong demand for successful business leaders capable of working in a global environment raises two main questions:

1. How to prepare leaders for their role in the global business?
2. What practical skills must future managers have to succeed?

However, as in many other professions, leaders' personality is closely related to the degree of success. Christopher Earley and Soon Ang (2003) introduced the most popular concept describing the desirable set of all personal characteristics and skills needed to adapt successfully to culturally diverse settings. They use the new term "Cultural Intelligence" or "Cultural Quotient (CQ)".

Certainly, in some aspects, cultural intelligence is a natural ability; however, many elements of cultural intelligence, such as knowledge and understanding of elements related to some cross-country environment, ability to observe and understand the behaviours of others from different cultural backgrounds can be successfully trained. Thus, it is essential for globally oriented companies to develop strategies and make long-term plans, improving leaders' adaptability in different social situations.

In their work, Earley and Ang (2003) described four cross-related dimensions of cultural intelligence (Figure 3.3), namely, metacognitive intelligence, cognitive intelligence, motivational intelligence, and behavioural intelligence.

Metacognition and cognition are closely related. Metacognition is the ability to control our thinking and consequently represents the ability to control planning and monitoring business processes. In comparison, cognition by itself represents the process of thinking. Earley and Ang described motivational intelligence as efficacy and confidence and the ability to be persistent to be aligned with personal values. Behavioural intelligence is more about the capability to adapt behaviour in different situations and environments.

In order to test the four-dimensional model (metacognition, cognition, behaviour, and motivation) in a real-life global economy, Ang et al. (Lee & Fuller, 2016) implemented measurable outcomes to the cultural dimensions describing them as cultural judgement and decision-making, cultural adaptation, and task performance in culturally diverse settings.

Many researchers also discuss the connection between cultural, social, and emotional intelligence. Emotional intelligence and social intelligence are closely associated with global leadership skills (Alon and Higgins, 2005) and global mindset development (Chen and Lovvorn, 2011; Van der Zee and Brinkmann, 2004).

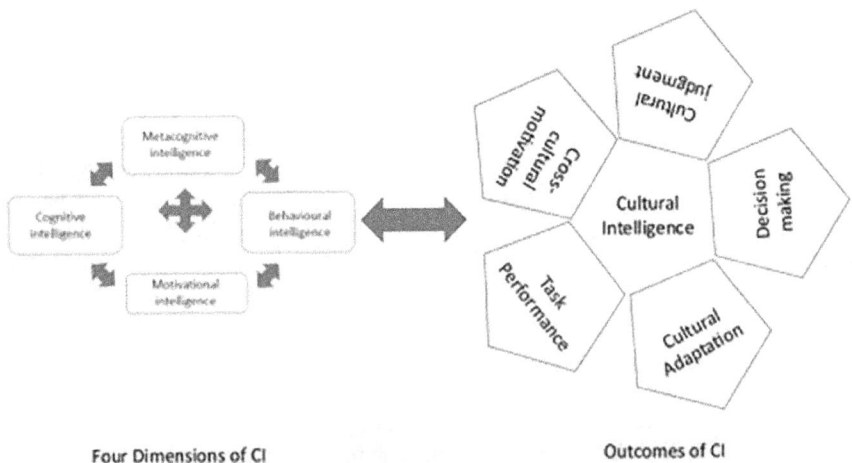

Four Dimensions of CI

Outcomes of CI

FIGURE 3.3 Four dimensions of cultural intelligence and measurable outcomes. (Adapted from Earley, P. C., and Ang, S. *Cultural Intelligence: Individual Interactions across Cultures.* Stanford University Press, Stanford, CA, 2003; Ang et al., 2007. With permission.)

From the leadership point of view, emotional intelligence can be defined as a leader's ability to feel and express emotion, with a good understanding of emotional knowledge as well as a highly adaptive ability to control emotions and to be able to regulate the emotions of others (Salovey et al., 2003).

Emotional intelligence is also related to the various components of leadership skills, such as teamwork unity (Rapisarda, 2002), attitude (Carmeli, 2023), performance, and, finally, work productivity (Akerjordet and Severinsson, 2008).

3.4.3   Cultural Intelligence Training

In some aspects, cultural intelligence is a natural ability. However, elements of cultural intelligence such as:

- Strong motivation to overcome differences in culture.
- Cultural knowledge of elements related to some social group.
- Ability to implement cultural knowledge into strategical planning at the global level.
- Ability to adapt behaviour in different situations.

All these skills can be significantly improved by proper education, training, and personal development of managers and leaders.

It is accepted that there is no general winning strategy in leadership training and that the best practice of training is by working in a foreign country and practising. In the training process, many authors state that cultural knowledge has a particular role that is crucial for resolving potential conflicts. Acquiring substantial cultural knowledge from various sources could significantly improve outcomes.

Cultural competence is another widely used term in the training of global leaders. Williams's (Strange, 2020) and Martin and Vaughn's (2007) studies discuss the attributes that could assist in a better understanding of the components of cultural competency. These attributes will guide you in developing cultural competence:

- Self-knowledge and awareness about one's own culture.
- Awareness of one's own cultural worldview.
- Experience and knowledge of different cultural practices.
- Attitude towards cultural differences.

The process of developing successful leadership with the ability to solve and overcome the problems evolving due to different cultural backgrounds needs the description of the cultural characteristics of some population, country, or specific group of people of interest.

Even if social studies of different cultures existed for a long time, the first serious attempt to relate culture with global process in business has been published after the WWII. One of the most used measures for a description of cultural characteristics is given by Hofstede (1980), known as GLOBE model (Clark et al., 2016), Arendt et al. (2019). Hofstede analyses the following dimensions:

- Individualism-collectivism.
- Uncertainty avoidance.

- Power distance (strength of social hierarchy)
- Masculinity-femininity (task-orientation versus person-orientation).

GLOBE model includes even nine dimensions.

1. Uncertainty avoidance – The extent of uncertainty is avoided by relying on established social norms.
2. Power distance – It is related to different distributions of power in cultures.
3. Institutional collectivism – It is related to the collective distribution of resources.
4. In-group collectivism – It is related to the individuals and their loyalty to the group.
5. Gender egalitarianism – It is related to the degree of differences between the treatment of genders.
6. Assertiveness – It is related to the degree of assertive, confrontational, and aggressive in social relationships.
7. Future orientation – It is related to the degree of engagement of the society in future planning.
8. Performance orientation – It is related to the importance of the rewards for performance improvements.
9. Humane orientation – It is related to the reward for being fair, altruistic, friendly, and kind.

However, this definition is raising many questions and discussions over last 40 years. Not everyone agrees. Some unanswered questions are *how to conceptualise and measure the impact of culture and define the main components of the culture.* Many authors state that the components of the culture are not static independent variables, and they change in time and depend on business activity (Brannen and Salk, 2000). However, during this time, both models have been refined by many authors. Including cultural intelligence in strategic planning and training already has a beneficial effect on the global economy.

3.4.3 FUTURE WORK ON CULTURAL INTELLIGENCE

However, there are numerous directions for improvements in training and planning. For example, there is a need for a better understanding of which factors must be included in the measurement of cultural intelligence to make training programmes on cultural intelligence more effective. Moreover, a better understanding of how individual characteristics such as personality, self-efficacy, and general intelligence relate to cultural intelligence (Liao and Thomas, 2020) would lead to a higher level of leadership.

Another direction for future work related to cultural intelligence is the development of customised training programmes that will be focused on both the personality of the leader and the particular environment.

And finally, one of the topics for future researchers will be a challenge to find a way to construct cultural intelligence at the team level and organise proper training of the teams.

REFERENCES

Akerjordet, K., & Severinson, E. (2008). "Emotionally Intelligent Nurse Leadership: A Literature Review Study." *Journal of Nursing Management, 16*(5): 565–577.

ldakkhelallah, A., Todorovic, M., & Simic, M. (2022). "Investigation on the Acceptance of Autonomous Vehicles." In *Autonomous Vehicle Technology Conference - APAC21, 21st Asia Pacific Automotive Engineering Conference*, Melbourne, 3–5 October 2022, SAE Australia, p. 6.

Alon, I., & Higgins, J. M. (2005). "Global Leadership Success through Emotional and Cultural Intelligences." *Business Horizons, 48*(6), 501–512.

Arendt, J. F. W., Pircher-Verdorfer, A., & Kugler, K. G. (2019). "Mindfulness and Leadership: Communication as a Behavioral Correlate of Leader Mindfulness and Its Effect on Follower Satisfaction." *Frontiers in Psychology, 10*, 34–56, Article 667.

Australian Government, Department of Agriculture, Fisheries and Forestry. *Snapshot of Australian Agriculture 2022.* [Online] Available: https://www.agriculture.gov.au/about/contact/our-offices.

Brannen, M. Y., & Salk, J. E. (2000). "Partnering Across Borders: Negotiating Organizational Culture in a German-Japanese Joint Venture." *Human Relations, 53*, 451–487.

Carmeli, A. (2003). "The Relationship between Emotional Intelligence and Work Attitudes, Behavior and Outcomes: An Examination among Senior Managers." *Journal of Managerial Psychology, 18*, 788–813.

Chen, J-S., & Lovvorn, Al S. (2011). "The Speed of Knowledge Transfer within Multinational Enterprises: The Role of Social Capital." *International Journal of Commerce and Management, 21*(1), 46–62.

Clark, J. et al. (2016). "GLOBE Study Culture Clusters: Can They Be Found in Importance Ratings of Managerial Competencies?" *European Journal of Training and Development, 40*(7), 534–553.

Earley, P. C., & Ang, S. (2003). *Cultural Intelligence: Individual Interactions across Cultures.* Stanford, CA: Stanford University Press.

Hofstede, G. (1980). "Culture and Organizations." *International Studies of Management and Organization, 10*(4), 15–41.

Johnson, C. (2003). "Enron's Ethical Collapse: Lessons for Leadership Educators." *Journal of Leadership Education, 2*(1), 11. [Online]. Available: https://journalofleadershiped.org/wp-content/uploads/2019/02/2_1_Johnson.pdf.

Lee, A., & Fuller, K. R., (2016). *Ang Lee: Interviews.* University Press of Mississippi.

Liao, Y., & Thomas, D. C. (2020). *Cultural Intelligence in the World of Work Past, Present, Future.* 1st ed. Cham: Springer International Publishing. https://link-springer-com.ezproxy.lib.rmit.edu.au/book/10.1007/978-3-030-18171-0.

Liu, G. M. (2022). *Adam Smith's America: How a Scottish Philosopher Became an Icon of American Capitalism.* American Philosophical Society.

Martin, M., & Vaughn, B. (2007). "Cultural Competence: The Nuts and Bolts of Diversity and Inclusion." *Strategic Diversity & Inclusion Management, 1*(1), 31–38.

OECD. "OECD Convention on Combating Bribery of Foreign Public Officials in International Business Transactions." https://www.oecd.org/corruption/oecdantibriberyconvention.htm (accessed 28.01.2023).

Rapisarda, B. A. (2002). "The Impact of Emotional Intelligence on Work Team Cohesiveness and Performance." *The International Journal of Organizational Analysis, 10*(4), 363–379.

Reddy, A. N. R., Marla, D., Simic, M., Favorskaya, M. N., & Satapathy, S. C. Eds. (2020). *Intelligent Manufacturing and Energy Sustainability* (Smart Innovation, Systems and Technologies Series 169). Singapore: Springer, pp. XLI, 842. [Online]. Available: https://link.springer.com/book/10.1007/978-981-16-0598-7.

Rio Tinto. (2022) "Jadar Project Update." https://www.riotinto.com/en/operations/projects/jadar (accessed 29.01.2023).

Rockstuhl, T. et al. (2011). "Beyond General Intelligence (IQ) and Emotional Intelligence (EQ): The Role of Cultural Intelligence (CQ) on Cross-Border Leadership Effectiveness in a Globalized World." *Journal of Social Issues*, *67*(4), 825–840.

Shandwick, W. (2022). "Reputation Accounts for 63 Percent of a Company's Market Value." *Weber Shandwick*. https://www.prnewswire.com/news-releases/reputation-accounts-for-63-percent-of-a-companys-market-value-300986105.html (accessed 28.01.2023).

Salovey, P., Mayer, J. D., Caruso, D., & Lopes, P. N. (2003). "Measuring Emotional Intelligence as a Set of Abilities with the Mayer-Salovey-Caruso Emotional Intelligence Test." In S. J. Lopez & C. R. Snyder (Eds.), *Positive Psychological Assessment: A Handbook of Models and Measures* (pp. 251–265). American Psychological Association.

Shin, S. J., Morgeson, F. P., & Campion, M. A. (2007). "What You Do Depends on Where You Are: Understanding How Domestic and Expatriate Work Requirements Depend Upon The Cultural Context." *Journal of International Business Studies*, *38*, 64–83.

Shultz, G. P., Taylor, J. B., & Friedman, M. (2020). *Choose Economic Freedom: Enduring Policy Lessons from the 1970s and 1980s*.

Strange, W. C. (ed.). (2020). *The Economics of Agglomeration*. Edward Elgar Publishing.

Todorovic, M., & Simic, M. (2019a). "Feasibility Study on Green Transportation." *Energy Procedia*, 160, 534–541. doi: 10.1016/j.egypro.2019.02.203.

Todorovic, M., & Simic, M. (2019b). "Managing Transition to Autonomous Vehicles Using Bayesian Fuzzy Logic." In Y.-W. Chen, A. Zimmermann, R. J. Howlett, & L. C. Jain (Eds.), *Innovation in Medicine and Healthcare Systems, and Multimedia* (pp. 409–421). Singapore: Springer Singapore.

Todorovic, M., & Simic, M. (2019c). "Transition to Electrical Vehicles Based on Multi-Attribute Decision Making." In *2019 IEEE International Conference on Industrial Technology (ICIT)*, 13–15 February 2019, pp. 921–926. doi: 10.1109/ICIT.2019.8755210.

U.S. Securities and Exchange Commission. (2004). "The Foreign Corrupt Practices Act - Prohibition of the Payment of Bribes to Foreign Officials." *U.S. Securities and Exchange Commission*. https://www.investor.gov/introduction-investing/general-resources/news-alerts/alerts-bulletins/investor-bulletins/foreign-0 (accessed 28.01.2023).

Van der Zee, K. I., & Brinkmann, U. (2004). "*Information Exchange Article*: Construct Validity Evidence for the Intercultural Readiness Check against the Multicultural Personality Questionnaire." *International Journal of Selection and Assessment*, *12*(3), 285–290.

Worldometer. (2023). "Countries in the World by Population (2023)." *Worldometer, US*. [Online]. Available: https://www.worldometers.info/world-population/population-by-country/.

4 Complex Evolving Systems and Iterative Approach to Solving Complex Problems

Lucy Lunevich
RMIT University

4.1 INTRODUCTION

This chapter includes the definition of the complex evolving system and its application to the Engineering Management discipline, as the discipline demands Engineering Manager's professional capabilities, and conceptual and interpersonal skills to translate business complexity into simplicity in order to be effective manager and leader.

This chapter explains:

1. Why systemic approaches are needed;
2. Explaining the distinctions between closed, partially open, and open systems;
3. Giving an introduction to autopoietic systems and key elements involved in the process of evolution.

It offers the definition of simple, complicated, complex, and complex evolving systems (CES) and provides some useful examples of them. Throughout, the language is largely non-technical, at the same time this chapter extensively quotes the relevant standard texts. Some fundamental properties of CES – such as the Second Fundamental Theorem of Thermodynamics – are discussed as well.

Definitions of cyber-physical-social systems (CPSS) have been also included to assist Engineering Manager to navigate increasing complexity and complex projects environment. For the first time, the Engineering Management academic handbook includes the concept of social learning. This is to assist Engineering Managers to effectively deal with various stakeholders, shareholders, and indigenous communities across the globe, where the project required careful consideration of local culture and social systems.

The social learning process described in this chapter outlines the stages of social learning and the appropriate actions required at each stage in order to facilitate positive outcomes of the negotiation. Engineering Manager frequently involves in community consultations, negotiations, mediation, and issues resolutions. Examples of these are negotiations with landowners about land access, access to other resources,

DOI: 10.1201/9781003374879-4

and negotiations with local councils and/or business owners. The process of negotiation could take years and can cause a significant project delay if not followed the
steps outlined in this chapter.

This chapter intends to take the Engineering Management discipline to a new
higher level so that Engineering Managers can meet the demands of the 21st-century
digital economy. It covers a deficit in understanding the complexity that not only is
glaring but also – given the current state of the world – has become patently unacceptable. Furthermore, this chapter is intended to assist higher education in designing new postgraduate programmes, which cover a deficit in the understanding of
complex project environments, which ultimately link to two fundamental concepts
of complex evolving system and social learning.

4.2 SIMPLE, COMPLICATED, AND COMPLEX SYSTEMS

4.2.1 Simple Systems

Understanding the difference between complex and complicated systems is becoming
important for many aspects of management, engineering management, and policy-mak-
ers. Each system is better managed with different leadership, tools, and approaches.

A major breakthrough in how to manage complex multi-stakeholder situations
and programmes has come through the field of systems theory. System theory or
system thinking is a way of helping people to see the overall structures, patterns, and
cycles in systems, rather than seeing only specific events or elements. It allows the
identification of solutions that simultaneously address different problem areas and
leverage improvement throughout the wider system, for instance, an organisation.

It is useful, however, to distinguish between different types of systems. In this
chapter, different systems have been described with some examples relevant to the
Engineering Management discipline and practical situations Engineering Managers
face in complex business environment.

Simple system or simple problem (such as following a recipe or protocol) may
encompass some basic issues of technique and terminology, but once these are
mastered, following the "recipe" carries with it a very high assurance of success.
Examples of this include engineering functional specifications, design, engineering
standards, engineering reports, data, risks assessment reports, modelling data, etc.
For instance, the example of a flow diagram in Figure 4.1 is a typical simple system.
Even pumps, water tank, reactor, valve, and pipes are required hydraulic calculations, and the system is a simple one and predictable.

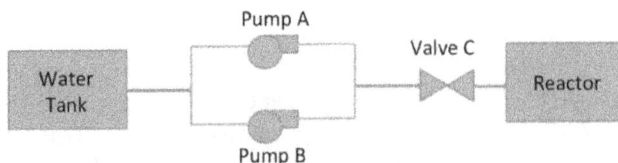

FIGURE 4.1 Flow diagram.

Other examples can be hydraulic calculations or geotechnical reports. If there is a question about a geotechnical report, an Engineering Manager can call a geotechnical engineer to explain and clarify it. If hydraulic calculations are not clear, an Engineering Manager can ask an hydraulic engineer to confirm it. If there is doubt about design standards (for instance, some industry has very specific or highly rigged design standards), Engineering Managers call on an engineering design manager and the client to discuss what level of reliability is required. All of these problems or situations are considered as simple systems or simple problems (Agnew et al., 1996; Ahmad et al., n.d.). Different academic schools have slightly different descriptions of simple systems and key components; however, simple systems are predictable, and something can go wrong if one do not follow the instruction or standards or established proven process.

4.2.2 COMPLICATED SYSTEMS

Complicated system or complicated problem (like sending a rocket to the moon) are different. Their complicated nature is often related not only to the scale of the problem but also to their increased requirements around coordination or specialised expertise. Specialised expertise is a common problem on big engineering projects, which have to be done in remote locations. If this is the case, Engineering Manager will need to rely on own professional network, national and/or global. However, rockets are similar to each other, and because of this following one success, there can be a relatively high degree of certainty of outcome repetition. Some specialised knowledge could be also developed over time dividing them into simple systems, or simplified problems and then consolidating them into new complicated systems.

According to Allen (2023), different leadership styles for different systems are required, for instance, to lead a team operating in complicated systems:

- Role defining – setting job and task descriptions
- Decision-making – find the "best" choice
- Tight structuring – use chain of command and prioritise or limit simple actions
- Knowing – decide and tell others what to do
- Staying the course – align and maintain focus

Complicated systems are all fully predictable. These systems are often engineered (Allen, 2023). These systems can be understood by taking them apart and analysing the details (Allen, 2023). From a management point of view, these systems can be created by first designing the parts and then putting them together. However, a complex system (CS) cannot be built from scratch and expect it to turn out exactly in the way that we intended. CSs are made up of multiple interconnected elements and are adaptive in that they have the capacity to change and learn from experience – their history is important. For instance, the construction of a piece of infrastructure whether it is a bridge, submarine, or airplane. It involves the development and approval of business case, feasibility study, concept design, detailed design, estimation of cost, procurements, commissioning, testing, quality assurance, data transfer, training, and defect validation – all of these interconnected elements, which in fact Engineering Manager must be across with confidence and understanding of each element and its role and impact on the project and project team.

4.2.3 COMPLEX SYSTEMS

In contrast, complex systems are based on relationships, and their properties of self-organisation, interconnections, and evolution. Complex system is constantly evolving and changing. As a result, dealing with a complex system requires developing capability and learning experience as you go. It is not simply experimental learning, which is the process of learning through experience, and it is more narrowly defined as "learning through reflection on doing", but overcoming old beliefs, moving from the comfortable place, and acquiring new capability fast.

Progressive interventions into the complex system require new capability and new knowledge in order to progress to the next level. Next, a new level won't be achieved if no new higher level of capability achieved, no obstacles to overcome. This interactive process, however, allow both parties to adjust, improve, increase resilience, and find new ways of interacting as new problems arise. Problems must be considered as an opportunity to accumulate new higher capabilities as only in this way the challenges can be converted into benefits or project outcomes. Example of this can be a company that joined an alliance (a mega project), but does not have the capacity to integrate into the project team and deal with new financial systems, business culture, or country's culture.

Research into complex systems demonstrates that they cannot be understood solely by simple or complicated approaches to evidence, policy, planning, and management. Some compare complex systems with raising a child or marriage. Formulas have limited application. Raising one child provides experience but no assurance of success with the next. Having unsuccessful marriage relation does not prevent one to have the next successful one. Experience can contribute, but is neither necessary nor sufficient to assure success. Every child is unique and must be understood as an individual. Every relationship is unique and must be understood as an individual case for a specific time and place. A number of interventions can be expected to fail as a matter of course. Uncertainty about the outcome remains. The most useful solutions usually emerge from discussions within the wider family and involve elaboration on values. Some members of the family might change their view and some might learn about the state of problems.

Management implications and processes involved in this content are significant. These differences have important implications for management and, in particular, for engineering management because it requires a fundamentally different approach to risk, people, communication management, and the business model approach to deliver successful projects.

Examples of complex systems include ourselves (human beings), the stock market, ecosystems, immune systems, and any human social-group-based endeavour in a cultural and social system and project team and organisation. It applies to Joint Ventures, Strategic Alliances, and Partnerships, frequently developing relationships; developing new business models as a result of alliances has strategic importance for companies, especially, one working on a global scale, across cultures. In fact, the data obtained from the last 20 years from the different Strategic Alliances and Joint Ventures indicate that learning experience, know-how is more important than temporary difficulties business faces while working within the Alliance.

CS defies attempts to be created in an engineering effort, and the components in the system co-evolve through their relationships with other components. But some understanding can be achieved by studying how the whole system operates, and the system can be influenced by implementing a range of well-thought-out and constructive interventions.

Getting people to work collectively in a coordinated fashion in areas such as poverty alleviation or catchment management, mega project environment, or joint venture setting is therefore better seen by agencies as a complex, rather than a complicated problem. In fact, many managers are happy to acknowledge it, but somehow this acknowledgement does not translate into different management practice and leadership styles. In most cases, the functional leadership style is required to navigate a complex environment, to lead the teams and to deliver key performance indicators (KPI).

Indicators of progress in managing a complicated system are directly linked through cause and effect. However, indicators of progress in a complex system are better seen as providing a focus around which different stakeholders can come together and discuss, with a view to potentially changing their practices to improve the way the wider system is trending. In many cases, people continue to refer to the system they are trying to influence as if it were complicated rather than complex, perhaps because this is a familiar approach, and there is a sense of security in having a blueprint, and fixed milestones. Furthermore, it is easier to spend time refining a blueprint than it is to accept that there is much uncertainty about what action is required and what outcomes will be achieved if other people see problems from very different perspectives. Put it simple, it is called "growth on job".

Allen suggests (1988) different leadership styles in order to manage and lead people in complex evolving environment and compares them with complex adaptive systems:

- Relationship building – working with patterns of interaction
- Sense making – collective interpretation
- Loose coupling – support communities of practice and add more degrees of freedom
- Learning – act/learn/plan at the same time
- Notice emergent directions – building on what works (Allen, 1988).

SOME REFLECTIONS FOR YOU:

1. What is the difference between complex and complicated systems?
2. Would you consider a business process improvement is a simple problem, or complicated problem or complex problem?
3. Do you think different management and leadership styles are required to successfully lead these various processes? How would you know which one is right for a situation?

4.3 SYSTEM THEORY

Systems theory starts by distinguishing between systems and non-systems. This is for the purpose of logic. The actual evolution of life on Earth since the origin of the Solar System, is a process in which unordered matter has organised itself. Chaos systems have been re-evolving into ordered structures of mutual interactions (Weis, 2008; Arthur, 1990, 1994). The Genesis (Emergence) of Systems – of increasingly complex organised entities – occurs under specific conditions, i.e., by way of specific processes occurring in unordered (chaotic) non-systems (Allen, 1988). A simple example for the Genesis, Emergence, or Origin, of a System is the Formation of a Sand Pile out of a continuously increasing set of individual Sand Corns. Let us take an experimental setting in which there is a circular plane and a device that permits dropping individual sand corns into the centre of this plane from a pre-determined height. Initially, the first few sand corns dropping on the plane will disperse and settle in various positions without touching one another – a single corn of sand does not make a sand pile, and several of them only form a set. (Ayres, 1994; Weis, 2008). With a continuous increase in the number of elements in this set, they will enter into ever closer relation with one another (Baccini, 1991; Bak, 1991). Once the number of sand corns reaches a critical density physical forces (gravity, frictional resistance) generate spontaneous interactions and relations among the elements which generate the formation of a sand pile, i.e., a particular spatial structure (Vester, 1983; Weis, 2008).

Originally, the elements of sets (which may potentially also be systems themselves) are separate from one another. Once a rising number of these elements begin to enter into close mutual relations of cause and effect, this may trigger the emergence, or genesis, of a new system of higher order (Vester, 1983; Weis, 2008). Thus, individual particles (atoms) may form a molecule; cells form an organism; and the interaction of animals, plants, and microbes generates an eco-system (Weis, 2008). When many small parts, elements, or systems, come together, they may either generate a Set – in which they remain separate from, or side by side with, one another – or some larger System (Vester, 1983; Weis, 2008).

Some argue that humans and the artificial systems they generate on this planet (such as roads, settlements, factories, mines, and land used for agriculture) were relatively spaced from one another for a long period of time (Vester, 1983; Weis, 2008). With small populations distributed over a vast terrain initially, and for a prolonged period of time, there was but little interaction among these systems far enough away from one another. With increasing population and density, however, these artificial systems have come into a close range from one another (Weis, 2008). This, in turn, has generated a wide variety of physical, chemical, energetic, and social interactions among them, between them and human populations, and between them and the biosphere (Vester, 1983; Weis, 2008). These mutual interactions have generated new systems overarching them, the system of human civilisation on Earth (Weis, 2008). According to Weis (2008), such a system need not be stable, i.e., sustainable, by necessity – the individual parts may affect one another in ways that may eliminate the system and all partial systems that are linked to one another within it. In close analogy with this, the evolution of human civilisation on Earth has generated new systems of mutual relations of cause and effect – systems that are characterised by exponential growth and, by necessity, increasing density and an increasingly global network of interactions. Any system needs stability to function, and a system always interacts with the surrounding and will evolve continuously.

As long as the relations and interactions among the elements within a set are negligible, the entity is NOT a system (Vester, 1983; Weis, 2008). Supersaturation of elements leads to a transition to a system. This occurs when a certain critical state is reached (Weis, 2008), in which the mutual interactions among the elements lead to a process, in which the set of elements in question organise themselves as a whole, in the form of a new entity. An ordered structure of interactions becomes a system. This new, or emergent, systemic order behaves totally different from the way in which these elements did before: Within the emergent system, they have become parts, or components, of the system (Weis, 2008):

> The components and parts of a system are linked with one another in a web of interactions, depend on one another and, in doing so, form a complex unified Whole or a new Entity. Each such combination of parts into a new whole, an individual entity, not only possesses certain Collective.

(Weis, 2008; Vester, 1983)

Weis (2008, p. 32) states:

> properties which result from the sum of its components. Instead, a System exhibits completely New or so-called Emergent Properties. These properties are specific Systemic Properties and Behavioural Characteristics, which DO NOT result from the properties and the behaviour of the individual parts of the system.

According to Weis (2008, p. 32) "a system is always an entity, or an integrated Whole, the properties of which cannot be reduced to the properties of smaller parts". Therefore "the behaviour of a system, cannot be explained by studying its individual parts, or by the collective sum of the individual properties of these parts" (Weis, 2008). And that, in turn, does not imply anything else but that a system – while it may consist of many parts – is a separate individual (Weis, 2008).

All real systems are more or less hierarchically organised entities that exist on various – more or less complex – levels of organisation or degrees of complexity (Vester, 1983; Weis, 2008).

> In hierarchically ordered or organized systems, each level subsumes all lower levels within itself, at the same time that the parts or components of a system may be systems themselves.

> Hierarchically organized systems are, from their very beginning, themselves systemic components of the total spectrum of existing systems and, hence, parts of higher ranking systems with which they are connected.

(Weis, 2008)

Thus, all systems exist within higher-ordered systems which, themselves, are once again parts of more comprehensive systems (Weis 2008). Each level of organisation has its own specific, distinctive, and generally valid, characteristics. This means that not all properties of a more highly organised system can be deduced from the properties of systems of a lower degree of organisation (Weis, 2008).

It is not possible to predict the properties of water from the molecular properties of hydrogen or oxygen. Similarly, the specific properties of ecosystems cannot be predicted from the knowledge one may have of isolated populations within it.

(Weis, 2008)

Weis (2008, p. 33) continues that "systems which emerge from the combination of components or component entities (systems) are entities which exist at a higher level of organization or complexity than did the individual components before their combination". On each new and higher level of complexity or organisation, systems display completely new or emergent properties that either did not exist at the previous lower level of organisation or were inconspicuous (Weis, 2008).

Such newly emergent properties of particular levels, or entities, of organization result from the functional interaction of their components.

(Weis, 2008)

By studying isolated or detached components, without taking their mutual interaction into account, it is impossible to predict the specific properties of the more highly organised entity (Weis, 2008; Karcanias, 2020). In similar way, by studying a single department within the organisation, it is impossible to define a specific improvement need for the department to perform better, more productive, and efficient.

It is in the nature of systems that they cannot be described by the sum of individual properties.

Cartesian science argued that, with each complex system, the behaviour of the whole system could be analysed by way of analysing the properties of its individual parts.

(Weis, 2008)

However, an intervention in this kind of system can produce a new system depending on the scale of intervention. For example, injection of substantial cash into the specific community can result in a different community? In this case, the result of intervention(s) can be assessed using various methods and public policy. The methods include impact evaluation, randomised controlling trials (RCT), quasi-experimental designs, statistical significance, theory of change, logic models, and process evaluations.

4.3.1 SYSTEMS SCIENCE

Systems science determines that systems cannot be understood by way of analysis.

The properties of the parts of a system are not properties which inhere in themselves, but can only be understood within the context of the larger whole of which they are part.

(Weis, 2008)

According to Weis (2008, p. 34) and others, the system problem is essentially the problem of the limitations of analytical procedures in science.

Analytical procedure means that an entity investigated be resolved into, and hence can be constituted, or reconstituted from, the parts put together, these procedures understood both in their material and conceptual sense.

(Weis, 2008)

Systems have characteristic patterns of organisation, i.e., a specific network of (self-) organising relations of their components, a specific configuration of ordered processes or relations that are mutually linked to one another. The pattern in which these processes and relations are organised is characteristic, and specific, for a particular class of systems at each level of organisation (Dickerson, 1978; Vester, 1983).

Systems possess, therefore, characteristic Patterns of Relations or Organization, a specific Network of Mutual Interactions, and Mutual Relations of Cause and Effect, which are characteristic for the Structure of the system in questions and inhere in it.

(Weis, 2008)

4.3.2 SELF-ORGANISED SYSTEMS

These systemic patterns or organisations are either determined exogenously or organised by the system itself (self-organised) (Weis, 2008), (Bak, 1991). They are different, and specific, for each individual system (Bak, 1991). Therefore, systems differ from one another by the specific habits and styles in which they are organised, or in which they organise themselves.

Systemic properties are properties of a specific pattern – this pattern is destroyed when the system gets dismantled into elements that are isolated from one another, so there is no point to study specific parts (Weis, 2008).

While the components of the original system are still there, in such a case, the specific configuration of the relations among them – the pattern of their (self-) organization – is destroyed, and therefore the system dies.

(Bak, 1991)

According to Weis (2008) and others systems essentially differ from non-systems because they are (self-) organised, because their parts, or components, are linked in a Web of Relations and are organised in a specific configuration. System has a specific internal structure in which these components form an ordered structure of mutual interaction (Bak, 1991; Barnet, 1975). They differ from non-systems in that their components exhibit specific patterns of integration (interconnectedness, networking) and organisation, specific structures, or configurations of cohesion (coherence, mutual relations, contiguity), both internal and external reciprocal interactions (interdependencies) and organising relation (Baccini, 1991; Bernstein, 1981).

SOME REFLECTIONS FOR YOU:

1. Consider examples of systems and non-systems?
2. Would a pile of tomatoes represent systems or non-systems?
3. Would Joint Venture, Strategic Alliance, or Partnership represent systems or non-systems?

4.4   COMPLEX EVOLVING SYSTEMS

4.4.1   CLOSED AND PARTIALLY OPEN, SYSTEMS

In a closed system, a given quantity of free Energy in a particular form transforms itself (Bak, 1997).

> In the process of its transformation, irreversibly into an equal quantity of bound, but disordered Energy – free Energy "dissipates" into the total system within which it was transformed. In a closed system, this process irreversibly increases Entropy, the share of energy within the system which is no longer freely available but bound and disordered.
>
> **(Bak, 1991)**

In the long run, any such system must, by necessity, tend towards a thermodynamic equilibrium, and disintegrate: The so-called Entropy of an isolated system can only increase to the point at which the system has reached its thermodynamic equilibrium (Bertalantein, 1950; Bertalanffy, 1968). The concept originates from Ancient Greek and implies as much as "Self-Creation", Self-Production, or Self-Regeneration and Self-Rejuvenation (Bertalanffy, 1968; Weis, 2008).

> Self-organization is only possible when the distance of the system from equilibrium passes certain critical thresholds.
>
> **(Weis, 2008)**

It only occurs when systems are in states that are far from equilibrium, and its occurrence is bound up with discrete transitions (Bak, 1991; Weis, 2008). Processes of self-organisation are frequently made up of sequences of kinetic transitions which, with increasing distance from equilibrium, occur under certain parameter values (Vester, 1983). This implies analogies to phase transitions from one particular state of equilibrium to another (Weis, 2008).

4.4.2   THE QUALITY OF ENERGY

It may suffice here to explain "the complex concept of entropy by defining it as a measure fort that part of energy which is not freely available and cannot be converted into a directed flow of energy, or work" (Bertalanffy, 1968). Entropy is a measure of the quality of the energy within a system. In distinction to a mechanical description, this introduces irreversibility (non-reversibility) or directedness of temporal processes as a characteristic of such systems (Weis, 2008). According to Weis (2008) each future macroscopic state of an isolated system can only display equal or higher entropy; every past state must be characterized by equal or lower entropy than the current state. A reversal of any particular state is impossible (Bertalantein, 1950; Bilsky, 1980).

All irreversible processes generate entropy, as a result the energy level of any bounded physical or chemical system decreases with time as the system loses energy to its surroundings (Bak, 1997). An example of this is the organisation or country. If an organisation does not employ new staff, from outside or the country does not have sustainable immigration law, then this system loses energy and stops renewing itself. In other words, such a system spontaneously changes from a higher to a lower energy state

(Bilsky, 1980). For example, all physical and chemical processes lead to transformations that release energy; those that require energy are highly unlikely. The oxidation of a carbohydrate – for example, the burning of a piece of paper – releases energy in the form of light and heat, and the products of this oxidation (carbon dioxide and water, in this case) contain less energy than the reactants (oxygen and carbohydrate) (Brown, 2006). Physical systems also dissipate energy, some of which are transferred to their surroundings.

4.4.3 ISOLATED OR CLOSED SYSTEMS OR CONSERVATIVE SYSTEMS

Isolated or closed system is a system without an environment. This type of physical system is called equilibrium systems.

> An equilibrium system is defined as a system which either has already reached a thermodynamic equilibrium (maximum entropy, disorganization, and disorder) and, therefore, is in equilibrium, or is as yet on its path towards that state.

> **(Weis, 2008)**

In the latter case, "the dynamics of the system are already oriented towards the equilibrium to be reached – equilibrium systems irreversibly move towards such a thermodynamic equilibrium" (Bilsky, 1980). This state of the system is called devolution, which is contrary to evolution.

> Devolving, or Equilibrium Systems are called Conservative Systems. Systems which Conserve their Structure.

> **(Weis, 2008)**

Such systems are distinguished from the class of the so-called evolving systems that include all biological systems (also can be countries, financial markets, or economies) (Butzer, 1996; Clark, 1986; Cohen, 1995). This is a reason why re-reorganisation of the old traditional systems is necessary. The process, however, is painful. Nevertheless, the process of creative destruction, which is universal, does just this – ruthlessly eliminates low intelligent systems with higher intelligence systems:

> The more realistic case is that of a partially open system under conditions which are such that it tends to its equilibrium in a similar way (the materially closed, yet energetically open, system of a sand clock, or the disintegration of an-organic and organic structures under the influence of physical and chemical environmental influences).

> **(Weis, 2008)**

Once it has reached that state, any exchange with the environment ceases (Clark, 1986; Cohen, 1995; Daly, 1976). In contrast to systems maintaining a given structure, evolving systems are open and, therefore, "far from thermodynamic equilibrium – such systems are so-called Nonlinear Disequilibrium Systems" (Weis, 2008).

4.4.4 OPEN SYSTEMS

Open System is an alive system. Open Systems are capable to continuously import free energy (in the form of light or other forms of potential energy, such as biomass, electricity, or fossil fuels) from their environment (Vester, 1983). Figure 4.2 shows

FIGURE 4.2   Open system.

an indicative model of an open system. It is capable of exchanging energy with its surroundings. This enables them to be an alive system.

At the same time, during this work process, these systems transform free energy into other forms of energy, all the while increasing entropy within the system (Bak P., 1997; Bak P. C., 1991). Natural systems such as organisms, populations, or ecosystems, however, are capable of generating and maintaining a high degree of internal order (and, therefore, a state of low entropy) (Bak, 1991). They do so by exporting energy forms that can no longer be used and, therefore, are no longer available or disposable, as potential energy within the system by way of "respiration" (Weis, 2008). In contrast to isolated systems, therefore, entropy within the system need not increase by necessity: It may remain stationary or may decrease, with the adjustment process being attained by way of exchange with the environment (Bak P., 1997; Bak P. C., 1991). In this case, what applies is the general extension of the Second Fundamental Theorem of Thermodynamics, according to which the change of entropy within a given system, dS, is the sum of entropy produced by irreversible processes within the system, diS, and the flow of entropy induced by exchange with the environment,

$$deS : dS = diS + deS.$$

The theorem maintains that the internal component diS – just like with an isolated system – can only be either positive or zero, but never negative (diS ≥ 0). The change of the flow of entropy between the system and its environment deS, however, may be either positive (import of entropy from without, or "immissions") or negative (export of entropy, or "emissions"). The total change in entropy within the system, therefore, may be positive (diS $\geq$ 0, deS > −diS), remain stationary (diS $\geq$ 0, deS = −diS), or may diminish (diS $\geq$ 0, deS $\leq$ −diS). If the system, as a whole, is to remain in equilibrium, imports and exports must be balanced (deS = 0), and at the same time, total entropy within the system must remain constant (dS = 0). This, however, is only possible if the production of entropy within the system itself stops (diS = 0). For this to happen, the system must be in thermodynamic equilibrium, i.e., must have stopped "working", and no longer maintains any transformation processes: Strictly speaking, such a system is "dead" (Daly, 1993; Bilsky, 1980; Weis, 2008).

This implies, by necessity, "that open systems can only be maintained, on a continuing basis, in states that are far from thermodynamic Equilibrium or in Disequilibrium, they must maintain relations of exchange with their environment" (Vester, 1983). At the same time, exchange with the environment can only be sustained, "if an internal state of Disequilibrium is sustained" (Vester, 1983). In a state of thermodynamic equilibrium, all processes end.

4.4.5 No Open System

No open system – "hence, no organism, and no single biological system" – is capable of existing by itself or without its environment (Bak, 1991). Open systems are systems which, by necessity, must continuously maintain relations of exchange with their environment, and the systems continuously regenerate themselves (Baccini, 1991).

Closed systems do not exist in reality – in reality, all systems are open and cross-linked to others. Closed systems only exist as a theoretical possibility and are usually used for the purpose of research and learning. In a theoretical field, frequently assumptions will be made that the system is a closed system for the purpose of study of specific parameters of this system. This allows the isolation of some variables and analysis of theoretical data obtained as a result of the study and development of a hypothesis. A hypothesis then can be studied to better understand the system itself.

> Living systems are, therefore, first of all, open Systems which maintain relations of mutual exchange, and develop – on each particular level of organization – characteristic Functional Systems.

> **(Weis, 2008)**

All biotic components of the biological spectrum only become real, living, biological systems, or bio-systems only because they take in, and process, abiotic components, material, and energy (Daly, 1976). The same is true for all anthropogenic social systems – they only become the real living systems which they are, by continuously maintaining relations of mutual exchange with their environment.

Living beings are complex evolving systems that can only keep themselves alive by maintaining a continuous inward flow of material and energy from their environment (Diamond, 2005; Daly, 1976). They cannot live by themselves or without their environment (Dickerson, 1978), they are inseparably connected with their environment, and they influence one another (Bak, 1997). Living beings or complex living systems, and therefore, they maintain some form of metabolism (Dickerson, 1978) (Edson, 1981).

Human beings, other living systems, and all complex living systems, therefore, are
systems which continuously transform.

(Vester, 1983)

Sometimes, changes are not easily observable within the period of human life.
Hence, they are not static or unchanging structures of components that are
arranged in some spatial order or structure, and they do not maintain relations of
mutual interaction. Instead –as already mentioned – "they are in fact Structures
of Processes in which certain forms of energy are transformed into other forms:
Process-Structures" (Weis, 2008).

4.4.6 AUTOPOIESIS AND EVOLUTION

As has been demonstrated above, "human beings – just like any other living biological systems – and anthropogenic social systems are open dynamic systems, disequilibrium systems (or disequilibrium structures)" (Weis, 2008). In order to survive,
they must, by necessity, "maintain continuous interaction with their environment"
(Weis, 2008; Vester, 1983), including cultural and social interactions.

The preliminary condition for the continuous dynamic existence of such
disequilibrium structures is that they are "partially open with respect to their
environment" (Weis, 2008) and that they maintain some macroscopic systemic
state which is far from equilibrium. According to Weis (2008) "thermodynamic
equilibrium is equivalent to cessation, standstill, shut-down, and death". The high
degree of disequilibrium – which is required to sustain the self-organising processes at work within the system (as well as between the system and its external
environment) – is maintained by the sustained maintenance of the exchange of
material and energy with the external environment – in other words, by way of
metabolism (Edson, 1981; Fischer-Kowalski, 1993).

> The dynamic of such a globally stable, but never inactive, structure was called Autopoiesis
> (Self-Production or Self-Regeneration). An autopoietic system strives, in the first instance,
> not after producing some form of output, but after continuously maintaining and regen-
> erating itself in the same process structure.

(Bak, 1997)

Under certain circumstances, such systems also generate new, or emergent process
structures (Vester, 1983). In such systems, "constitutive (anabolic) and decompos-
ing (catabolic) processes are continuously at work simultaneously" (Vester, 1983).
In doing so, these systems not only dissolve their evolution, but also their temporary existence within a particular structure, into processes (Weis, 2008). In the
sphere of life, there is little that remains solid and unchanged (Bak, 1991; Weis,
2008). An autopoietic structure is a structure of transformation processes, or a
process structure or a result of many processes (Bak, 1991).

Nevertheless:

> autopoietic systems (or, structures) are not only geared do reproducing their partic-
> ular given structure and, by way of doing so, reproducing themselves: Under certain

circumstances, they are also capable of rebuilding (reconditioning, regenerating, reintegrating, remaking, renewing, renovating, replacing, reviving) themselves.

(Vester, 1983)

They are able to change, to evolve and "to spontaneously generate new (emergent) properties or process structures in this process" (Bak, 1991).

4.4.7 COMPLEX EVOLVING SYSTEM

Open systems which are in states far from "thermodynamic equilibrium, and which evolve through an open sequence of structures therefore, are logically called complex evolving systems" (Bernstein, 1981). "They are coherent systems, the structure of which does not remain unchanged, but changes in a coherent way. All biological organisms, communities, and ecosystems, are dynamic, self-organizing, autopoietic, coherent, or complex, evolving systems" (Vester, 1983) – in the very same way as the complex anthropogenic systems so beloved in economics, sociology, or political science: Households, firms, governments, oligopolies, networks, markets, economic systems (regimes), and other niceties. Complex evolving systems are open disequilibrium systems of a particular kind which – in contrast to conservative systems geared to conserve a particular structure – maintain so-called dissipative self-organization, and are generally called dissipative systems (or structures) (Weis, 2008; Vester, 1983).

As dissipative structures, complex evolving systems produce entropy which, however, does not get accumulated within the system. Instead, such entropy is part of a continuous exchange of energy with the external environment (Bak, 1997; Weis, 2008). According to Weis (2008), by maintaining this continuous exchange of material and energy (metabolism), the system maintains its internal disequilibrium – and this very same disequilibrium maintains the processes of exchange that it requires to survive. In doing so, dissipative structure continuously regenerates themselves, and maintains a specific dynamic regime (Alexander, 2022), a globally stable space-time-structure. Such structures seem to be exclusively concerned with their own identity and self-regeneration (Bak, 1997; Weis, 2008):

> Hence, dissipative systems are not characterized by the static measure of the amount of entropy which, at a particular moment, is within the system: Instead, what is decisive is the dynamic measure of the rate at which entropy is being produced within the system, and the rate at which it maintains exchange with its environment.

(Weis, 2008)

Thus, the crucial parameter characterising dissipative systems is the intensity of their throughput, as well as turnover, of energy.

Dissipative structures display two kinds of behaviours: When they get close to a state of (thermodynamic) equilibrium, their internal order gets destroyed (just like that of closed, or isolated, systems) (Weis, 2008). When they are in states that are far from equilibrium, they maintain ordered structures by way of instabilities and fluctuations (exogenous, or endogenous, shocks), out of which new (emergent) order may evolve (coherent behaviour) (Bertalantein, 1950).

Dissipative structures that, at first, emerge spontaneously and transcend the original thermodynamic order do not constitute the ultimate end of evolution (Weis, 2008). As long as they maintain exchange of energy with their environment, and as long as the fluctuations which occur (emerge, appear, arise, develop) are absorbed within the limits (scope, bounds) of the dynamic regime in question, the structure is stable, in principle (Karcanias, 2020; Weis, 2008). Though, no structure of a disequilibrium system is stable in and by itself. Each system may be forced, or driven, over a point of instability, into a new regime, once the fluctuations exceed certain critical thresholds (Karcanias, 2020). This, in turn, corresponds to a qualitative change of the dynamic regime of the system in question.

> The transition to a new dynamic regime renews the capacity of the system in question to produce entropy – a process which may be associated with life in the widest sense of the term. The spontaneous creation of new forms of order, or emergent (process) structures occurs only under specific conditions.
>
> **(Weis, 2008)**

> In the case of the most simply hydro-dynamic or chemical dissipative structures, their evolution can be precisely specified and formalized – as a matter of principle, they equally apply to all such structures that are more complex. The transition from a laminar flow to a turbulent flow when one turns up a faucet, the emergence of new macroscopic phenomena of order such as the Bénard-instability in certain liquid systems, or dynamic phenomena in certain chemical reaction systems of the Belousov-Zhabotinsky (BZ) type, requires openness toward the exchange of energy and material with the environment, a state far from (thermodynamic) equilibrium, and both auto- or cross-catalytic processes, and/or auto-catalytic self-augmentation of certain process stages.
>
> **(Weis, 2008)**

Thus, there are two factors that are decisive for the evolution of systems: (1) The intensity of their throughput of energy, material, and information must increase and transcend certain critical thresholds; and (2) The systems must have auto- and/ or cross-catalytic components and/or processes – this implies that either the systems at large grow, or they have certain components which grow (Weis, 2008; Vester, 1983).

Living, or complex evolving systems continuously transform energy and are characterised by chains or networks of processes that transform energy. In doing so, they work or produce in certain ways (Weis, 2008). The development of highly ordered structures is always dependent on high-grade forms of energy with part of the imported energy always transformed into some specific high-grade form (Weis, 2008). Such systems, therefore, are not static, or invariant, structures of components that are configured in a specific spatial order and do not maintain interaction with one another. Instead, as already mentioned, they are structures of processes, in which specific forms of energy are transformed into other forms – hence, they are process structures (Weis, 2008; De Santo, 2020). The evolution of such systems, therefore, implies the transformation of existing structures of processes in which energy is being transformed (De Santo, 2020).

Such spontaneous, self-organising transitions to new process structures are dynamic processes that vary among systems depending on their degree of organisation and complexity (Weis, 2008). The most elementary case is the case of the so-called equilibrium phase transitions taking place in physical systems (Weis, 2008; Zattoni, 2020). Water changes its aggregate state (frozen-liquid-gaseous), depending on temperature, and on reaching certain critical thresholds – something that is valid for a host of chemical elements and groups of materials. Hydrodynamic systems change their structure, depending on the quantity of throughput (laminar-turbulent flow), or temperature (Bénard-Convection), chemical systems of the BZ type evolve because of auto- and cross-catalytic processes that are triggered by the addition of new elements that participate in the reaction (Weis, 2008; Zattoni, 2020; Karcanias, 2020).

In dissipative physical and chemical systems, the (continuous, or discontinuous) transitions from one macroscopic state of order to another one occurs because of changes in the throughput of energy and material which are exogenously caused (Weis, 2008; Karcanias, 2020). In contrast to this, the process structures of dissipative physical and chemical systems (d)evolve, when certain parameters of order which are specific to the system in question exceed (or fall below) certain critical thresholds (Weis, 2008; Karcanias, 2020). This is caused by an increase, or decrease, of the throughput of energy, material, and/or information (Heylighen, 1997; Weis, 2008; Zattoni, 2020; Georgiev, 2022).

Kauffman (1993, 1995) and others, who have extensively studied, and documented, the role of natural selection and the spontaneous emergence of order in self-organising systems (Kaufmann, 1993, 1995). In this approach, the evolution of Life is conceived as a continuous evolution of increasingly more complex process structures which occurs spontaneously, and by way of self-organisations. Darwin reduced the sources of the overwhelming and beautiful order which graces the living world to a single singular force: natural selection (Georgiev, 2022; Weis, 2008). This single-force view fails to notice, fails to stress, and fails to incorporate the possibility that simple and complex systems exhibit order spontaneously (Zattoni, 2020). That spontaneous order exists, is hardly mysterious (Weis, 2008; Cang, 2017; Justice, 2012). The non-biological world is replete with examples, and no one would doubt that similar sources of order are available to living things:

> What is true for dissipative physical and chemical systems is also true for biological systems– whether it is human beings, complex anthropogenic systems (such as cities, markets, or other complex evolving systems, the (continuous or discontinuous) transition from one macroscopic state of order to another one occurs by way of changes in the throughput of energy or matter.

Such changes may result from random exogenous shocks as well as endogenous causes (Weis, 2008; Heylighen, 1997).

> The evolution of process structures of biological systems occurs, when they are driven over a threshold into some new dynamic regime by fluctuations which exceed (or fall below) certain critical reference values.

(Weis, 2008)

This is the case, when certain parameters of order – "which are specific to the system in question – exceed (or fall below) critical thresholds as a result of an increase (or decrease) in the throughput of energy, matter, and/or information" (Vester, 1983). Under such circumstances, the system in question is incapable of absorbing any further instability, shocks, or fluctuations (caused exogenously or endogenously).

> The fluctuations which we are talking about here in no way refer to concentrations, or other macroscopic parameters, but to fluctuations in the mechanisms which result in modifications of kinetic behaviour (such as rates of reaction, or diffusion).

(Weis, 2008)

Such fluctuations may hit the system, more or less at random, from outside by way of adding new participants in reactions, or by changing the quantitative relations within the existing original reaction system (Weis, 2008). On the contrary, they may be generated within the system itself, by way of positive feedbacks, which – in this case – is called evolutionary feedback: (1) Instability formation of a new dissipative structure; (2) Increase in the production of entropy critical threshold; and (3) Instability formation of a new, emergent, dissipative structure (Weis, 2008; Zattoni, 2020).

As a general rule, the evolution of complex evolving biological systems involves the transition to more complex dissipative regimes characterised by higher rates of energy throughput, production of entropy, increasing complexity, and increasing volume of metabolic processes (increase in the intensity of work) (Weis, 2008):

> The spontaneous formation of new forms of order, or emergent (process) structures in dissipative systems results from fluctuations and instabilities which can no longer be absorbed, and are caused by random exogenous shocks, or endogenous growth (auto-, and cross-catalytic processes, positive feedback).

(Weis, 2008)

The decisive parameters of control for the particular systems in question, therefore, are:

> (i) the Growth rates of the mass of energy, matter, and information, which these systems convert in their metabolism, and (ii) the growth in the extent (volume, turnover) of these systems (by way of increasing numbers of systemic component, participants in reactions, or of organisms in populations).

(Weis, 2008)

4.5   CYBER-PHYSICAL-SOCIAL SYSTEMS

It is the overriding trend of the present-day world that traditional systems and mobile devices are currently transforming into intelligent systems and smart devices. Against this backdrop, cyber-physical systems (CPSs) and Internet-of-Things (IoT) emerge as the times require. To achieve the parallel interactions between the human world and the computer network in real time, IoT along with wireless mobile communication and computing opens up some future opportunities as well as challenges for constructing a novel cyber-physical-social system (CPSS) that takes human factors into account during the system operation and management.

Many novel technologies and cyber innovations applied to physical control system infrastructures such as power, water, wastewater, and dams are about to change the world, not just for short-term profits but also for broader societal interest. As a result, society will be more productive.

The Grand Ethiopian Renaissance Dam (GERD) is an example of constructing one of the largest physical control system infrastructures today, creating a reservoir on the Blue Nile to supply electricity to Ethiopia and support the country's societal development (Scalco, 2022). GERD has also created discord with neighbouring downstream countries Egypt and Sudan, whose economies depend on the Blue Nile (Scalco, 2022). Internet Protocol (IP)-connected control systems of the GERD infrastructure make operations vulnerable to cyber failure. Uncertainty of agreement among professionals about cybersecurity and system security engineering coupled with unpredictable social and political developments and uncertainty of the technologies themselves make these systems and societies interacting with the systems more vulnerable to disruption. Clearly, as society becomes more technologically advanced, risk will also increase dramatically. In this content, it must be stated that the more advanced society is the higher risk of failure of the control systems.

There are many ways advances in systems sciences and digital engineering converge with the systems engineering (SE) discipline (Scalco, 2022). Such advances require cultural and systems engineering process and methodology changes and advance to support the digital transformation of physical systems into higher level technologically sustainable systems. Societal benefits and vulnerabilities to these advances illustrate the value of social systems engineering models and methodologies for measuring the uncertainty of agreement for achieving control physical-social system multi-concern assurance (Scalco, 2022). New approach to risk assessment is required to assess holistically the level of integration of the CPSs.

4.6 SYSTEM THINKING AND SOCIAL LEARNING PROCESS

4.6.1 System Thinking as Approach for Integration

According to Gharajedaghi (2012) and others, systems thinking is an approach to integration or managing conflict between different parts of the system, which might have different understandings of matters and priorities. This is based on the belief that the component parts of a system will act differently or event against each other when isolated from the system's environment or other parts of the system. Moreover, systems thinking concerns with an understanding of a system by examining the linkages and interactions between the elements that comprise the whole of the system (Justice, 2012; Gharajedaghi, 2012).

Systems thinking in practice encourages us to explore inter-relationships (context and connections), perspectives (each actor has their own unique perception of the situation), and boundaries (agreeing on scope, scale, and what might constitute an improvement) (Justice, 2012; Lind, 1988).

Systems thinking is particularly useful in addressing complex or wicked problem situations, managing chaos, or translation complexity into simplicity and developing new capabilities. These problems cannot be solved by any one actor, any more than a complex system can be fully understood from only one perspective (Justice, 2012; Heylighen, 1997). It requires a team of people who can learn and adapt as they go. This is where the social learning process explains the steps necessary to be taken. Moreover, because complex evolving systems are continually evolving, systems thinking is oriented towards organisational and social learning.

4.6.2 SOCIAL LEARNING

Social learning is the process whereby people becomes actively involved in the developing of mutually acceptable solutions to a problem or decisions that affect their community or company (Justice, 2012). The social learning approach can be broken into two components – cognitive enhancement and moral development. Cognitive enhancement involves participants gaining technical competence and learning about collective values and preferences (Justice, 2012) (Diamond, 2005). The second component, moral development, involves the ability of individuals to make judgements about right and wrong and setting outside self-interest; the ability to take on the perspective of others; developing moral reasoning and problems-solving skills; and learning how to integrate new cognitive knowledge into your own opinion. Learning how to cooperate with others to solve common problems (Justice, 2012; Diamond, 2005).

The social learning process consists of the following steps and is applicable to complex evolving situations, for instance, community learning about the state of environmental problems, project team learning about specific challenges the project face (budget, time, scope, risks), organisational restructure involving several departments, etc.) (Justice, 2012; Karcanias, 2020). According to Justice (2012), the social learning approach to people participation can be broken into two components – cognitive enhancement and moral development (Justice, 2001). Cognitive enhancement involves participants gaining technical competence and learning about collective values and preferences including (Justice, 2012):

- Learning about the state of problem
- Learning about the possible solution
- Learning about other people's interests or groups in the problem
- Acknowledging your own interest in the problem
- Learning about the communication methods required to achieve agreement with the group.
- Practicing integrated thinking about the problem (incorporating all of the above).

The second component, moral development, involves the ability of individuals to make judgements about right and wrong and setting aside self-interest (Justice, 2001). This would involve:

- Developing a sense of self-respect and responsibility to self and others, regardless of how these may affect one's own personal interests or values, and acting accordingly
- The ability to take on the respective of others

- Developing moral reasoning
- Developing problem-solving skills
- Developing a sense of solidarity with the group
- Learning how to integrate new cognitive knowledge into your own opinion
- Learning how to cooperate with other to solve common problems (Justice, 2012).

The above looks a simple, but it is not. In the project environment, it might take a time to navigate these processes if one is not aware of many channels of communications between people (Justice, 2012).

When Engineering Manager encounters situations, which are complex and messy, then systems thinking can help Engineering Manager understand the situation systemically and apply the correct approach, especially in a multiple project environment where time and resources are limited (Diamond, 2005). This helps Engineering Manager to see a big picture and take a balcony view – from which he/she may identify multiple leverage points that can be addressed to support constructive change or move project or team forward from the point of conflict and confrontations to effective working relationships. It assists Engineering Manager to see the connectivity between elements in the situation, so as to support joined-up actions (Justice, 2012; Lind, 1988).

4.7 ITERATIVE APPROACH TO SOLVING COMPLEX PROBLEMS

4.7.1 INTRODUCTION

Professor Matt Andrews at Harvard University developed the idea of iterative, experimental processes of finding and fitting solutions to complex problems (Lind, 1988; Harvard, 2022). This approach is similar to adaptive environmental management, which involves learning and building capability as the team goes.

His observation is that many processes within the organisation are based on best practice ideas that are seen to foster success in other places (Andrews, 2021). But "best practice" is only one type of idea we should think of working within looking for solutions to our problems or want to improve the efficiency of the existing processes, for instance, what is the best process to facilitate strategic business plan development, its consultations, approvals, and endorsement? (Andrews, 2021) Andrew recommends the problem-driven interactive approach to solving complex problems where the problem is broken into several small parts (Andrews, 2021; Australia, 2017).

The process involves building a team or organisational capability, learning from each step of process and building capability for the next step. A step-by-step approach helps a team or an organisation involved (1) break down problems analysing their root causes, (2) identify entry points, (3) search for possible solutions, (4) take action and reflect upon what have been learned, (5) adapt and then (6) act accordingly; when a solution is clear, the team gains the capability it needs to deal with the problem (Andrews, 2021).

The concept is not completely new, as mentioned above, a similar approach is used in the adaptive environmental management when uncertainty is too high and the problem is too costly to solve. What is important to understand is that this is a dynamic process with tight feedback loops that allow the team to build a specific solution to the problem that fits into a specific project context. According to Andrews (2021), the approach rests on four principles:

1. Local solution to local problem
2. Pushing problem-driven positive deviance
3. Try, learn, adapt
4. Scale through diffusion (Andrews, 2021; Harvard, 2022). Free download of the PDIA process guide is available from https://bsc.cid.harvard.edu.

4.7.2 Initial Problem Analysis

Professionals, especially engineers and managers, and project teams frequently prefer a solution-driven approach, which could lead to ignoring of a real problem. Furthermore, part of the identification of the problem is required to build company or team capability. Why bother to go through complex discussions if a solution seems to be there, normally a best practice. This regularly happened on mega projects where engineering teams face several potential scenarios to solving a problem. While it is clear that people learn on projects, there is simply not enough time and management capability to step back and consider that teams must learn, discover something new, and gain new capabilities before the next step in the project can be made. It is highly likely that decision on preferred scenario can be only evaluated with new knowledge.

PDIA, developed by Prof Andrews is about building capability to solve problems through the process of solving good problems (Andrews, 2021). A good problem is one that cannot be ignored, motivates, and drives change. It can be broken down into small causal elements, allow real, sequenced, strategic responses (Andrews, 2021). A good problem is locally driven, where local actors define, debate, and refine the problem statement through shared consensus (Harvard, 2022; Andrews, 2021). By doing so, new knowledge will be created and shared between team members. In this case, new capability will be gained by the team and company:

I. The first step in this process is to construct the problem. The PDIA guide prompts to ask the following questions: What is the problem? Why does it matter? To whom does it matter? Who needs to care about? What will the problem look like when it is solved? For this step, many quality management tools like decision – trees, decision – matrix analysis, brainstorming, mapping can be used to learn about the state of problem. It is better to use various visual tools, data presented in colour with meaning to the project outcomes. This is a process of deconstruction of the problem. The art of deconstruction is the process of taking any problem and breaking it down into a set of smaller problems (Andrews, 2021). Simplified version of this concept is that if one can solve all of the smaller problems, then the big problem will be solved at least partly.

Complex problems are intractable and the right solutions are hard to identify. This often leads reformers to push for preferred best practice solutions that they know will not build real capability but will at least offer something to do (Andrews, 2021).

II. To mitigate this risk, the problem needs to be broken down into smaller, more manageable sets of coal points for engagement, that are open to localised solution building. This is the second step of PDIA process called the deconstruction step. To deconstruct the problem, various quality management techniques could be used. Andrews (2021) suggests fishbone diagram and "5-why technique", which allow users to identify multiple root causes and to further break down each cause into its sub-causes. Fishbone diagrams can be used quite effectively in this step (see Figure 4.3).

It is critical to involve professionals from different backgrounds (design, operation, maintenance, management, customer service). Professionals can bring wealth of their experience to deconstruct and test problems from various perspectives, thus allowing for a more robust deconstruction of the problem. At this stage, it is vital to have evidence data, collect new data, and identify which data are required to verify the problem. The answers to the questions should be informed by data/evidence to convince others of their validity (Andrews, 2021; Harvard, 2022). It is essential to distinguish between the perception of the problem and the problem itself.

Most deconstructed problems take the form of meta-problems and raise questions like:

- Where do I begin to solve the problem?
- What do I do?
- How do I ensure that all causal strands are addressed?

III. Solving the problems requires multiple interventions that allow for multiple entry points for change. Each cause and sub-cause of the fishbone diagram is essential a separate point of engagement, and offers different opportunities for change. The PDAI process refers this opportunity as the "space for change". This change in space is contingent on contextual factors commonly

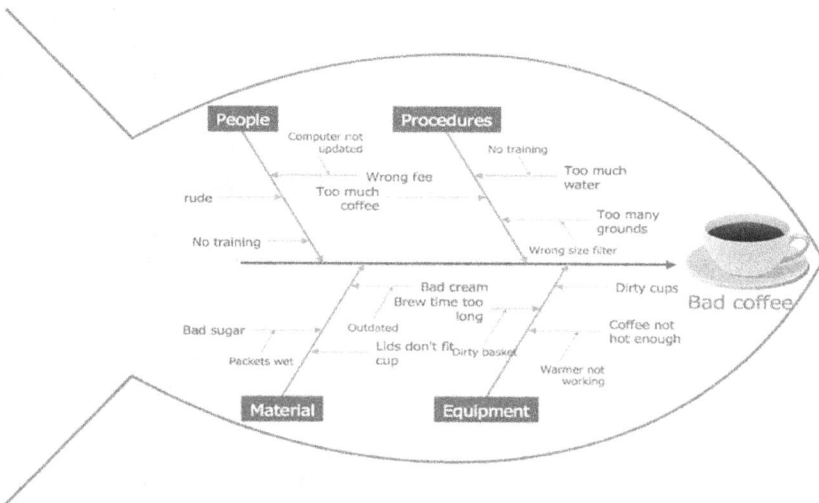

FIGURE 4.3 Example of fishbone diagram to deconstruct the problem of bad coffee.

found to influence problem–solution success, shaping what and how much one can do in any policy or reform initiative at any time. This is the third step sequencing, which is critical in helping the project team with the timing and staging of the engagement process (Andrews, 2021; Harvard, 2022).

The goal is to make as good an estimate as possible, in a transparent fashion as possible, so that the team allows itself to progressively learn about the context and turn uncertainty into clear knowledge (Andrews, 2021; Harvard, 2022).

4.7.3 IDENTIFY ACTION STEPS

The deconstruction and sequencing processes help the team and Engineering Manager who leads the team to think about where the team should act. The challenge still remains "what to do"? This is a serious challenge when dealing with complex problems, given that the solutions are usually unclear. If Engineering Manager can admit that not knowing what to do is common when facing a complex problem, then the process can take a different, maybe more effective shape.

According to Andrews (2021), the "what" answer to complex problems does exist and can be found but must emerge through active iteration, experimentation, and learning and discussion, sharing ideas. This means that answers cannot be pre-planned or developed in a passive or academic fashion by specialists applying knowledge from other contexts (Andrews, 2021). Andrews points out that a real solution to complex problems comes in the form of many small solutions to the many casual dimensions of the problem (2021).

Crawling the design space, the fourth step in doing PDIA, assists the team and Engineering Manager to look for data and experiment with multiple alternative solutions. In this stage, the project team learn to identify alternative solutions that might inform the next strategy to deal with the complex problem. According to Andrews (2021), the process yields positive and negative lessons from each idea, with no individual idea provided to be "the solution". The lessons usually lead to the emergence of new hybrids, or locally constructed solutions that blend elements from all of the ideas (Andrews, 2021). Table 4.1 demonstrates the options where ideas can come from.

4.7.4 TAKE ACTION, BUILDING AND MAINTAINING AUTHORISATION

Engineering Manager needs authority to undertake any initiative aimed at building project team capability. However, it is not easy to build authority to act, even when Engineering Manager has the authority to lead the project. Authorisation environments are commonly fragmented, and difficult to navigate (Andrews, 2021). Programmes, projects, and other strategic opportunities typically cross multiple authority domains in which many different agents and processes act to constrain or support behaviour. People who made decisions are frequently not in your office or country. Authorising structure often vertical as well, with agents at different levels of an organisational structure enjoying control over different dimensions of the same process (Andrews, 2021; Justice, 2012).

Informality often reigns in these challenges as well, manifest in personality and relationships-authority structures. These structures are seldom well known, which makes it extremely difficult to know what really authorises what in any context.

TABLE 4.1
The Design Space: Where Do We Get Ideas From?

Existing Practice	Latent Practice	Positive Deviance	External Practice
These is always some existing practice or capability which provides an opportunity to learn about what works in the local context, what does not work, and why. Common tools to help in this process include gap analysis, programme evaluation, site visits, immersions, and inspections.	This is the set of potential ideas that are possible but require some focused attention to emerge. Rapid results type interventions where groups of people are given a challenge to solve a focal problem in a defined period with no new resources is an example. The ideas that emerge from these rapid initiatives can also become the basis of permanent solutions to existing problems.	Positive deviance relates to ideas that are already being acted upon in the change context and that yield positive results, but are not the norm (hence the idea of deviance). Finding these positive deviants, codifying them and broadly diffusing; the core principles of their success are crucial.	The final read of opportunity in the design space is often the first set of ideas the team and Engineering Manager suggest and look at. These are often multiple external good/best practice ideas to learn from and the find and fit process should start by identifying a few of these – rather than setting for one prematurely. These ideas need to be translated into the local context.

Whatever formal or informal, authority structures are often fickle and inconsistent (Justice, 2012; Andrews, 2021). Authorisers will sanction new activities for many reasons and may lose interest or energy for many reasons, which no one will explain. This means that one is never guaranteed continued support from any authoriser for any period of time, no matter what promises are made (Andrews, 2021). Therefore, authority needs to be treated as a variable and not as something fixed. It is dynamic and with well-structured strategies, and it can be influential in expanding your change space (Andrews, 2021; Harvard, 2022).

Table 4.2 summarises the key aspects to secure support from various senior management groups in the company.

TABLE 4.2
A Basic Triple-A Change Space Analysis

Authority	Acceptance	Ability
• Who has the authority to engage: Legal, Procedural, Informal? • Which of the authoriser(s) might support engagement now? • Which of them would probably not support engagement now?	• Which agents (person/organisation) have an interest in this work? • For each agent, in a scale if 1–10 m think about how much they are likely to support engagement? • What proportion of strong acceptance agents do you have? • What proportion of "low acceptance" agents do you have?	• What is your personal ability? • Who are the key agents you need to work on any opening engagement? • How much time would you need to engage? • What is your resource ability? • How much money would you need to engage? • What other resources do you need to engage?

SOME REFLECTIONS FOR YOU:

Please think about some problems you had in the past:

1. What authority do you need and where will you look to find it?
2. What is your communication and persuasion strategy to convince your authorisers?
3. Does the authoriser agree that you have a problem?
4. What would make the authoriser care about the problem?
5. Does the authoriser support the experimental iteration you propose?
6. What could convince the authoriser that you as Engineering Manager needs an experimental iteration approach?

4.7.5 Reflect on Action and Building Capability

Trying a number of small interventions in short rapid cycles help to assure common risks in reforming and learning processes, of either appearing too slow in responding to a problem or of leading a large and expensive capacity-building failure (Andrews, 2021). This is because each step offers a quick action that is relatively cheap and open to adjustment, and with multiple actions at any one time. According to Andrews (2021), there is an enhanced prospect of early success (Andrews, 2021).

The small steps help to flush out contextual challenges, including those that emerge in response to the interventions themselves (Andrews, 2021). Facilitating such positive deviations and contextual lessons is especially important in uncertain and complex contexts where reforms are unsure of what the problems and solutions actually are and often lack confidence in their ability to make things better (Andrews, 2021; Harvard, 2022).

Designing your first iteration is a key step in doing PDIA where multiple solutions ideas are identified and put into action, iteration is a key step that progressively allows locally legitimate solutions to emerge and fosters adaptation to the idiosyncrasies of the local context. It is essential to begin by trying in your context to become a little bit more functional. And then learning from that experience, getting some legitimacy from the quick wins, iterating again, with maybe a bigger step the next time around, learning again and getting legitimacy again, working a way up, step by step – learning by doing and creating new capability. Please also refer the PDIA guide and some examples from www.bsc.cid.harvard.edu. (Harvard, 2022).

There is a similar approach in the engineering field calls "Value Engineering Workshops".

In PDIA, there is no separation between the design and the implementation phase of solving complex problems. This is a simulation process that occurs via embedding experiential learning into the iteration process. The idea of iteration around specific steps instead of taking big jumps is so the team can stop and learn from their experiences (Harvard, 2022). Check-in points offer opportunities to ask what was learning as the team tries to address the challenge, and especially learn new knowledge – that is not codified or written down but is based on what the team did in taking the steps.

In social science, it is recognised that knowledge can be constructed as a result of interactions between people or members of the team. This is also called tacit knowledge, which is the key knowledge required to capture and build on when working on complex problems. In the context of the digital economy, Engineering Managers increasingly face problems of navigating complex environment and the ability to translate complexity into simplicity.

According to Andrews (Andrews, 2021), each iteration has five dimensions: (1) it is time bound (with a short period at first) in which (2) the project team identifies multiple ideas; (3) it acts upon ideas; (4) it stops to take stock of your experience and test the validity of your assumptions in specific contexts; and (5) it revises your ideas to try again. In this process, Engineering Manager and the project team are both the course and user of emergent knowledge, as compared to many other approaches where the learner is a passive recipient of knowledge. It is believed that active discourse and engagement are vital in complex change processes and must therefore be facilitated through the iterations (Andrews, 2021) (Figure 4.4).

4.7.6 ADAPT AND ITERATE

According to Andrews (2021), doing PDIA is hard, because Engineering Manager and the team should be under no illusions that the problems the team confronts, the forces arrayed against real reform or problem-solving, the incumbent systems in which they are embedded, and the seemingly modest starting points from which PDIA begins, can all combine to make the challenge before us seems daunting and overwhelming – and on a bad day impossible. It can be a painful process. Many things taken for granted, in fact, took hundreds of years to invent, test, evaluate, re-design, recognise the mistakes, errors made, rectify, and validate. It took many generations of engineers and sciences to adopt and implement the British Standards.

One day, perhaps, something like PDIA will be normal and normative way of engaging with complex development challenges, but only a committed global social movement of citizens and development professionals will bring it about (Andrews, 2021; Harvard, 2022).

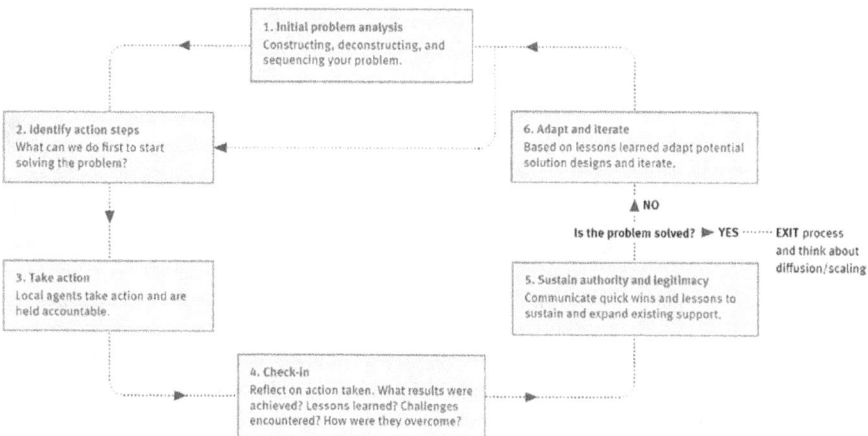

FIGURE 4.4 The iterative process. (From Andrews, M., *PDIA, Interactive Process Approach*, Harvard University Press, Boston, MA, 2021. With permission.)

FIGURE 4.5 Complex problem is solved using the PDIA process.

4.7.7 LEARNING PROCESSES

The PDIA process, which is similar to adaptive environmental management practices, has been used widely in the natural resource sector by companies such as Shell Global, ExMobil, and British Gas during the development of mega projects. The PDIA process offers a template to follow and explains all steps needed to gain some significant benefits from all of these invested times by the team and Engineering Manager. The problem will be solved no matter how difficult it was at the beginning of the journey. The last bit to solve a problem will be found if the PDIA process is used (Figure 4.5).

REFERENCES

Agnew, J., et al. (1996). *Human Geography. An Essential Anthology*. Oxford: Blackwell.

Ahmad, Y. J., et al. (n.d.). *Environmental Accounting for Sustainable Development. A UNEP-World Bank Symposium* (pp. 23–45). Washington, DC: The World Bank.

Alexander, E. R. (2022). Complexity, Institutions and Institutional Design. In E. R. Alexander, *Handbook on Planning and Complexity* (pp. 19–33). London: Springer.

Allen, P. (1988). Dynamic Models of Evolving Systems. *Systems Dynamics Review*, 4, 109–130.

Allen, W. (2023, Jan 09). *Complicated or Complex – Knowing the Difference Is Important*. Retrieved from Learning for Sustainability: https://learningforsustainability.net/post/complicated-complex/.

Andrews, M. (2021). *PDIA, Interactive Process Approach*. Boston, MA: Harvard University Press.

Arthur, W. B. (1990, Feb). Positive Feedbacks in the Economy. *Scientific American*, 262, 92–99.

Arthur, W. B. (1994). *Increasing Returns and Path Dependence in the Economy*. Boston: The University of Michigan Press.

Australia, E. (2017). *The Digital Economy: Engineers Australia Submission*. Sydney: Engineers Australia.

Ayres, R. U. (1994). Industrial Metabolism: Theory and Policy. In R. U. Ayres, *Industrial Metabolism: Restructuring for Sustainable Development* (pp. 3–20). Tokyo: United Nation University Press.

Baccini, P. B. (1991). *Metabolism of the Anthroposhere*. Berlin: Springer Verlag.

Bak, P. (1997). *How Nature Works: The Science of Self-Organised Criticality*. Oxford: Oxford University Press.

Bak, P. C. (1991). Self-Organised Criticality. *Scientific American*, 264(1): 26–33.

Barnet, H. (1975). Pressures of Growth upon the Environment. In K. Boulding, *Environment Quality in Growing Economy* (pp. 15–20). Baltimore, MD: Johns Hopkins University Press.

Bernstein, B. (1981, Dec 30). Ecology and Economics: Complex Systems in Changing Environments. *Annual Review of Ecology and Systematics*, 12, 309–330.

Bertalanffy, L. (1968). *General Systems Theory: Foundations, Development, Applications*. New York: Braziller.

Bertalantein, B. (1950). The Theory of Open Systems in Physisc and Biology. *Science*, 111, 23–29.

Bilsky, L. J. (1980). *Historical Ecology*. London: Kennicat Press.

Brown, L. (2006). *Plan B 2.0: Rescuing a Planet under Stressand a Civilization in Trouble*. New York: W.W. Norton & Company.

Butzer, K. W. (1996). Civilisations: Orgnisms or Systems. In J. L. Agnew, *Human Geography. An Essential Anthology* (pp. 268–281). Oxford: Blackwell Publishers.

Cang, H. R. (2017). *Invasion Dynamics*. London: Oxford University Press.

Clark, W. C. (1986). Sustainable Development of Biosphere: Themes for a Research Program. In W. M. Clark, *Sustainable Development of Biospere* (pp. 5–48). Cambridge, MA: Cambridge University Press.

Cohen, J. E. (1995). *How Many People Can the Earth Support?* London: W.W. Norton & Company.

Daly, H. E. (1976). On Economics as a Life Science. *Journal of Political Economy*, 76, 392–406.

Daly, H. E. (1993). *Steady-State Economics*. London: Earthscan.

Daly, H. E. (1994). *For the Common Good: Redirecting the Economy toward Coomunity, the Environment and a Sustainable Future*. Boston, MA: Beacon Press.

De Santo, G. N. (2020). Nephorology a Discipline Evolving into Complexity: Between Complex Systems and Philosophy. *Journal of Nephrology*, 33, 1–4.

Diamond, J. (2005). *How Societies Choose to Fail or Succeed*. New York: Viking.

Dickerson, R. (1978). Chemical Evolution and the Origin of Life. *Scientific American*, 239, 70–86.

Edson, M. (1981). Emergent Properties and Ecological Research. *The American Naturalist*, 118, 593–596.

Fischer-Kowalski, M. H. (1993). Metabolism and Colonisation: Modes of Production and the Physical Exchange between Socieities and Nature. *Innovation and Social Sciences Research*, 6(4), 415–442. doi:10.1080/13511610.1993.9968370

Georgiev, G. Y. (2022). *Efficiency in Complex Systems*. London: Springer.

Gharajedaghi, J. (2012). *Systems Thinking Managing Chaos and Complexity: A Platform for Designing Business Architecture*. Amsterdam: Elsevier.

Harvard. (2022, Oct 9). *The Building State Capability Team*. Retrieved from https://bsc.hks. harvard.edu/publications/building-state-capability-evidence-analysis-action/

Heylighen, F. (1997). Publications on Complex, Evolving Systems: A Citation-Based Survey. *Complexity*, 2, 31–36.

Justice, M. (2012). The Roles of Procedural Justice and Social Learning in Improving Self Organizing Capabilities of Local Communities for Sustainable Development in Decentralized Indonesia. *OIDA International Journal of Sustainable Development*, 3(10), 73–90.

Karcanias, N. L. (2020). Complex Systems and Control: The Paradigm of Structure Evolving Systems and System of Systems. In E. P. Zattoni, *Structural Methods in the Study of Complex Systems* (pp. 3–39). Switzerland. doi:10.1007/978-3-030-18572-5_1

Kaufmann, S. (1993). *The Origins of Order: Self-Organization and Selection in Evolution*. New York: Oxford University Press.

Kaufmann, S. (1995). *At Home in the Universe. The Search for Laws of Self-Organization and Complexity*. New York: Oxford University Press.

Lind, E. T. (1988). *The Social Psychology of Procedural Justice*. New York: Plenum Press.

Scalco, A. P. (2022). *Social Systems Engineering for Achieving Cyber Physical-Social System Multi-Concern Assurance*. First published: 13 September 2022. Retrieved from https:// incose-onlinelibrary-wiley-com.ezproxy.lib.rmit.edu.au/doi/full/10.1002/iis2.12893.

Vester, F. (1983). *Ballungsgebiete in der Krise: Vom Verstehen and Planen menschlicher Lebensraume*. Munchen: Deutscher Taschenbuch Verlag.

Weis, E. (2008). *Fundamentals of Complex Evolving Systems: A Primer*. Vienna: Socialecologyvienna.

Zattoni, E. P. (2020). *Structual Methods in the Study of Complex Systems*. Springer Nature Switzerland AG.

5 Leadership, Group Leadership, Functional Leadership

Sebastian Salicru

This chapter presents the Engineering Managers' Leadership Capability Framework (EMLCF). Despite the many efforts made around the world to define engineering leadership, design suitable curricula for engineering leadership educational programmes, and develop engineering leadership development programmes, no unified leadership framework for engineering managers exists to date. This is mostly due to the fact that leadership development practices have traditionally been narrowly focused and remain rooted in old paradigms (Salicru, 2017). To remedy this deficiency, the EMLCF draws on and integrates previously neglected leadership research areas, such as adult leadership and adult development (Kegan, 1982; Kegan & Lahey, 1984); adaptive leadership (Heifetz et al., 2009); ethical leadership (Brown & Treviño, 2006; Den Hartog, 2015; Yukl et al., 2013); various forms of pluralistic leadership (Denis et al., 2012; Sergi et al., 2012) – collective leadership (Edwards & Bolden, 2022), distributed leadership (Bolden, 2011), and shared leadership (Crevani et al., 2007); global mindset (Clapp-Smith & Lester, 2014; Javidan & Walker, 2012); holistic leader and leadership development (Clerkin & Ruderman, 2016; Dhiman, 2017; Quatro et al., 2007); and emotion, networking, creativity, innovation, and spirituality (Pearce, 2007). The EMLCF is, arguably, the first holistic Leadership Capability Framework ever developed for engineering managers, introduced by Salicru (2023).

5.1 INTRODUCTION

Engineering management (EM) relates to the application of scientific, engineering, and managerial efforts (Sage & Rouse, 2014). It is a specialised type of management concerned with the application of engineering principles to business practices. EM integrates technological, savvy, and engineering problem-solving with organisational, legal, administrative, planning, and operational performance management capabilities to enable complex enterprises, from conception to completion. Examples of EM application include areas such as product development, manufacturing, construction, design engineering, industrial engineering, technology, production, or any other field that employs people who perform an engineering function. Engineering managers are responsible for feasibility, planning, and implementing engineering projects from business case development to commissioning and handover. This includes providing supervision and guidance to other engineers, managers,

DOI: 10.1201/9781003374879-5

and project managers. Engineering managers work across multiple industries and disciplines – namely, chemical, civil, electrical, and mechanical engineering – each covering a wide range of fields.

Since the mid-1970s, the EM discipline has continued in a pattern of explosive growth (Kocaoglu, 1991). Societies have experienced rapid progress and unprecedented population growth (Paul et al., 2018), and as artificial intelligence (AI) is rapidly becoming more critical to the digital world (Murgai, 2018), the global economy is undergoing pervasive digital technological changes across most aspect of human enterprise (Gluckman, 2018). More recently, as the world rapidly transitions to a lower carbon economy, new opportunities are emerging to pursue green development (Adams, 2019; Hausmann, 2022; Hickel & Kallis, 2020). This includes engineering green growth (Juma, 2013) with new industries, markets, and technologies creating new paths to prosperity for those who act early. As a result, engineers face both challenges and opportunities to solve increasingly complex problems of a magnitude never seen previously.

> Engineers in the past were able to have successful careers based on their technical merit and management skills.
>
> **(Flowers, 2002, p. 16)**

The new digital and green economy, coupled with the fact that "people are the real key to digital transformation" (Kane, 2019, p. 44), calls for new capabilities to create an urgency for digitisation and the leadership required to drive this vision forward (Kohnke, 2017). As a result, engineering managers need to possess critical non-technical capabilities to enable them to understand and successfully navigate the various social, political, economic, cultural, environmental, and ethical aspects of the technical projects on which they are working (Killgore, 2014). Such non-technical capabilities include knowledge and skills in the areas of management and leadership. In fact, "leadership must be a key element advancing for the engineering profession to remain relevant and connected in an era of heightened outsourcing and global competition" (Farr & Brazil, 2009, p. 3), and leadership has been referred to as "the essential of engineering management" (Giegold, 1981, p. 49).

The need to educate 21st-century engineers with stronger leadership skills and identity has been acknowledged extensively (Amirianzadeh et al., 2011; Bayless & Robe, 2010; Cerri, 2016; Dunwoody et al., 2017; Farr & Brazil, 2012; Fasano, 2011; Flowers, 2002; Giegold, 1981; Hinkle, 2007; Klassen et al., 2020; Mawson, 2001; McCuen, 1999; National Academy of Engineering, 2004; Paul & Falls, 2015; Paul et al., 2018; Rottmann et al., 2015; Schell et al., 2022; Walesh, 2012; Weingardt, 2000). Engineering industry leaders such as General Electric, Lockheed Martin, NASA, National Instruments, Northrop Grumman, and Raytheon offer leadership development programmes to their newly qualified engineering employees. Such programmes focus on facilitating the transition from working with an academic mindset to a corporate one and entail leadership training, career development, and rotational assignments (Compton-Young et al., 2010). Despite the compelling recognition of the needs outlined above, there is lack of consensus on the definition of leadership in engineering, and graduates are still missing the acquisition of leadership capabilities through their education (Paul et al., 2018).

The engineering literature of the last decade, however, reveals that significant effort has been made to close this gap by proposing various definitions, competencies, and models of leadership in engineering (AlSagheer & Al-Sagheer, 2011; ASCE, 2008;

Bennett & Millam, 2013; Compton-Young et al., 2015; Crawley et al., 2014; Daniels, 2009; Farr et al., 1997; Gordon, 2021; Hartmann & Jahren, 2015; Hensey, 1999; Hess, 2018; Ivey, 2002; Parkin, 1997; Prieto, 2013; Sabatini & Knox, 1999; Schuhmann, 2010; Shaw, 2002; Wellington, 2009). Paul et al. (2018), for example, analysed 163 definitions of engineering leadership and identified the following four main themes comprising a total of ten categories:

1. Leading and influencing others (three categories: lead others, influence others, and be a role model);
2. Personal effectiveness (two categories: excellence and get things done);
3. Engineering competency (three categories: solve problems, project management, and engineering ethics); and
4. Collaboration (two categories: work with others and listen to others).

Other themes that emerge from the above-mentioned studies include the creation of vision; initiative; confidence; personal drive; communication; execution; character; emotional intelligence; ethics; self-realisation; facilitation of the work of others; organisation of effort production of deliverables; meeting schedules and customer quality requirements; assessment of risk; interpersonal interactions; engagement; building trust; team dynamics and team building; creativity; and innovation. Mawson (2001) summarises these themes by stating: "To be successful as leaders, engineers must achieve personal growth on at least four levels, emotional, imaginative, cognitive and behavioral" (p. 44). More specifically, this means taking a holistic perspective of human development.

Holism is both a philosophical perspective and a practical approach, and its underlying premise is that the whole is greater than the sum of its individual or specialised parts or views that contribute to it (Haynes, 2009). This holistic perspective integrates research areas that traditionally have either been ignored or received less scholarly attention than conventional or mainstream leadership research. Examples of such areas include adult constructive-developmental theory (McCauley et al., 2006), adult holistic development (Rogers et al., 2006), moral development (Lapsley & Narvaez, 2004), consciousness (Vincent et al., 2015), and spiritual leadership (Fry, 2003, 2005).

> Holistic leaders are adept at operating in the analytical, conceptual, emotional, and spiritual domains of leadership practice.
>
> **(Quatro et al., 2007, p. 427)**

Hence, holistic leader development requires practices such as mindfulness, social connections, body-based practices, and the use of positive emotions (Clerkin & Ruderman, 2016). In turn, such practices assist to counter the effects of stress, overload, and exhaustion by promoting and maintaining mental health and well-being and encouraging leaders to become fully human. This holistic perspective on leadership development is in line with Petrie's (2014) notion of "vertical leadership development" (p. 1), as opposed to horizontal leadership development – increasing technical skill sets and building leadership competencies. While such skills are essential and necessary, they are no longer sufficient in today's world. Vertical leadership development, on the contrary, goes deeper and entails developing more sophisticated and complex ways of thinking capability, clearer insights, and greater wisdom to tackle today's complexity and challenges (Petrie, 2015).

The areas emerging from the engineering literature outlined previously are referred to, interchangeably, as skills, competencies and, occasionally, capabilities. While the terms "competence" and "capability" both denote manifestations of human abilities and skills, and are thus inherently closely related to each other, there are differences between the two. "Competency is the possession of the skills, knowledge and capacity to fulfil current needs" (Nagarajan & Prabhu, 2015, p. 7) and has been traditionally used to capture the knowledge, skills, and abilities (KSA) to perform in a technical domain. Other authors have defined competency as a "capability or ability" (Boyatzis, 2009, p. 750). Over time, however, the competency paradigm has become somewhat outdated, as it only enables employees to master the skills and knowledge they require to perform a particular job. Capability-building models, on the contrary, enable individuals to integrate knowledge and skills needed to innovate and adapt to the dynamic changes of any industry or work environment (Hirt, 2020). Capability refers to the integration of knowledge, skills, and personal qualities used in response to varied, familiar, or unfamiliar circumstances, which requires competence, communion, creativity, and coping (Stephenson, 1994). The application of capabilities entails the creation of innovative learning experiences (Graves, 2013; Stephenson & Weil, 1992).

> Capable people are those who: know how to learn; are creative; have a high degree of self-efficacy; can apply competencies in novel as well as familiar situations; and work well with others.
>
> **(Hase & Davis, 1999, p. 2)**

As opposed to competency, the term capability is broader and more holistic, as it includes attributes, attitudes, and behaviours and other ranges of resources (e.g., psychological states) used to describe an ability to achieve certain outcomes in the future. To lead effectively in a digital and uncertain world, Ancona (2005), for example, proposes four key capabilities of effective leadership: (1) Sensemaking – the act of discovering new terrain as it is invented and the process of mapping new terrain as it is being created; (2) Visioning – developing a vision about something exciting and important; (3) Relating – building trusting and collaborative relationships with others and creating coalitions for change; and (4) Inventing – creating processes and structures to turn the vision into reality by implementing the steps needed to achieve the vision of the future.

In relation to leadership development, "the competency approach to leadership could be conceived of as a repeating refrain that continues to offer an illusory promise to rationalize and simplify the processes of selecting, measuring and developing leaders, yet only reflects a fragment of the complexity that is leadership" (Bolden & Gosling, 2006, p. 147). The stance to move beyond the competency movement, which originated in the late 1960s with the changing economic and political context – along with the concept of "managerial competency" – has been endorsed by prominent leadership researchers and practitioners (Boyatzis, 1993; Carroll et al., 2008; Conger & Ready, 2004; Ladkin, 2010; Ruderman et al., 2014).

Competency-based models are highly structured. Leadership in real life is less clear-cut, being much more fluid, dynamic, and chaotic. Accordingly, competencies limit the full picture of leadership as a truly social and relational phenomenon (Carroll et al., 2008; Ruderman, et al., 2014). From this perspective, accepting the colonisation of leadership by such a distinctly managerial concept as the competency paradigm is particularly problematic, inappropriate, and misplaced (Carroll et al., 2008). In line with this, and given that leadership constitutes a constellation of cognitions, skills, behaviours, behavioural repertoires, and competencies, this chapter adopts the term capability (i.e., the ability to do something, or execute a specific action, or to achieve certain outcomes). From this perspective, capabilities could be considered as "meta-competencies" (Tubbs & Schulz, 2006) or "abilities that underpin or allow for the development of competencies, as well as characteristics that individuals will need in addition to competencies such as motivation and key cognitive abilities" (Van der Merwe & Verwey, 2007. p. 35). More specifically, the term "global leadership capabilities" will be used to integrate the above findings, along with the extant leadership literature, in a contemporary and futuristic framework. The term "global" is adopted because it denotes more than just the geographic location or reach of business operations, and it includes working across cultures either face-to-face or via virtual teams. It also captures the inherent value of a cultural diversity of people, which is now the norm in most organisations.

Building on the extant leadership in engineering literature, and informed by theoretical and empirical findings in leadership research, this chapter presents the Engineering Managers' Leadership Capability Framework (EMLCF) – a contemporary and futuristic framework of leadership education and development for engineering managers. To this end, this chapter is organised as follows. To begin, a broad overview of the leadership construct and some preliminary conceptualisations are outlined. Next, the main types of leadership, along with their linkages and convergences, are presented. An overview of the evolution of leadership theories from a historical perspective, which is classified in four waves, follows. Further, five key distinctions related to the study of leadership are made in order to develop a common and consistent language. The current context for leadership, and the new demands imposed by such context upon leaders and organisations, is explained next. This is followed by a comprehensive inspection of the EMLCF. Finally, concluding remarks are provided.

5.1.1 LEADERSHIP: BROAD OVERVIEW AND PRELIMINARY CONCEPTUALISATIONS

Leadership is a highly sought-after and highly valued commodity.

(Northouse, 2019, p. 32)

This assertion is not at all surprising given that leadership has been linked to organisational performance (Bass, 1985; Day & Lord, 1988; Knies et al., 2016; Smith et al., 1984; Thomas, 1988; Xenikou & Simosi, 2006), creativity and innovation (Amabile & Khaire, 2008; Houghton & DiLiello, 2010; Hughes et al., 2018; Mainemelis et al., 2015; Lee et al., 2020; Reiter-Palmon & Illies, 2004), and the economic growth of nations (Jones & Olken, 2005; Easterly & Pennings, 2020). Clearly, the social impact of leadership is huge. In fact, "the success of all economic, political, and organizational systems depends on the effective and efficient guidance of the leaders of these systems" (Barrow, 1977, p. 231). The study of leadership is therefore exciting. Notwithstanding, it can also be confusing and frustrating. This is because, as acknowledged by Riggio (2019), "the study of leadership is both immensely fascinating and enormously complex" (p. 1). Entering the study of leadership can be confusing because of the many paradoxical and controversial viewpoints that exist on the topic. As a complex social activity, leadership involves numerous interconnected psychological and interpersonal processes (Goethals et al., 2014). Not surprisingly, leadership has been defined as "one of the world's oldest preoccupations" (Bass, 1990, p. 3).

Paradoxically, despite the fact that it has been observed and studied for centuries, leadership is a very contested construct that has been described as one of the "least understood phenomena on the earth" (Burns 1978, p. 2). Hence, the first challenge related to the study of leadership is being able to define it. Almost 50 years ago, Stogdill (1974) stated that "there are almost as many different definitions of leadership as there are persons who have attempted to define the concept" (p. 259). Rost (1991), for example, identified over 200 different definitions for leadership written from 1900 to 1990. According to Kellerman (2012, cited in Volckmann, 2012), around 1,400 different definitions of the terms "leader" or "leadership" exist. In relation to classification systems of taxonomies used to study leadership, Fleishman et al. (1991) identified up to 65 different systems that have been generated to define the dimensions of leadership (e.g., personality, behaviour, skills, and processes perspectives).

As a result of the above, the study of leadership could be even more confusing for students of engineering. This is because, as an applied science, engineering relies on and resembles an exact science more than leadership does. Engineering is "the study of using scientific principles to design and build machines, structures, and other things, including bridges, roads, vehicles, and buildings" (Cambridge Dictionary, 2023). As a result, the content of engineering learning is embodied in an established system of pure sciences with "correct" answers (e.g., mathematics or physics). This is despite the fact that "as a creative and scientific activity that transforms nature to serve the needs and wants of large numbers of people, engineering has both physical and human dimensions" (Auyang, 2006, p. 2).

Leadership, on the contrary, is a socially constructed phenomenon (Meindl et al., 1985), with ambiguous and confusing meanings and a multitude of definitions (Kjellström et al., 2020), which has been informed predominantly by the humanities and social sciences – mostly psychology, sociology, education, management, and public administration (Riggio, 2019). Hence, the content of leadership learning is embodied in a multisystem of subjective and conflicting viewpoints, which don't have "right" or "wrong" answers. From this perspective – and using Elder and Paul's (1997) analogy, like a judge in a court of law – the study of leadership requires sound judgement and critical thinking; that is, the ability to analyse and evaluate information critically. More specifically, critical thinking is about being: sceptical without being cynical; open-minded without being naive; decisive without being stubborn; evaluative without being judgemental; and forthright without being opinionated (Facione, 2011).

A useful way to deal with this array of complexities, fragmentations, and contradictions surrounding the study of leadership is via integration. In this case, integration refers to the synthesis of relevant knowledge from diverse theories and models of leadership. In relation to defining leadership among the myriad of existing definitions, for example, Chemers (2014) integrates multiple perspectives and provides the following definition: "Leadership is a social process in which a person is able to enlist the aid of others in the accomplishment of a common task" (p. 1). Such a definition is broad enough that it most likely would be accepted by most theorists and researchers. Similarly, Yukl (2006), identifying that leadership is a process that occurs in groups, involves influence and the achievement of common goals, defines leadership as "the process of influencing others to understand and agree about what needs to be done and how to do it, and the process of facilitating individual and collective efforts to accomplish shared objectives" (p. 8). Further, and more succinctly, Northouse (2019) states that "leadership is a process whereby an individual influences a group of individuals to achieve a common goal" (p. 43). The next section offers an overview of the main types of leadership and their corresponding convergence.

5.1.2 Main Types of Leadership

Following the integration rationale mentioned above, and considering the convergence of leadership research, Table 5.1 lists alphabetically the main 18 types of leadership found in the literature. This includes at least one definition, related concepts and/or leadership types, underlying or related theories, and their source.

5.2 THE EVOLUTION OF LEADERSHIP: THE FOUR WAVES OF LEADERSHIP

Leadership – like history – is constantly evolving. Over a century ago, leadership became a topic of academic research, and definitions, models, and perspectives have evolved continuously since then. This section provides a broad overview of the evolution of leadership theories over the last two centuries.

The evolution of leadership theory can be divided into four main waves or generations as follows. The first wave includes leader-centred approaches to or theories of leadership (Taits and Skills). The second wave encompasses Situational and Contingency theories. The third wave, or New Leadership Era, constitutes Transactional and Transformational theories. Finally, the fourth wave represents Post-heroic, Emergent, or Contemporary approaches to leadership. Table 5.2 captures these four waves in more detail.

As illustrated above, post-heroic models of leadership place less emphasis on the importance of individual leaders by paying more attention to the agency of followers and collective action. Thus, recognising that leadership can occur at any level in an organisation. Such approaches emerged as a response to the evolving nature of work by leaving behind individualistic leadership models from the industrial era defined by mechanistic thinking, command and control, and authoritarian systems. Thus, giving rise to new and more democratic approaches better suited to the current knowledge-intensive economy. The next section outlines five key distinctions related to leadership in order to develop a common and consistent language and facilitate the learning of leadership.

5.3 FIVE KEY LEADERSHIP-RELATED DISTINCTIONS

This section outlines five distinctions related to leadership, which are particularly relevant when studying it. They are designed to develop the ability to think critically, and learn deeply and intentionally about leadership.

TABLE 5.1
Main Types of Leadership

Leadership Type	Definition	Related Concepts and/or Leadership Types, Underlying or Related Theories, Sources
1. Adaptive leadership	"The practice of mobilizing people to tackle tough challenges and thrive" (Heifetz et al. 2009, p. 14).	Adaptive leadership theory (DeRue, 2011; Heifetz, 1998; Heifetz & Laurie, 2001).
2. Authentic leadership	"A pattern of leader behavior that draws upon and promotes both positive psychological capacities and a positive ethical climate, to foster greater self-awareness, an internalized moral perspective, balanced processing of information, and relational transparency on the part of leaders working with followers, fostering positive self-development" (Walumbwa et al., 2008, p. 94).	Authentic leadership (Avolio et al., 2004; Gardner et al., 2005, 2011).
3. Creative leadership	"Creative leadership is the ability to deliberately engage one's imagination to define and guide a group toward a novel goal – a direction that is new for the group" (Puccio et al., 2010, p. xviii). "Leading others towards the attainment of a creative outcome" (Mainemelis et al., 2015, p. 393). "Adaptability is driven by organizational creativity, which has been defined as a continuous process of thinking innovatively, or finding and solving problems, and implementing new solutions" (Basadur, 2004, p. 104).	Creative leadership (Basadur, 2004; Mainemelis et al., 2015; Puccio et al., 2010; Sternberg et al., 2003). Work environment for creativity (Amabile et al., 1996).
4. Distributed leadership	"Leadership is probably best conceived as a group quality, as a set of functions which must be carried out by the group" (Gibb 1954, cited in Gronn 2000, p. 324). "A dynamic, interactive influence process among individuals in groups for which the objective is to lead one another to the achievement of group or organizational goals or both. This influence process often involves peer, or lateral, influence and at other times involves upward or downward hierarchical influence" (Pearce & Conger, 2003, p. 1).	Multiple forms of "Pluralistic Leadership" (Denis et al., 2012), namely: • Emergent leadership • Democratic leadership • Collaborative leadership • Collective leadership • Participative leadership • Shared leadership Bolden (2011), Cullen et al. (2012), Edwards and Bolden (2022), Fairhurst et al. (2020), Friedrich et al. (2016), Yammarino et al. (2012).

(Continued)

TABLE 5.1 (Continued)
Main Types of Leadership

Leadership Type	Definition	Related Concepts and/or Leadership Types, Underlying or Related Theories, Sources
5. Ethical leadership	"The demonstration of normatively appropriate conduct through personal actions and interpersonal relationships, and the promotion of such conduct to followers through two-way communication, reinforcement, and decision-making" (Brown et al., 2005, p. 120).	Behavioural ethics (De Cremer & Moore, 2020; Mitchell et al., 2017; Treviño et al., 2006). Values-based leadership (Copeland, 2014; Rao, 2017).
6. Followership	"Implicit followership theories (IFTs) are defined as individuals' personal assumptions about the traits and behaviors that characterize followers" (Sy, 2010, p. 73). "The social construction of followership involves the emergence of a leadership relationship that occurs when (1) a potential leader perceives or infers a group of individuals to be his or her followers or (2) when individuals in a group begin to view themselves as members of a larger group led by a leader" (Shondrick & Lord, 2010, p. 9). "Followership is a relational role in which followers have the ability to influence leaders and contribute to the improvement and attainment of group and organizational objectives. It is primarily a hierarchically upwards influence" (Carsten et al., 2010, p. 559). "In contrast to the traditional approach to leadership development, we argue that followers should also be included in leadership development efforts in order to prepare them to exercise responsible self-leadership and to effectively utilize shared leadership. This need is especially important in the case of team-based knowledge work" (Pearce & Manz, 2005, p. 130).	Implicit followership theories (IFT), and follower-centred perspectives (Chaleff, 1995; Kellerman, 2008; Kong et al., 2019; Lord et al., 2020). Upward leadership (Carsten et al., 2010). Self-leadership and Shared leadership (Pearce & Manzo, 2005).

(Continued)

TABLE 5.1 (*Continued*)
Main Types of Leadership

Leadership Type	Definition	Related Concepts and/or Leadership Types, Underlying or Related Theories, Sources
7. Functional leadership	Functional leadership focuses on the leaders' relationship to their teams (Fleishman et al., 1991), and postulates that the main role of leaders is to ensure the critical group's needs (task accomplishment and maintenance or relational functions) are adequately met (Hackman & Walton, 1986). "Functional leadership is centered on goal oriented leadership activities that may promote team processes which are likely to drive team effectiveness. Moreover, this team leadership approach is centered on improving the effectiveness of task performing groups in organizational contexts, rather than addressing any abstract context. Furthermore, functional leadership theory has a high potential to be used as a framework for leaders' training programs" (Santos et al., 2015, 471). Views leadership behaviour as representing "a form of organizationally-based problem solving", "a social problem-solving syndrome involving many cognitive capacities in the generation, selection, and implementation of influence attempts" (Fleishman et al., 1991, p. 259).	Team leadership (Burke et al., 2006; Hackman & Walton, 1986; Morgeson et al., 2010; Travis Maynard et al., 2017; Zaccaro et al., 2001).
8. Global leadership	"A process of influencing the thinking, attitudes and behaviors of a global community to work together synergistically toward a common vision and common goals" (Osland et al., 2006, p. 204).	Global mindset (Beechler & Javidan, 2007; Maznevski & Lane, 2004; Pucik, 2005).

(*Continued*)

TABLE 5.1 (Continued)
Main Types of Leadership

Leadership Type	Definition	Related Concepts and/or Leadership Types, Underlying or Related Theories, Sources
9. Group leadership	Views leadership is a relational property of groups.	Group leadership (Hoyt et al., 2003).
	Emphasise a team-centric view of leadership (Kozlowski et al., 2016).	Team leadership (Burke et al., 2006; Hackman & Walton, 1986; Kozlowski et al., 2016; Morgeson et al., 2010; Travis Maynard et al., 2017; Zaccaro et al., 2001).
	Leaders cannot exist without a group of followers and followers cannot exist without leaders.	Leaders' self-efficacy, and collective efficacy and effectiveness (Chemers, 2001).
	"Leadership identifies a relationship in which some people are able to influence others to embrace, as their own, new values, attitudes, and goals and to exert effort on behalf of and in pursuit of those values, attitudes, and goals. The relationship is almost always played out within a group—a small group such as a team, a medium-sized group such as an organization, or a large group such as a nation" (Hogg et al., 2005, p. 991).	Leader-Member Exchange (LMX) Theory (Graen & Uhl-Bien (1995).
	The process of providing focus and direction to a specific group of people. It involves facilitating and guiding the actions of group participants as well as accepting responsibility for the outcome of the group's efforts.	Social identity theory of leadership (Hogg et al., 2005).
10. Leadership-as-practice	The leadership-as-practice (L-A-P) movement views leadership occurring as a practice or activity rather than through the traits, charisma, and heroic acts of individual actors (Raelin, 2011).	Carroll et al. (2008), Raelin (2011, 2016, 2018, 2020).
	Leadership-as-practice has a collective orientation because it is less about what one person thinks or does, and more about what people may accomplish together. Hence, it is concerned with how leadership emerges and unfolds through day-to-day experience (Raelin, 2020).	
11. Relational leadership	Conceptualises three interactive domains of leadership: the leader, the follower, and their relationship.	Leader-Member Exchange (LMX) Theory (Graen & Uhl-Bien (1995).
	Leadership is as a process centred on the interactions between leaders and followers.	
	LMX theory proposes that leaders form different dyadic exchange relationships with different subordinates.	

(Continued)

TABLE 5.1 (Continued)
Main Types of Leadership

Leadership Type	Definition	Related Concepts and/or Leadership Types, Underlying or Related Theories, Sources
12. Self-leadership	"Self-leadership is a process through which individuals control their own behavior, influencing and leading themselves through the use of specific sets of behavioral and cognitive strategies" (Neck & Houghton, 2006, p. 270). "Self-leadership is a process through which people influence themselves to achieve the self-direction and self-motivation necessary to behave and perform in desirable ways" (Houghton & Neck, 2002, p. 672). Process through which individuals influence their own behaviour to achieve the self-direction and self-motivation necessary to perform, empower themselves, and achieve personal excellence.	Self-leadership (Harari et al., 2021; Houghton & Neck, 2002; Manz, 2015; Neck et al., 2012; Neck & Manz, 2012). Self-influence and self-regulation (Carver & Scheier, 1981). Self-control (Thoresen & Mahoney, 1974). Self-management (Andrasik & Heimberg, 1982).
13. Servant leadership	"The Servant-Leader is servant first.… It begins with the natural feeling that one wants to serve, to serve first. Then conscious choice brings one to aspire to lead" (Greenleaf, 1977, p. 7). Going beyond one's self-interest is a core characteristic of servant leadership. Servant-leaders empower and develop people; they show humility, are authentic, accept people for who they are, provide direction, and are stewards who work for the good of the whole. Focuses on the humble and ethical use of power as a serving leader, cultivating a genuine relationship between leaders and followers, and creating a supportive and positive work environment (van Dierendonck, 2011).	Servant leadership (Graham, 1991; Hunter, 2004; Parolini et al., 2009). Authentic leadership (Avolio et al., 2004; Gardner et al., 2005, 2011). Transformational leadership (Burns, 1978).

(Continued)

TABLE 5.1 (*Continued*)
Main Types of Leadership

Leadership Type	Definition	Related Concepts and/or Leadership Types, Underlying or Related Theories, Sources
14. Situational leadership	Situational leadership, or leadership in context, means that leadership depends on the situation at hand. Situational Leadership is based on an interplay among: (1) the amount of direction or task behaviour leaders provide; (2) the amount of relationship behaviour or socioemotional support leader use; and (3) the level of readiness exhibited by followers (maturity) on any specific task, function, activity or objective that leaders are attempting to accomplish through individuals or groups – followers (Hersey & Blanchard, 1969, 1993) The maturity of the follower determines both the style of leadership likely to have the highest probability of success, and the power base leaders should use to induce compliance or influence behavior (Hersey et al., 1979).	Hersey and Blanchard (1969, 1993), Hersey et al. (1979).
15. Spiritual leadership	"Spiritual leadership theory (SLT) is a causal leadership theory for organizational transformation designed to create an intrinsically motivated, learning organization. Spiritual leadership comprises the values, attitudes, and behaviors required to intrinsically motivate one's self and others in order to have a sense of spiritual survival through calling and membership—i.e., they experience meaning in their lives, have a sense of making a difference, and feel understood and appreciated" (Fry et al., 2005, p. 835).	Spiritual leadership (Benefiel, 2005; Dent et al., 2005; Fry, 2003, 2005). Ethical leadership (Brown & Treviño, 2003; Brown & Mitchell, 2010).
16. Strategic leadership	"Strategic leadership, in its simplest form, is leadership that manifests at the highest level of an organization" …. Strategic leadership is inherently grounded in digital transformation, innovation, and the upper echelons, with a growing footprint that spans across basic management and organizational activities" (Singh et al., 2023, p. 1). "A person's ability to anticipate, envision, maintain flexibility, think strategically, and work with others to initiate changes that will create a viable future for the organization" (Ireland & Hitt, 2005, p. 63).	Strategic leadership (Samimi et al., 2022). Capacity to learn, the capacity to change, and managerial wisdom (Boal & Hooijberg, 2000). Human and social capital (Hitt, & Ireland, 2002). Organisational learning (Vera & Crossan, 2004).

(*Continued*)

TABLE 5.1 (*Continued*)
Main Types of Leadership

Leadership Type	Definition	Related Concepts and/or Leadership Types, Underlying or Related Theories, Sources
17. Transactional leadership	Transactional leadership means that leaders rely on praise, rewards, and punishment, or the avoidance of disciplinary action, to achieve optimal followers' job performance or compliance (Bass et al., 2003).	Managerial leadership (Yukl, 1989).
18. Transformational leadership	"Transformational leadership is a process that changes and transforms people. It is concerned with emotions, values, ethics, standards, and long-term goals. It includes assessing followers' motives, satisfying their needs, and treating them as full human beings. Transformational leadership involves an exceptional form of influence that moves followers to accomplish more than what is usually expected of them. It is a process that often incorporates charismatic and visionary leadership" (Northouse, 2019, p. 263).	• Charismatic leadership • Inspirational leadership • Values-based leadership • Ethical leadership • Visionary leadership Bass (1985), Bennis and Nanus (1985), Burns (1978), Conger (1999), Hester (2012), Kouzes and Posner (2017), Mhatre and Riggio (2014), Shamir and Howell (1999).

TABLE 5.2
The Four Waves of Leadership

Wave	Period	Theory/Approach	Focus	Source
First wave (leader-centred theories)	1840 onwards (19th-century)	Great Man theory (heroic leadership)	Innate characteristics of leaders.	Carlyle (1907), King (1990), Organ (1996).
	1910s–1940s	Trait theory	Emphasises the personal attributes of leaders. Physiological, demographic, personality, intellectual, personality traits (e.g., The Big Five model of personality traits)	Colbert et al. (2012), DeRue et al. (2011), Judge and Bono (2000), Judge et al. (2002), Lord et al. (1986).
	1950s	Skills approach	Shift from a focus on personality characteristics, (usually are viewed as innate and largely fixed) to an emphasis on skills and abilities that can be learned and developed. Three basic personal skills: technical, human, and conceptual.	Katz (1955), Mumford et al. (2000).
Second wave	1940s–1960s	Behavioural theories • McGregor's theory X and theory Y • The Ohio state two-factor model • Blake-Mouton managerial grid	Behaviour and styles of leaders. Focus on what leaders actually do, rather than on their qualities. This area has attracted most attention from practising managers.	Blake and Mouton (1964), Fleishman (1953), Fleishman and Harris (1962), Halpin and Winer (1957), Likert 1961), McGregor (1960).

(Continued)

TABLE 5.2 (Continued)
The Four Waves of Leadership

Wave	Period	Theory/Approach	Focus	Source
	1960s–1990s	Situational and contingency theories: • Hersey-Blanchard situational leadership model • Fiedler's contingency theory • House's path-goal leadership theory • Vroom-Yetton-Jago decision-making model of leadership • Tannenbaum & Schmidt's leadership continuum	Situational theory views leadership as specific to the situation in which it is being exercised, describes leadership style, and stresses the need to relate the leader's style to the maturity level of the followers. Contingency approaches are the refinement of the situational approach and focus on identifying the situational variables which best predict the most effective leadership style to fit specific circumstances.	Fiedler (1967, 1971), Hersey and Blanchard (1969), House (1971, 1996), Tannenbaum and Schmidt (1958), Vroom and Yetton (1973), Vroom and Jago (2007).
		Relational theories Leader–Member Exchange (LMX) Theory	Conceptualises leadership as a process that is centred on the interactions between leaders and followers.	Graen and Uhl-Bien (1995).
Third wave (new leadership era)	1990s onwards	Transactional theory • Managerial leadership (authoritative)	Focuses on the exchanges that occur between leaders and followers by emphasising extrinsic rewards, avoiding unnecessary risks, and focusing on improving organisational efficiency. Uses bureaucracy, policy, power, and authority to maintain control.	Bass (1985, 1990, 2009), Bass et al. (2003), Burns (1978), Yukl (1989).
		Transformational theory • Authentic leadership • Charismatic leadership • Ethical leadership	Raising followers' level of consciousness about the importance and value of achieving desired outcomes, and the methods of reaching those outcomes. The process whereby leaders engage with others and create connections that raises the level of motivation and morality in both leaders and followers.	Bass (1985, 1990, 2009), Bass et al. (2003), Bennis and Nanus (1985), Burns (1978), Greenleaf (1977), Kellerman (2008), Kouzes and Posner (2017), Northouse (2019).

(Continued)

TABLE 5.2 (*Continued*)
The Four Waves of Leadership

Wave	Period	Theory/Approach	Focus	Source
		Post-heroic approaches • Shared leadership • Collaborative leadership • Servant leadership • Follower-based leadership (followership)	"Followership is a process whereby an individual or individuals accept the influence of others to accomplish a common goal" (Northouse, 2019. p. 439). Focus on engaging followers. Focus on followers leading each other. Focus on the whole system of an organisation	Bolden (2011), Cullen et al. (2012), Denis et al. (2012), Edwards and Bolden (2022), Fairhurst et al. (2020), Friedrich et al. (2016), Pearce and Conger (2003), Yammarino et al. (2012).
Fourth wave (post-heroic, emergent, or contemporary)	Contemporary	Leadership-as-practice (L-A-P) movement Leaderful practice	"Rather than refer to leadership as occurring through the traits or behaviors of particular individuals, the leadership-as-practice movement looks to leadership as occurring as a practice" (Raelin, 2011, p. 196). "A practice is a coordinative effort among participants who choose through their own rules to achieve a distinctive outcome. Leadership as-practice has a markedly collective orientation because it is less about what one person thinks or does and more about what people may accomplish together" (Raelin, 2020, p. 481). L-A-P focuses on the everyday practice of leadership including its moral, emotional, and relational aspects, rather than its rational, objective, and technical ones (Carroll et al., 2008). "L-A-P is a process model that cannot be reduced to an individual or even to discrete relations" (Raelin, 2011, p. 201). Leaderful practices are collective, concurrent, collaborative, compassionate, and co-creative (5Cs, Salicru, 2020).	Carroll et al. (2008), Raelin (2011, 2016, 2018, 2020), Salicru (2020).

5.3.1 Implicit and Research Theories of Leadership

Implicit leadership theories (ILTs), also referred to as "people's naive conceptions" of leadership (Offermann et al., 1994, p. 43), have been defined as images that people have of "a leader in general or of an effective leader" (Schyns & Meindl 2005, p. 21). They relate to the individuals' schemas or mental structures of leaders (Lord, & Maher, 1993), or idealised images or representations they hold and associate with the term leader and leadership (Keller, 1999). ILTs, then, include the individuals' preconceived notions, personal assumptions, or perceptions about leadership and the characteristics that constitute an ideal leader (e.g., traits, qualities, behaviours). Such everyday theories or mental representations are like prototypes or stereotypes stored in people's memory, which become activated when the person meets an individual that matches such characteristics (Schyns & Riggio, 2016). ILTs are socially shared among members of a particular culture or society and can be categorised into eight prototype dimensions, namely, sensitivity, dedication, tyranny, charisma, attractiveness, masculinity, intelligence, and strength (Epitropaki & Martin, 2004). This explains, for example, the disproportionate low representation of certain minorities as leaders (Sy, 2010).

Implicit theories (or mindsets), in general, refer to the fundamental core beliefs that individuals hold about various aspects of the human condition, which they use to understand the world, and to guide their behaviour, and in turn affect their learning (Dweck, 2012). Implicit theories reflect the tacit knowledge we all have learned about the world. As children, we all acquired opinions from our parents and teachers, which rarely have been explicitly articulated. As a result, most people are not fully conscious of these sets of beliefs or assumptions and find it difficult to put them into words. Nevertheless, such implicit theories or beliefs are manifested via personal opinions or expectations, and have an enormous impact on how individuals interpret, think about, and act in everyday situations.

Over time, implicit theories build individuals' meaning system that in turn set their learning trajectories and prime specific learning behaviours. Domain-specific implicit theories have a strong impact on learning about any specific topic or domain (e.g., leadership). Similarly, implicit followership theories (IFTs) exist (Junker & van Dick, 2014; Lord et al., 2020) and relate to "individuals' personal assumptions about the traits and behaviors that characterize followers" (Sy, 2010, p. 73).

Implicit theories can be conceptualised along a continuum that ranges from a fixed mindset – which assumes intellectual abilities as relatively fixed and unchangeable – to a growth mindset, which maintains that intellectual abilities can be developed (Yeager & Dweck, 2020). Due to the automatic and unconscious nature of implicit theories, it is important to pay conscious attention to these mechanisms of action in order to cultivate a growth mindset.

Scientific theories, on the contrary, are testable via research. Hence, they need to be made explicit in order to formulate testable propositions or hypotheses in ways that other scientists can replicate them (Runco, 1999). Given the power of implicit theories for learning in different educational contexts (Karlen & Hertel, 2021), when learning about leadership, individuals are likely to become exposed to new information that contradicts their ILTs. Therefore, being aware of this distinction matters, in order to reduce the bias deriving from ILTs. As pointed out by Elliott (2023), the study of leadership requires a continuous critical scrutiny of the sociohistorical conditions that have shaped leadership, in order to develop responsible practices for equitable organisations. Accordingly, this chapter explores the leadership construct by examining theoretical and empirical findings from the research literature.

5.3.2 LEADERSHIP AND MANAGEMENT

Leadership and management are considered as two distinct constructs (Kotterman, 2006; Lunenburg, 2011; Maccoby, 2000; Sarros, 1992; Sutton, 2010; Toor, 2011; Toor & Ofori, 2008; Zaleznik, 1997). Despite the fact that both terms are often used interchangeably in some of the literature, and in everyday life, they not synonymous (Bass, 2009). This distinction was first put forth by Zaleznik (1997) by arguing that while both leaders and managers make a valuable contribution to the firm, their contributions are different. Managers promote stability and the status quo by embracing the process, seeking stability, order, and control, and trying to solve problems quickly – sometimes before they understand problems fully. Leaders, in contrast, drive corporate success by focusing on inspiration, vision, and human passion by championing new approaches and change, tolerating chaos and lack of structure, and being willing to delay closure, to understand the issues more thoroughly. According to Zaleznik (1997), leaders are more artists, scientists, and other creative thinkers than managers. In a nutshell, managers are concerned about getting things done, and leaders are concerned with what things matter and mean to people. According to Giegold (1981), the distinction between leadership and management is important to the field of engineering management because engineering managers face different and more difficult challenges than other managers. This includes the fact that engineering professionals are very sensitive to their manager's leadership style and that they become demotivated by insensitive or unskilled leaders.

Management entails carrying out position responsibilities and exercising authority. It relates to the administrative function of planning, organising, budgeting, controlling, and monitoring, which is performed to achieve stated goals and objectives (Yukl, 1989). Leadership, on the contrary, relates to the human side of the enterprise. Leading is about influencing the commitment of people. Hence, leadership is about people, relationships, and change. Leadership entails motivating, inspiring, building trust and relationships, and coaching people (Maccoby, 2000). Management entails controlling a group or a set of entities to accomplish goals, and counting value; leadership refers to the ability to influence, motivate, and enable others to contribute towards the success of the organisation and creating value (Nayar, 2013). Management focuses on the standardisation of products and services, predictably achieving consistency on budgets and quality, day after day and week after week. Leadership is completely different, as it relates to the vision and future of the organisation by finding opportunities, empowering people, and facilitating lasting change (Kotter, 2013). Bennis and Nanus (2007) captured this distinction by stating: "Managers do things right, while leaders do the right things" (p. 12). Table 5.3 summarises the main differences between management and leadership, according to different authors.

TABLE 5.3
Management and Leadership

Management	Leadership	Source
• Building competence, control, and appropriate balance of power • Promoting stability and the status quo • Embracing process • Seeking stability, order, and control • Resolving problems quickly – sometimes before they are fully understood	• Focusing on inspiration, vision, and human passion, as drivers of corporate success • Championing new approaches and change • Tolerating chaos and uncertainty • Lacking of structure • Willing to delay closure to understand the issues more fully	Zaleznik (1997)
Carrying out position responsibilities and exercising authority	Influencing commitment	Yukl (1989)
Administrative function: • Planning • Organising • Budgeting • Controlling • Monitoring • Controlling a group or set of entities to accomplish a goal • Counting value	People, relational and change function: • Motivating • Inspiring • Building trust and relationships • Change • Coaching	Maccoby (2000)
	Influencing, motivating, and enabling people to contribute Creating value	Nayar (2013)
Produces order and consistency • Predictability • Standardisation • Allocation of resources	Produces change and movement • Sets vision and future direction • Empowers people • Facilitates change	Kotter (2013)
Present focus	Future focus	Sarros (1992)
Doing things right	Doing the right things	Bennis and Nanus (2007)

The above-outlined distinctions warrant some further explanation and cautionary notes. First, as noted by Sarros (1992), leadership and management are two distinct and complementary systems of action. Hence, organisations need both functions, as well as people who are effective at both leading and managing, in order to be competitive. From this perspective, it should not be assumed that leadership is better or more desirable than management, or vice versa. Contemporary organisations need leaders to challenge the status quo, and to inspire and lead change and innovation. However, they also need managers to set, implement and monitor process and systems to develop and maintain smooth day-to-day functioning operations. Second, given that exercising leadership or management functions depends on the incumbents themselves, and that such functions are not mutually exclusive, they can overlap – although they don't always do. According to Yukl (1989), for example, "a person can be a leader without being a manager, and a person can be a manager without leading" (p. 253). Similarly, Knight (2005) contends that all leaders manage but not all managers lead.

5.3.3 LEADERSHIP, POWER, AND AUTHORITY

Leadership is generally accepted in the literature as the ability to influence individuals towards the achievement of goals (Yukl, 1989). The three key elements of this broad definition are people, influence, and goals. Power relates to the desire to have an impact on others (McClelland, 1995), or influence of other's behaviours (Mintzberg, 1983). More specifically, power is the potential ability of a person to change or control the values, needs, attitudes, opinions, objectives, and behaviour of others (Rahim, 1989). Power, then, relates to leadership as part of its influencing process between leaders or followers (Northouse, 2019).

Authority is the institutionalised power between a superior (e.g., manager) and a subordinate that ensures compliance with the superior's wishes because s/he is the boss (Munduate & Medina, 2004). Authority, therefore, is the formal power individuals hold by virtue of their position in the organisational hierarchy (Gibson et al., 2012).

Leadership, therefore, can be conceptualised as involving power relationships and processes of influence between leaders and followers, in which the leaders exercise greater influence over the followers to achieve collective goals. It's important to realise that there is a difference between exercising social pressure and being genuinely persuaded. Hence, power is essential to leadership to influence group, team, and organisational members, but insufficient by itself for leadership. Power, then, is not the same as leadership, although it's often seen as a feature of it. According to Hollander and Offermann (1990), "leadership clearly depends on responsive followers in a process involving the direction and maintenance of collective activity" (p. 179). The key difference between power and leadership is that power is the ability to control others' behaviours, while leadership is the ability to influence others' behaviours. Influence is the force that leaders exert to induce a change in their followers (French & Raven, 1959). Table 5.4 summarises the main differences between power, authority, and leadership.

Understanding power and how to used it within the context of leadership is critical, as it relates to the understanding of ethical leadership (Ciulla, 2003), and the dark side of leadership, destructive leadership, or how leaders use their leadership in toxic and destructive ways to achieve their own personal interests (Krasikova et al.,

TABLE 5.4

Summary of Differences Between Leadership, Authority, and Power

Power	Authority	Leadership	Source
Potential ability of a person to change or control the values, needs, attitudes, opinions, objectives, and behaviour of others.	Institutionalised power between a superior (e.g., manager) and a subordinate that ensures compliance with the superior's wishes because s/he is the boss. The formal power individuals hold by virtue of their position in the organisational hierarchy.	Ability to influence, persuade or inspire others towards action without using power or force.	Gibson et al. (2012), Hollander and Offermann (1990), French and Raven (1959), McClelland (1995), Munduate and Medina (2004), Rahim (1989)

2013). The literature identifies different types of power according to two dimensions, namely, source and base. The two main sources of power are position and personal. Position power emerges from the formal position held in the organisation structure. This source of power generates employee compliance as a type of social influence. Personal power results from the leaders' personal attributes and the type of relationship established with their subordinates. This source of power generates employee internalisation and identification as types of social influence (Munduate & Medina, 2004). Internalisation relates to the followers' internalised believe of the same values of their leader, as necessary for the effectiveness of their work. Identification relates the perceived oneness with another individual (e.g., the leader), "where one defines oneself in terms of the other" (Ashforth et al., 2016. P. 28). Thereby, followers deliberately select a leader who is accountable, trustworthy, and displays desirable values and behaviours, from whom they can feel proud of working with. Thus, generating a high level of satisfaction in the relationship with their leader.

In relation to the bases of power, French and Raven's (1959) original power fivefold typology is arguably the most accepted. This typology comprises five types of power: legitimate, reward, coercive, expert, and referent power. Subsequently, Raven (1965) distinguished informational power as a sixth power type. Henceforth, in line with the literature, the terms managers and leaders, and employees, subordinates, team members, and followers will use interchangeably. Table 5.5 summarises the six main types of power, including their source, type of social influence, definitions, examples, and respective expected outcomes.

As depicted in Table 5.3, only legitimate power shares the three types of social influence (compliance, internalisation, and identification). Leaders/managers possess legitimate power when their followers/subordinates believe they have a legitimate right to, and expect from them, exert influence over them. Hence, followers/subordinates willingly accept this influence from their leaders/managers. Reward and coercive power rely on followers believing that their leader can punish them or provide them with their desired rewards. Relying on these forms of power only, eventually, will very likely generate limited follower loyalty and compliance. There is overwhelming research evidence demonstrating that coercive, authoritative, or forcing power styles are likely to engender passive compliance, strong resistance (disengagement), poor performance and productivity, and negative effects on long-term organisational health. Contrastingly, expert power and referent power are effective power bases that elicit employees' enthusiasm and commitment (engagement), and high levels of performance and productivity (Singh, 2009; Yukl, 1989).

Other typologies of power that can be found in the literature include Morgan's (1997) 14 sources of power, which have similarities with French and Raven's (1959) taxonomy; Salancik & Pfeffer's (1977) strategic-contingency model of power, which distinguishes between political and institutionalised power; and Kipnis et al.'s (1980) eight means of influence in the workplace (assertiveness, ingratiation, rationality, sanctions, exchange, upward appeals, blocking, and coalitions).

Given that leadership entails empowering others, and that empowerment has been defined in terms of the transfer of power or authority to employees (Bennis & Nanus, 1985; Burke, 1986), the concept of empowerment is briefly defined next.

TABLE 5.5

Six Types of Power: Their Source, Type of Social Influence, Definitions, Examples, and Expected Outcomes

Power Type, Source, and Type of Social Influence	Definitions	Source
1. Legitimate Power Source: Position Type of social influence: Compliance, internalisation, and identification	Legitimate power (also referred to as 'formal authority' or 'bureaucratic power') is the power assigned to a given position within an organisational structure. This power comes with the position and is assigned to the person who occupies a specific position within the organisation. Legitimate power stems from the justifiable right to request compliance from another organisational member. *Examples* Managers have the right, considering their position and job responsibilities, to expect their subordinates/team members to comply with legitimate requests. Managers have the authority to give subordinates/team members tasks or assignments. Subordinates/team members comply with their managers' request simply because their managers have legitimate rights or authority to ask them to do their work in certain ways. *Expected outcomes* Legitimate power can be effectively for some time. Continued reliance on it, however, may create dissatisfaction, resistance, and frustration among employees. If legitimate power does not match expert power, there may cause negative effects on productivity. Dependence on legitimate power only may lead to minimum compliance and increased employee resistance.	Elias (2008), French and Raven (1959), Munduate and Medina (2004), Pfeffer (1992)
2. Reward Power Source: Position Type of social influence: Compliance	Reward power is the power whose basis is the ability to reward. It is when powerholder promise some form of compensation to employees in exchange for compliance. Admittedly, reward power is inherent within the organisational structure (e.g., salaries). Managers' ability to influence employee's behaviour by providing them with things they want to receive (e.g., pay raises or bonuses, promotions, favourable work assignments, greater responsibility, new or special equipment, praise, or recognition). *Examples* Managers offer special rewards or benefits to subordinates/team members, as they find it advantageous to trade favours with them. *Expected outcomes* Reward power can influence the frequency of employee-performance behaviours initially. Prolonged us, however, can lead to a dependent relationship in which employees feel manipulated and become dissatisfied. Hence, causing decreases in performance.	Pfeffer (1992), French and Raven (1959), Munduate and Medina (2004)

(Continued)

TABLE 5.5 (Continued)

Six Types of Power: Their Source, Type of Social Influence, Definitions, Examples, and Expected Outcomes

3. Coercive Power	Definitions	French and Raven (1959),
Source: Position	Coercive power relates to the use of threat, punishment, or recommend punishment, in order to gain compliance.	Munduate and Medina (2004),
Type of social influence: Compliance	Coercive power is predicated upon fear.	Singh (2009)
	Examples	
	Managers threaten subordinates with termination of employment, withholding or depriving pay increases, or complaining about them to higher levels of management, should they not comply with certain requests.	
	Managers make things difficult for their subordinates, who comply to avoid getting into trouble due to their managers' anger and punitive behaviour.	
	Expected outcomes	
	Coercive power may lead to temporary compliance by employees. However, it can generate undesirable side effects such as frustration, fear, revenge, or alienation. This in turn may lead to dissatisfaction, poor performance, and employee turnover. Hence, coercive power may also be associated with conflict.	
4. Expert Power	Definitions	French and Raven (1959),
Source: Personal	Expert power is when managers rely on their superior knowledge, skills, or abilities in order to gain compliance.	Munduate and Medina (2004)
Type of social influence: Internalisation	*Examples*	
	Manager have the knowledge, experience, and proven ability to perform, earn respect, and defer to their judgment in certain matters.	
	Managers have the expertise to make sound decisions related to the work at hand.	
	Subordinates/team members follow the advice or instructions of their managers because they perceive them as possessing a high-level expertise in their field.	
	Expected outcomes	
	Expert power generates a climate of trust, which generates influence that can be internalised as employee motivation. This internalised employee motivation then requires less managers' surveillance of employees, and less reliance on using reward or coercive power.	

(Continued)

TABLE 5.5 (Continued)

Six Types of Power: Their Source, Type of Social Influence, Definitions, Examples, and Expected Outcomes

5. Referent Power Source: Personal Type of social influence: Identification	**Definitions** Referent power relates to the followers' identification with or the desire to be associated with the leader. Referent, or charismatic, power is the power of managers to influence employees by force of character or personal charisma. Referent power is when subordinates/team members comply with the requests of their managers because they recognise them as powerholders and influencing agents. Employees identify with their managers (e.g., personality identification, shared identity, hero worship, shared culture, or idealisation are some other source). *Examples* Employees comply with their managers' requests because they wish to move up the ladder or organisational hierarchy since they wish a similar position as that of their manager in the future. Managers have personal qualities that make them easy to be liked (e.g., have an attitude of enthusiasm and optimism that is contagious). Employees like their managers and enjoy doing things for them. *Expected outcomes* Referent power can lead to enthusiastic and unquestioning trust, loyalty, commitment, and compliance from employees. Like expert power, considerably less surveillance or supervision of employees (or use of reward or coercive power) is required.	French and Raven (1959), Munduate and Medina (2004), Singh (2009)
6. Informational Power Source: Position and Personal Type of social influence: Compliance	**Definitions** Informational power is driven by a powerholder's superior knowledge and information. The capacity to influence others based on the leader's knowledge of facts relevant to a specific situation. Power is derived from the ability to be able to access privilege information, as well as share or withhold it. It can be used to help others, to hurt others, or as a bargaining tool. *Examples* The leader has access to information not available to the followers, and this information convinces them the leaders is right. The leader has information team members need to do their work effectively. A project manager has all the information for a specific project. Nonetheless, it is hard for the manager to keep this power for too long, as eventually this information will be released. This is not an effective long-term strategy. *Expected outcomes* Because informational power is related to positional power, given that access to information often (but not exclusively) relates to the position the manager holds in the organisation, like coercive power may lead to temporary compliance by employees. However, can generate undesirable side effects such frustration, fear, revenge, conflict, or alienation. This is a short-term type of power that will not necessarily influence positively or build leader credibility.	Munduate and Medina (2004), Raven (1965)

5.3.3.1   Empowerment

As we look ahead into the next century, leaders will be those who empower others.

(Bill Gates, 2015, p. 167)

Psychological empowerment in organisations is "the perception by members that they have the opportunity to help determine work roles, accomplish meaningful work, and influence important decisions" (Yukl & Becker, 2006, p. 201). According to Thomas and Velthouse (1990), psychological empowerment entails the intrinsic task of motivating others by providing a sense of control and active orientation to their work role, which manifests in four cognitions: meaning, competence, self-determination, and impact. Robbins et al. (2002), assert that the most critical step in the process of empowering employees is the creation of a work environment within an organisational context that provides both an opportunity to exercise employees' full range of authority and power (empowered behaviours), and the intrinsic motivation within employees to engage in that type of behaviour (psychological empowerment). Table 5.6 summarises the four cognitive dimensions of psychological empowerment,

The four components described above are essential prerequisites for individuals' motivation to engage in empowered behaviour at work. More specifically, employees must want to do the task by feeling that it is worthwhile (meaning). They also must believe they are competent to engage in the behaviours required to do the by the environment (competence), must perceive they can make their choices (self-determination), and believe that their actions will have a significant influence on what

TABLE 5.6
Psychological Empowerment

Dimension	Definition	Source
1. Meaning	Relates to the value of the work goal or purpose judged in relation to employee's own ideals and standards. The fit or alignment between the demands of employees' work role and their own beliefs, goals, values, and standards. That is, the extent to which employees care about their work.	Hackman and Oldham (1980), Robbins et al. (2002)
2. Competence	The belief employees hold regarding their capability to skilfully perform their work activities.	Bandura (1977, 1982)
3. Self-determination	The sense of choice concerning the initiation or regulation of one's actions Indicates an individual's sense of choice or autonomy in initiation and regulation of actions or work behaviours and processes.	Deci et al. (1989)
4. Impact	The belief that one can influence strategic, administrative, or operational activities and outcomes in one's work unit. Denotes an individual's perceived degree of influence over outcomes in one's work environment.	Ashforth (1989), Spreitzer (1995)

happens in their environment (impact). Finally, Tracy (1992) offers the following ten steps to empower others:

1. Tell people what their responsibilities are.	6. Provide them with feedback on their performance.
2. Give them authority equal to the responsibility assigned to them.	7. Recognise them for their achievements.
3. Set standards of excellence.	8. Trust them.
4. Provide them with the needed training.	9. Give them permission to fail.
5. Give them knowledge and information.	10. Treat them with dignity and respect.

5.3.4 ADAPTIVE LEADERSHIP (TECHNICAL PROBLEMS VS ADAPTIVE CHALLENGES)

The distinction between leadership and authority builds on the one presented above and relates to adaptive leadership – "the practice of mobilizing people to tackle tough challenges and thrive" (Heifetz et al., 2009, p. 14). Adaptive leadership is a distributed leadership approach, in that it assumes that leadership can be displayed by people across an organisation, not only by those in positions of authority or in management roles. From this perspective, while distinct, leadership and management complement each other within a broad system of action. Management is linked to a position with authority, which is used to address "technical" or "routine" problems – those that are easy to identify and well defined and can be solved by applying well-known solutions or the knowledge of experts.

Leadership, on the contrary, addresses "adaptive" challenges – murky and systemic problems with no easy answers (Heifetz & Laurie, 2001). Adaptive challenges are difficult to define, have no known or clear-cut solutions, and call for new ideas to bring about change in numerous places that involve many stakeholders (Heifetz et al., 2009). Hence, adaptive work is distressing and painful for the individuals going through it need to take unfamiliar roles and responsibilities and change their values, preferences, and ways of working (Heifetz, 1998). According to Heifetz and Laurie (2001), solutions to adaptive challenges are to be found within the collective intelligence of employees at all organisational levels.

Table 5.7 summarises the main differences between technical problems and adaptive challenges, along with some examples.

The single biggest cause of leadership failures within organisations is due to not being able to identify adaptive challenges, and therefore treating them like if they were technical problems (Heifetz et al., 2009). Due to its highly technical nature, this is highly relevant to engineering. As noted by Ludwig (2001), when confronted with such types of complex problems, the management paradigm fails. Postmodern organisations are adaptive systems that need to match the complexity of their environment to survive (Boisot & McKelvey, 2010).

An equivalent distinction of technical problems and adaptive challenges is that between "tame" and "wicked" problems (Churchman, 1967; Grint, 2005; Rittel & Webber, 1973). Tame problems are those that we have experienced before and for which we have a known solution (e.g., building a small bridge or a tunnel). Wicked problems, on the contrary – like adaptive challenges – are ill-formulated, always occur in a social context, have complex interdependencies and innumerable causes, are difficult to recognise,

TABLE 5.7
Differences between Technical Problems and Adaptive Challenges

Technical Problems (Management)	Adaptive Challenges (Leadership)	Source
• Relatively easy to identify	• Difficult to identify (easy to deny)	Heifetz and
• Often lend themselves to quick an easy, clear-cut, and well-known solutions	• Require changes in people's values, beliefs, priorities, roles, responsibilities, relationships, and approaches to work	Laurie (2001), Heifetz et al. (2009)
• Often can be solved by an authority or the knowledge or advice of an expert	• Needs to be solved by people with, or affected by, the problem	
• Require change in just one or a few places – often contained within organisational boundaries	• Require change in multiple places – usually across organisational boundaries	
• People are generally receptive to technical solutions	• People often resist even acknowledging adaptive challenges	
• Solutions can often be implemented quickly – even by edict	• "Solutions" require experiments and new discoveries; they can take a long time to implement and cannot be implemented by edict	

Examples

Take the required measurements for building a bridge	Complete a mega project (e.g., a very large infrastructure project, a new generation of submarines), which requires establishing an alliance between organisations with very different organisational cultures
Implement a new electronic system within an organisation	Ensure all parties involve use the system according to specified requirements and timelines
Increase the pressure of the pumping system in a mine site or oil rig	Negotiate the construction of open-cut mining or a nuclear plant in a land protected with a native title

change constantly, involve many stakeholders with different values and agendas, have no known solutions, and often are symptoms of other problems (e.g., tackling poverty, terrorism, public policy). An example of an adaptive challenge that engineering managers would be likely to find themselves on, would be working on a mega project (e.g., a very large infrastructure project and a new generation of submarines), which requires establishing an alliance between organisations with very different organisational cultures (e.g., government or asset owner, various private design, supply, and construction organisations). Another example would be negotiating the construction of open-cut mining or a nuclear plant in a land with protected with a native title.

In essence, adaptive or wicked challenges cannot be solved with the existing mindsets, or repertoire of skills or modus operandi. When organisations attempt to address such problems, they usually uncover a gap between their current capacity and that actually needed to do so effectively.

5.3.5   LEADER DEVELOPMENT AND LEADERSHIP DEVELOPMENT

The distinction between leader development and leadership development (LD) is an important one and relates to the distinction between human capital and social capital (Day & Dragon, 2015). Similarly, the related distinction between LD and management development (MD) needs to be highlighted as different (yet interrelated) concepts, just as the differences between leadership and management were highlighted previously. This is despite the fact that both literatures do indeed overlap (Day, 2000).

MD includes managerial education and training (Latham & Seijts, 1998), and emphases the acquisition of specific types of knowledge, skills, and abilities to enhance performance in management roles (Baldwin & Padgett, 1994). Another aspect of MD relates to the application of proven solutions to known (technical) problems, which gives it a training orientation (Day, 2000). LD, on the contrary, relates to expanding the collective capacity of organisational members to engage effectively in leadership roles and processes (McCauley et al., 1998). Leadership roles are those that come with and without formal authority. Contrastingly, MD focuses on formal managerial positions or roles within the structure of the organisations. LD is oriented toward building capacity to tackle unexpected problems or challenges that could not have been predicted, or that occur from the breakdown of traditional organisational structures and the associated loss of sensemaking (Weick, 1993). An example of this would the 2010 Deepwater Horizon disaster that killed 11 people and smothered the Gulf of Mexico following an explosion on a BP oilrig, which caused what has been considered the largest marine oil spill in history (Monnier, 2021). Hence, LD has an anticipatory orientation for the unknown, as opposed to MD which focuses on ensuring consistency and predictability.

5.3.5.1   Leader Development

Leader development is "the expansion of the capacity of individuals to be effective in leadership roles and processes" (Day & Dragon, 2015, p. 134). Its focus is on developing the knowledge, skills, and abilities of individuals within formal leadership roles, which progressively need to develop capabilities in three domains: leading oneself, leading others, and leading the organisation (McCauley et al., 2010). This in line with the traditional view conceptualises leadership as a skill at the individual level. Hence, organisation invests in training and developing employees to enhance and protect their human capital (Lepak & Snell, 1999). Human capital is the knowledge, skill, creativity, and health of the individual (Becker, 2002). From this perspective, development is assumed to occur mostly via training the individual primarily in intrapersonal skills and abilities (Neck & Manz, 1996; Stewart et al., 1996). This approach is aligned with the concept of self-leadership (Manz, 2015; Neck & Houghton, 2006; Neck & Manz, 1996; Neck & Manz, 2012). Self-leadership is "a process through which individuals control their own behavior, influencing and leading themselves through the use of specific sets of behavioral and cognitive strategies" (Neck & Houghton, 2006, p. 270). Through this process individuals influence themselves to achieve the self-direction and self-motivation necessary to perform (Manz, 2015). Examples of the type of capabilities to develop at the intrapersonal competence level associated with leader development include emotional intelligence (EI, Goleman, 1996, 2011; Goleman & Boyatzis, 2017) and psychological capital (PsyCap, Luthans & Youssef, 2004; Luthans et al., 2004; Luthans et al., 2007). EI or emotional quotient (EQ) has been defined slightly differently according to the various EI/EQ models proposed by their corresponding researchers and will be discussed in detail later in this chapter as one of the main capabilities of the EMLCF. In short, at the most general level, and according to Goleman (1996), EI relates to the ability to identify, recognise, and regulate emotions in ourselves and in others and has four major EI domains (self-awareness, self-management, social awareness, and relationship management).

PsyCap is

"an individual's positive psychological state of development and is characterized by: (1) having confidence (self-efficacy) to take on and put in the necessary effort to succeed at challenging tasks; (2) making a positive attribution (optimism) about succeeding now and in the future; (3) persevering toward goals and, when necessary, redirecting paths to goals (hope) in order to succeed; and (4) when beset by problems and adversity, sustaining and bouncing back and even beyond (resiliency) to attain success" (Luthans et al. 2007, p. 3).

(Luthans et al. 2007, p. 3)

PsyCap will also be discussed later in more detail as a component of Global Mindset – another key capability of the EMLCF.

The expected outcomes to be achieved in developing individuals from an individualistic leadership, personal development perspective, or intrapersonal domain, include changes in a leader's knowledge, skills, abilities, self-views, or schemas (Kjellström et al., 2020). Theoretically, the above capabilities and intended outcomes are linked to Kegan's (1980, 1982, 1994) and Kegan and Lahey (1984) constructive-developmental theory of adult development, which focuses on the growth and elaboration of individuals' ways of understanding the self and the world. Constructive-developmental theory builds on the seminal work of Piaget (1954), relates to the development of meaning and meaning-making processes across the lifespan, and has been used to advance the understanding of leadership and LD (McCauley et al., 2006). The main five pedagogical methods or processes used to teach and develop the above leader development capabilities include 360-degree feedback, training, coaching, and mentoring (Day, 2000). Such practices, however, ignore 50 years of research indicating that leadership is a complex interaction between a chosen individual (the leader) and the social and organisational environment (Fiedler, 1996).

5.3.5.2 Leadership Development

LD is "the expansion of a collective's capacity to produce direction, alignment, and commitment" (McCauley et al., 2010, p. 20). The term collective, within this definition, refers to any group of people sharing their work, such as work groups, teams, organisations, partnerships, alliances, or communities. LD, therefore, shifts the focus from the processes of developing individual leaders who influence their followers towards the achievement of shared goals, to viewing the collective as a single entity or unit of focus and measurement (team, organisation, or community) and producing direction, alignment, and commitment (DAC) for such collective. This implies adopting a collective leadership model (Cullen et al., 2012; Fairhurst et al., 2020; Friedrich et al., 2016; Hunter et al., 2012; Mumford et al., 2012; Ospina et al., 2020; Raelin, 2018; Yammarino et al., 2012), with a focus on non-hierarchical and collectivistic configuration structures. Such an approach requires a paradigm shift, and re-configuration of power-based structures, from traditional vertical, hierarchical leadership towards more horizontal, shared, or distributed forms of leadership (Gronn, 2002; Pearce & Conger, 2002; Pearce et al., 2008). From this perspective, LD focuses on the benefits deriving from the social resources embedded within work relationships or social capital (Brass & Krackhardt, 1999), as opposed to focusing on developing the individual knowledge, skills, and abilities of individuals (human capital). Social capital is created through relational or interpersonal exchange of durable networks of more or less institutionalised relationships (Bourdieu, 1986). By leveraging from these networked relationships, social capital enhances exchange of resources, cooperation, and collaboration to create innovation (Bouty, 2000) and organisational value (Tsai & Ghoshal, 1998). Therefore, social capital is grounded in a relational model of leadership and requires an interpersonal lens (Drath & Palus, 1994), and its effectiveness is based on commitments, and mutual obligations that are supported by reciprocated trust and respect (Brower et al., 2000). Commitments, trust, and respect correspond to Nahapiet and Ghoshal's (1998) three different aspects of social capital: structural, relational, and cognitive.

Examples of the type of capabilities to develop at the intrapersonal competence level associated with LD include the social awareness and relationship management clusters of EI; capabilities such as collaboration, teaming, and teamwork (Edmondson, 2012); team psychological safety (Edmondson, 1999); and climate for creativity and change (Ekvall, 1996), which comprises nine dimensions (challenge, freedom, trust/openness, idea time, playfulness/humour, risk-taking, idea support, debate, and conflict). Team psychological safety is "a shared belief that the team is safe for interpersonal risk taking "(Edmondson, 1999, p. 354). This, however, doesn't suggest carelessness or permissiveness by team members, but rather a sense of confidence that the team will not reject, embarrass, or punish any member for speaking up. This confidence derives from mutual respect and trust among members of the team. The expected outcomes to be achieved in developing this interpersonal domain include changes in the collective capacity for leadership in a group, team, or organisation (Kjellström et al., 2020); innovation (Horth and Buchner, 2014); and "organized complexity" (Gharajedaghi, 1999, pp. 92–93).

The theoretical foundations of such capabilities and intended outcomes include team coaching theory (Hackman & Wageman, 2005) and social network analysis (SNA, Freeman, 2004). Team coaching is the "direct interaction with a team intended to help members make coordinated and task-appropriate use of their collective resources in accomplishing the team's work" theory (Hackman & Wageman, 2005, p. 269). It is the collaborative endeavour of reflection and dialogue team leaders use to help their teams improve the processes that lead to achieved performance (Clutterbuck et al., 2016) and involves multiple techniques (Lancer et al., 2016). SNA refers to the set of theories, tools, and processes for understanding the relationships and structures of a network. It represents organisations as social groupings that show certain patterns of interaction evolving over time. SNA aims to identify these structures and patterns, as well as their evolving nature, causes, and consequences (Freeman, 2004). Network analysis, then, seeks to uncover various kinds of patterns and tries to determine the conditions under which those patterns arise and discover their consequences. The above is also in line with systems thinking theory (Bailey, 2005; Gharajedaghi, 1999; Monat et al., 2020; Reynolds & Holwell, 2010). Systems thinking views organisations as complex adaptive systems and advocates the understanding of reality by emphasising the relationships among the parts of a system, as opposed to focusing only on the parts themselves from a conventional reductionist thinking approach, which regards the organisation as a machine. From this perspective, the behaviours and structures of organisations emerge from the collective interaction (collective conversation) of their organisational members (Boal & Schultz, 2007).

The methods or processes used for LD include are range of integrated strategies such as: Action learning; action research; Creative Problem Solving; job assignments; networking; group facilitation; and team coaching. Their aim is assisting people understand how to relate to others, build commitments to effectively coordinate their actions, and develop extended social networks (Day, 2000). These relational and multilevel views of leadership include the networked patterns of social relationships linking individuals and teams to larger collectives. This leads to new approaches for network-enhancing leadership development to improve the leadership capacity of organisations. Table 5.8 summarises the main differences between management and leadership, according to seven dimensions of comparison (capital type; leadership model or perspective; domain; capability type; expected outcomes; theoretical foundation; and methods and practices).

As mentioned previously, leadership is a highly contextualised phenomenon. This means that leadership never occurs in a vacuum. As the context changes, to be effective, leadership needs to adjust to the new context (Antonakis et al., 2003; Fairhurst, 2009; Oc, 2018; Osborn et al., 2002; Osborn et al., 2014; Porter & McLaughlin, 2006; Shamir & Howell, 1999). The next section provides an overview of the current for leadership.

5.4 CURRENT LEADERSHIP CONTEXT

Leadership is a highly contextualised social phenomenon. As the context changes, to be effective, leadership also has to change and be embedded in its context (Osborn & Marion, 2009). More specifically, the context of leadership relates to the environment, conditions, or circumstances (e.g., physical, sociocultural, economic, political) in which leadership exists and is observed (Liden & Antonakis, 2009). Regrettably, as noted by Zaccaro and Klimoski (2002), "most theories of organizational leadership in the psychological literature are largely context free" (p. 12). This problematic because the current leadership context is very different than what it was just over two decades ago. Hence, leadership cannot be effectively without attending to such contextual changes.

According to Arthur (1996), five trends have been driving the new economy and redefining their corresponding domains: (1) globalisation has redefined the concept of space; (2) networking and connectivity have reorganised structures; (3)

TABLE 5.8

Summary of Differences between Leader Development and Leadership Development

Comparison Dimension	Leader Development	Leadership Development	Source
Capital type	Human capital	Social capital	Day (2000), Nahapiet and Ghoshal (1998), Kjellström et al. (2020)
Leadership model and perspective	• Individual • Self-leadership	• Collective • Distributed • Network Leadership • Relational • Shared	Drath and Palus (1994), Manz (2015), McCauley et al. (2010), Cullen et al. (2012), Fairhurst et al. (2020), Friedrich et al. (2016), Hunter et al. (2012), Mumford et al. (2012), Ospina et al. (2020), Raelin, (2018), Yammarino et al. (2012)
Domain	Intrapersonal	Interpersonal	Day (2000)
Capability type	Self-leadership: Self-direction and Self-motivation	Emotional Intelligence: Social awareness and Relationship Management	Manz (2015), Goleman (1996, 2011)
	Emotional Intelligence • Self-awareness • Self-management	• Collaboration • Teaming • Teamwork • Team psychological safety	Goleman and Boyatzis (2017), Edmondson (1999, 2012)
	Psychological Capital • Hope • Confidence (self-efficacy) • Resiliency • Optimism	Climate for creativity and change (challenge, freedom, trust/openness, idea time, playfulness/humour, risk-taking, idea support, debate, and conflict).	Luthans and Youssef (2004), Luthans et al. (2004, 2007), Luthans and Youssef-Morgan (2017), Ekvall (1996)

(Continued)

TABLE 5.8 (*Continued*)
Summary of Differences between Leader Development and Leadership Development

Comparison Dimension	Leader Development	Leadership Development	Source
Expected outcomes	Changes in a leader's knowledge, skills, abilities, self-views, and schemas (mindsets).	• Changes in the collective capacity for leadership in a group, team, or organisation. • Innovation Leadership • Organised Complexity	Horth and Buchner (2014), Kjellström et al. (2020), (Gharajedaghi (1999)
Theoretical foundation	Constructive-developmental theory of adult development	• Team Coaching Theory • Social Network Analysis • Systems Thinking	Freeman (2004), Hackman and Wageman (2005), Kegan (1980, 1982, 1994), Kegan and Lahey (1984), Bailey (2005), Monat et al. (2020), Reynolds & Holwell (2010)
Methods and Practices	• 360-dgree feedback • Training • Coaching • Mentoring	• Action Learning • Action Research • Creative Problem Solving • Job Assignments • Networking • Group Facilitation • Team Coaching	Clutterbuck et al. (2016), Cullen et al. (2014), Day (2000), Grayson and Baldwin (2011), Hawkins, 2004, 2021), Lancer et al. (2016)

dematerialisation of products into knowledge has increased the value of intangible assets; (4) speed has become a primary source of competitive advantage; and (5) increasing returns have redefined competition. Similarly, despite the enormous benefits derived from advances in digital technology, big data, artificial intelligence, and data-driven innovation, the risks of their misuse can lead to data workflows that bypass privacy and data protection laws, as well as the failure of ethical imperatives (Da Bormida, 2021). The above-outlined changes have had, and continue having, a profound impact across industries, who now must adapt by playing the new rules for business (Jaworski & Scharmer, 2000). Hence, as noted by Fullan (2001), "The more complex society gets, the more sophisticated leadership must become" (p. ix).

The new context is one of high velocity, complexity, turbulence, and social and economic unrest. High-velocity environments are those that become hypercompetitive due to continuously changing expectations caused by the disruption of new technologies and/or regulations, which quickly makes information inaccurate, obsolete, and where conventional approaches no longer work (Bogner & Barr, 2000). These environments are also inherently turbulent (Lichtenthaler, 2009). Turbulence refers "the amount of change and complexity in the environment of an industry" (Kipley, Lewis, & Jewe, 2012, p. 251) created by constantly changing economic conditions (Perrot, 2011). Related constructs to those mentioned above that are found in the literature include dynamic environments (Sirmon, Hitt, & Ireland, 2007), reliability-seeking organisations (Vogus & Welbourne, 2003), and clock speed (Nadkarni & Narayanan, 2007).

The acronym VUCA (volatility, uncertainty, complexity, and ambiguity), which originated in the early 1990s from the U.S. Army War College, also become popular to describe this new world (Horney, Pasmore, & O'Shea, 2010). The term "VUCA world" describes new environments characterised by volatility – the speed and turbulence of change; uncertainty – the fact that outcomes and familiar actions are less predictable; complexity – the enormity of interdependencies in globally connected economies and societies; and ambiguity – the multitude of options and potential outcomes resulting from them (Bennett & Lemoine, 2014). Consequently, this new environment has created unfamiliar, and often confusing, situations by posing new types of challenges referred to as adaptive challenges, as opposed to technical ones, that required adaptive leaders (Doyle, 2017). This is particularly relevant to technical professions like engineering. As a result, adaptive leadership (DeRue, 2011; Heifetz et al., 2009) will be discussed in more detail later in this chapter.

Further, in the current globalised economy, many organisations operate on a global scale. Hence, we now live and work in a global village. This means most organisations have diverse cultural, political and institutional systems to help them achieve their global ambitions while managing multiplicities, tackling huge challenges, grappling with instability and navigating ambiguity (Osland et al., 2012). Culture is pervasive and has multiple layers that can often be invisible to the untrained eye. It acts like a pair of glasses that colours our vision. Culture works like a powerful filter through which we perceive and experience reality. It is like the mental software that we use to decode, interpret, encode and send messages. Culture determines how people "do things around here", it is the "unwritten rules of the social game", and what we consider "normal" in any given society. It is the glue that holds societies together. Leadership beliefs, expectations and practices are not readily portable from one culture to another. Hence, applying them uniformly across geographies is a fool's errand, much as we'd like to think otherwise. All this makes culture a critical business risk (Salicru, 2017). As result, proving effective leadership within this global context, requires global leadership and having a global mindset (Clapp-Smith & Lester, 2014) – the ability to absorb information, understand traditions and cultural norms with openness and awareness of diversity, and to be able to exercises to affect change. Intercultural competence (Zheng, 2015) or cultural intelligence (CQ) – the ability to interact effectively in multiple cultures (Ang & Van Dyne, 2015; Crowne, 2008) is also necessary.

Adding to this complexity are the challenges mentioned thus far, are the challenges associated with the digital revolution that has transformed the economy and society. This includes: the need to introduce technological, as well employment and workforce management innovations (e.g., virtual teams); the transfer of tacit knowledge into explicit knowledge; cyber security; and the governance required to manage ethical concerns related to data collection and privacy (Flyverbom et al., 2019). In summary, the increasingly complex, dynamic, uncertain, socio-culturally, technologically, and ethically demanding context calls for new leadership capabilities.

5.5 THE ENGINEERING MANAGERS' LEADERSHIP CAPABILITY FRAMEWORK (EMLCF): LEADERSHIP CAPABILITIES FOR THE 21ST CENTURY AND BEYOND

The EMLCF is a holistic framework that integrates eight high-level capabilities or meta-competencies. Table 5.9 summarises these eight capabilities.

Next, each one of these eight capabilities will be discussed in more detail.

CAVEAT

It is important to notice that while the above are represented as discrete capabilities within the framework, there is a degree of overlap between some of them. For example, some components of emotional intelligence overlap with some aspects of cultural intelligence and ethical behaviour. Social awareness is a case in point (e.g., the ability to take the perspective of and empathise with others, including those from diverse backgrounds and cultures, to understand social and ethical norms for behaviour). From this perspective, the EMLCF should be viewed as an integrated or blended whole unified into a functional open system, which contains complementary and supplementary aspects that confirm and reinforce each other from various research standpoints, as opposed to a fixed or rigid boundary-less closed system. As an open system, the EMLCF allows interactions between its internal elements and the environment.

TABLE 5.9

The Engineering Managers' Leadership Capability Framework (EMLCF) – Eight Global Leadership Capabilities

Capability	Definition/Related or Interchangeable Constructs/Components/Benefits	Sources
1. Self-leadership and psychological capital	A process through which individuals influence their own behaviour to achieve the self-direction and self-motivation necessary to perform, empower themselves, and achieve personal excellence Meaningfulness, purpose, self-determination, competence, and self-efficacy Greater job satisfaction, lower stress levels, and transformational leadership Psychological Capital (PsyCap) relates to the state of development that influences individuals' levels of satisfaction and performance. PsyCap is the result of the powerful synergistic effect of four psychological states: hope, efficacy, resilience, and optimism (HERO).	Manz (2015), Neck et al. (2012), Neck and Manz (2012), Dolbier et al. (2001), Harari et al. (2021) Luthans and Youssef (2004), Luthans et al. (2004, 2007), Luthans and Youssef-Morgan (2017)
2. Contextual intelligence	Ability to recognise and diagnose the many contextual factors inherent in an event, and then intentionally and intuitively adjust behaviour to exert influence in that context Hindsight, foresight, and insight	Khanna (2014, 2015), Kutz (2008, 2017), Kutz and Bamford-Wade (2014), Oc (2018)
3. Sensemaking, framing, and storytelling	Sensemaking is about making sense of the world around us by structuring the unknown to be able to act in it. Sensemaking relates to contextual rationality, and is built on vague questions, muddy answers, and negotiated agreements that attempt to reduce confusion. It is mostly required in rapidly changing contexts where surprises and adaptive challenges emerge, for which people are confronted and unprepared. Framing re-organises experiences and produce new meanings. Frames are cognitive heuristic or mental shortcuts that people use to help make sense of complex information. Storytelling is a powerful way to explicitly or implicitly transfer both information and emotion that can move others to action. Storytelling is a strong motivational strategy in response to crises or during times of change, upheaval, and uncertainty.	Sensemaking (Ancoina, 2012; Aron & Leykum, 2022; Maitlis & Christianson, 2014; Weick, 1995, 1993, 2001, 2009, 2010; Weick et al., 2005) Framing (Cornelissen & Werner, 2014; Fairhurst, 2005, 2010; Fairhurst & Sarr, 1996). Storytelling (Denning, 2005; Mitroff & Kilmann, 1975; Snowden, 2000).

(Continued)

TABLE 5.9 (Continued)
The Engineering Managers' Leadership Capability Framework (EMLCF) – Eight Global Leadership Capabilities

Capability	Definition/Related or Interchangeable Constructs/Components/Benefits	Sources
4. Learning agility	Willingness and ability to learn new competencies in order to perform under first-time, tough, or different conditions. Comprises five factors: mental agility, people agility, change agility, results agility, and self-awareness.	De Meuse et al. (2011), Lombardo and Eichinger (2000)
5. Global leadership, global mindset, and cultural intelligence	Global Leadership: A process of influencing the thinking, attitudes, and behaviours of a global community to work together synergistically toward a common vision and common goals. Global Mindset: A set of attributes and skills that contribute to effective leadership in a global corporation; The ability to develop and interpret criteria for personal and business performance that are independent from the assumptions of a single country, culture, or context; and to implement those criteria appropriately in different countries, cultures, and contexts; The process of influencing individuals, groups, and organisations (inside and outside the boundaries of the global organisation) representing diverse cultural/political/institutional systems to help achieve the global organisation's goals; A highly complex cognitive structure characterised by an openness to and articulation of multiple cultural and strategic realities on both global and local levels, and the cognitive ability to mediate and integrate across this multiplicity; The capability to influence others unlike yourself – and that is the key difference between leadership and global leadership. Cultural Intelligence: The capability to function effectively in culturally diverse settings; A person's capability to adapt effectively to new cultural contexts; A person's adaptation to new cultural settings and capability to deal effectively with other people with whom the person does not share a common cultural background and understanding; Related/interchangeable constructs: cultural, intercultural, or cross-cultural competence, and cultural adaptability – the ability of an individual to effectively interact, work, and develop meaningful relationships with people of various cultural backgrounds.	Dorfman et al. (2012), Giddens (1999), House et al. (2004), Javidan et al. (2010), Osland (et al. 2006) Beechler and Javidan, 2007), Javidan and Walker (2012), Levy et al. (2007), Maznevski and Lane (2004), Pucik (2005) Ang et al. (2007), Earley and Ang (2003), Earley and Mosakowski (2004), Van Dyne et al. (2015)

(Continued)

TABLE 5.9 (*Continued*)

The Engineering Managers' Leadership Capability Framework (EMLCF) – Eight Global Leadership Capabilities

Capability	Definition/Related or Interchangeable Constructs/Components/Benefits	Sources
6. Emotional intelligence	The ability to perceive and express emotions, to use emotions to facilitate thinking, to understand and reason with emotions, and to effectively manage emotions within oneself and in relationships with others The capacity for recognising our own feelings and those of others, for motivating ourselves, and for managing emotions well in ourselves and in our relationships Self-awareness, self-Management or self-regulation, social awareness, and relationship management	Chemiss et al. (2001), Goleman (1995, 1998, 2011), Mayer and Salovey (1997), Mayer et al. (2008), Salovey and Mayer (1990)
7. Creative thinking and innovative behaviour	Creativity relates to the production of novel and useful ideas, or socially valued products or services. Innovation relates to the production or adoption of useful ideas and idea implementation, and is central to the long-term survival of organisations. Leadership is a chief predictor of creativity – the precursor of all innovation. Leaders establish work environments that are conducive to creative thinking and innovation. Leader behaviour shapes company culture and climate, and predicts innovative workplace behaviour – the behaviour that guides the initiation and intentional introduction of new and useful ideas, processes, products, services, or procedures. Transformational and participative or collaborative leadership, generates employees' intrinsic motivation, psychological empowerment, creative thinking, and innovative workplace behaviour (IWB). IWB entails: searching out new technologies, processes, techniques and/or concepts/ideas; generating new and creative ideas; promoting and championing new ideas to others; implementing new and useful ideas; and developing adequate plans and schedules for this implementation.	Amabile et al. (1996), Hennessey and Amabile (2010), Mumford and Gustafson (1988), Salicru (2017), Scott and Bruce (1994), Van de Ven (1986), Whitehurst (2016)

(*Continued*)

TABLE 5.9 (*Continued*)
The Engineering Managers' Leadership Capability Framework (EMLCF) – Eight Global Leadership Capabilities

Capability	Definition/Related or Interchangeable Constructs/Components/Benefits	Sources
8. Ethical behaviour and ethical leadership	Ethics is central to science and engineering, is at the heart of leadership, and has been recognised as an essential component in business success. The bottom line of business success always includes an ethics component. Ethical leadership is the demonstration of appropriate conduct through personal actions and interpersonal relationships by promoting such conduct to followers through two-way communication, reinforcement, and decision-making. Ethical leadership entails exercising influence in ways that are ethical in both means and in ends. Ethical leadership is essential to build a culture of corporate social responsibility. It also improves employee attitudes, job satisfaction, affective commitment, and work engagement, and reduces employee turnover intentions. Behavioural ethics explains why good people sometimes do bad things. It is an emerging discipline that studies business ethics scientifically by drawing on research from behavioural psychology, cognitive science, neuroscience, and evolutionary biology. Behavioural ethics focuses on how and why people make ethical – and unethical – decisions, with the aim to improve people's ethical decision-making and actions.	Ciulla (2014), Bazerman and Gino, 2012; Brown et al. (2005), De Cremer et al. (2010), De Cremer and Moore (2020), Mitchell et al. (2017), Thomas et al. (2004), Tanner et al. (2010), Treviño et al. (2003, 2006)

5.5.1 Self-Leadership and Psychological Capital

Self-leadership is a process through which individuals influence their own behaviour to achieve the self-direction and self-motivation necessary to perform, empower themselves, and achieve personal excellence (Manz, 2015; Neck et al., 2012; Neck & Manz, 2012). Self-leadership strategies have been found to facilitate empowerment by enhancing meaningfulness, purpose, self-determination, competence, and self-efficacy – an individual's belief in their capacity to act in the ways necessary to reach specific goals (Bandura, 1977).

Hence, people with high levels of self-efficacy are more likely to believe they can achieve what they want to accomplish. Self-leadership also derives greater job satisfaction and lower stress levels (Dolbier et al., 2001), and is positively associated with conscientiousness, openness, extraversion, and transformational leadership (Harari et al., 2021). Developing and building self-leadership entails both behavioural and cognitive strategies that fall into three main categories:

1. Behavioural-focused strategies that promote self-management (self-goal setting, self-observation, self-reward, self-punishment, and self-cueing);
2. Natural reward strategies to develop intrinsic motivation; and
3. Constructive thought pattern strategies, which involve visualising successful performance, self-talk, and evaluating beliefs and assumptions.

The revised self-leadership questionnaire (RSLQ, Houghton & Neck, 2002) is one of the most reliable and valid measures of self-leadership skills, behaviours, and cognitions. The RSLQ consists of 35 items in nine subscales. Table 5.10 unpacks the three dimensions of self-leadership.

The positive organisational behaviour (POB) movement (Cameron & Spreitzer 2012; Luthans, 2002) offers an alternative psychological capital model. POB is "the study and application of positively oriented human resource strengths and psychological capacities that can be measured, developed, and effectively managed for performance improvement in today's workplace" (Luthans, 2002, p. 59). This organisational science movement focuses on the dynamics that lead to extraordinary individual and organisational performance by developing human strengths (Cameron & Caza, 2004). From a POB perspective, psychological capital (PsyCap, Luthans & Youssef-Morgan, 2017) relates to the state of development that influences individuals' levels of satisfaction and performance. PsyCap is the result of the powerful synergistic effect of four psychological states: hope, efficacy, resilience, and optimism (HERO). The Psychological Capital Questionnaire (PCQ) has been recognised as the standard scale to measure PsyCap (Dawkins et al., 2013). Table 5.11 captures the HERO model of PsyCap, including definitions for each of the four construct and example items from the PCQ.

PsyCap is considered a vital factor for both leader and leadership development (Pitichat et al., 2018). The integration of authentic leadership and PsyCap fosters employees' creativity (Rego et al., 2012), and leader PsyCap promotes innovative behaviour in employees (Wang et al., 2021).

TABLE 5.10

Self-Leadership Dimensions – Focus and Examples

Dimensions	Focus	Sub-Scales (9)
Behaviour-focused	Aimed at increasing self-awareness, leading to the management of behaviours involving necessary but perhaps unpleasant tasks. Designed to encourage positive, desirable behaviours that lead to successful outcomes, while suppressing negative, undesirable behaviours that lead to unsuccessful outcomes.	Five sub-scales 1 Self-goal setting strategies 2 Self-reward 3 Self-punishment 4 Self-observation 5 Self-cueing
	Examples • I establish specific goals for my own performance. • I use written notes to remind myself of what I need to accomplish. • I pay attention to how well I am doing in my work. • I keep track of my progress on projects I'm working on.	
Natural reward strategies	Aimed at changing perceptions of an activity by focusing on the task's inherently rewarding aspects. Emphasise the enjoyable aspects of a given task or activity, which result when incentives are built into the task itself and a person is motivated or rewarded by the task itself. Foster feelings of increased competence, self-control, and purpose.	A single sub-scale 1. Focusing thoughts on natural rewards
	Examples • I try to surround myself with the objects and people that bring out my desirable behaviours. • I seek out activities in my work that I enjoy doing. • I find my own preferred way to do things.	
Constructive thought pattern strategies	Aimed at creating and maintaining functional patterns of habitual thinking. Include the evaluation and challenging of irrational beliefs and assumptions, mental imagery of successful future performance, and positive self-talk.	Three sub-scales 1 Visualising successful performance 2 Self-talk 3 Evaluating beliefs and assumptions
	Examples • I use my imagination to picture myself performing well on important tasks. • I purposefully visualise myself overcoming the challenges I face. • I often mentally rehearse the way I plan to deal with a challenge before I actually face the challenge.	

TABLE 5.11
The HERO Model of Psychological Capital

Component	Definition and Examples
Hope	The will and the way – one's desire, ambition, and expectation to persevere and, when necessary, to change direction to reach one's goals. A positive motivational state based on an interactively derived sense of successful: (1) agency (goal-directed energy); and (2) generation of pathways (planning to meet goals). *Examples* • I am energetically pursuing my work goals. • I have several ways to accomplish the work goals. • When I set goals and plan to work, I concentrate to achieve these goals.
Efficacy	The confidence to succeed – the belief in one's ability to take on and succeed at challenging tasks within a given context. The conviction about one's abilities to generate the motivation, and generate the cognitive resources or courses of action needed to successfully execute a specific task within a given context. *Examples* • I feel confident in analysing a long-term problem to find a solution. • I am confident in my performance that I can work under pressure and challenging circumstances. • I feel confident that I can accomplish my work goals.
Resilience	The capacity to bounce back from adversity, conflict, and failure to succeed, and adapt to changing and stressful demands. The capacity to rebound or bounce back from adversity, conflict, failure, or even positive events, to progress and increased responsibility. *Examples* • I usually manage difficulties one way or another at work. • Although my task has failed, I will try to make it succeed again. • Although too much responsibility at work makes me feel awkward, I can go through to work successfully.
Optimism	A positive explanatory style that attributes positive events to personal, permanent, and pervasive causes, and interprets negative events in terms of external, temporary, and situation-specific factors; this results in a generalised positive outlook that yields positive expectancies. In contrast to a pessimistic explanatory style that attributes positive events to external, temporary, and situation-specific causes, and negative events to personal, permanent, and pervasive ones. *Examples* • I'm optimistic about what will happen to me in the future as it relates to work. • At work, I always find that every problem has a solution. • If I have to face with a bad situation, I believe that everything will change to be better.

5.5.2 CONTEXTUAL INTELLIGENCE

Contextual intelligence (CI) "is the ability to quickly and intuitively recognize and diagnose the dynamic contextual variables inherent in an event or circumstance and results in intentional adjustment of behavior in order to exert appropriate influence in that context" (Kutz, 2008. p. 23). It relates to contextual leadership (Oc, 2018) and the need for leaders to understand the context in which they are required to lead. As discussed earlier, leadership never takes place in a vacuum; hence, the importance of understanding its contextual factors. Context relates to the nature of interactions and interdependencies among and between the multiple events and agents within a system; namely – cultures, people, ideas, values, experiences, and alliances. CI relates to the awareness of the dynamics among these events and agents, which ultimately informs behaviour in a given socially complex environment. This environment must be considered in light of an unpredictable future, while taking into consideration history and tradition (Kutz & Bamford-Wade, 2014). CI has three key abilities: hindsight; foresight; and insight. Hindsight relates to an intuitive grasp of relevant past events, for leaders to take full advantage of what they have learned in the past. Foresight entails acute awareness of the present context for leaders to clearly articulate what they wish to become, and clarify what they will do to reach their goals and aspirations. Insight is the convergence of hindsight and foresight. That is, informed by hindsight and inspired by foresight, leaders can gain the clarity and understanding to make appropriate decisions to exert influence within the context at hand (Kutz, 2017). In sum, CI relates to the understanding of the limits of knowledge, and to adapt that knowledge to a context different from the one in which it was acquired (Khanna, 2014).

5.5.3 SENSEMAKING, FRAMING, AND STORYTELLING

Sensemaking was first coined by Weick (1995) as "the making of sense" (p. 4), and relates to making sense of the world around us by structuring the unknown to be able to act in it (Waterman, 1990). This includes continuously understanding developments in your business or work environment, and interpreting their consequences for your organisation and industry. For example, how digitisation, AI, and new technologies will reshape your industry? "Sensemaking is the process by which people give meaning to an experience that is somehow at odds with expectations" (Aron & Leykum, 2022, p. 96). "Sensemaking involves coming up with plausible understandings and meanings; testing them with others and via action; and then refining our understandings or abandoning them in favor of new ones that better explain a shifting reality" (Ancona, 2012, p. 5). From this perspective, sensemaking is not concerned with accuracy or finding the "correct" answer, but rather about "plausibility" by creating a more meaningful picture that enables people to act.

According to Maitlis and Christianson (2014), sensemaking is the process that enables individuals to understand and make sense of experiences, events or issues that are confusing, ambiguous, or unexpected. Sensemaking then is mostly required in rapidly changing contexts where surprises and adaptive challenges emerge for which people are confronted and unprepared (Heifetz et al., 2009). The genesis of sensemaking, then, is chaos and confusion, as people attempt to answer the question, "what's the story?", by mapping the context and the ongoing unpredictable experiences thrown at them (Weick et al., 2005). It involves turning circumstances into a situation that is comprehended explicitly in words and that serves as a springboard into action (Weick et al., 2005). Therefore, sensemaking is a conversational and narrative process that entails multiple communication categories – written and spoken, formal and informal (e.g., rumours, gossip, negotiations, and exchange of stories). Hence, the importance of framing and storytelling for leaders as means to simplify complex and confusing situations.

Framing (Cornelissen & Werner, 2014; Fairhurst, 2005, 2010; Fairhurst & Sarr, 1996) relates to the ability shape the meaning of a subject [or situation], to judge its character and significance (Fairhurst & Sarr, 1996, p. 3). Frames are cognitive heuristic or mental shortcuts that people use to help make sense of complex information. They can significantly affect the intractability of a conflict by creating mutually inconsistent interpretations of events (Kaufman et al., 2003). Frames include definitions of situations that re-organise experiences and produce new meanings. They are as multidimensional and multi-layered. The art of framing includes five key language devises: metaphor, jargon or catchphrases, contrast, spin, and stories. They highlight how reality, truth, objectivity, and legitimacy manifest themselves linguistically and contribute to mixed messages (Fairhurst, 2005). This affords people adaptive sensemaking – the ability to frame, understand and respond to an evolving situation and mobilise to action (Cornelissen & Werner, 2014).

Storytelling is a powerful way to explicitly or implicitly transfer both information and emotion (Snowden, 2000) that can move others to action. Impactful stories appeal to the intellect and evoke emotion (Denning, 2005), and contextualise and encapsulate messages (Pink, 2005). Hence, it not surprising that leaders through history have used storytelling as a powerful motivational strategy in response to crises or during times of change, upheaval, and uncertainty (Forster et al., 1999). In organisations, storytelling serves multiple purposes, namely: problem solving and conducting action research (Mitroff & Kilmann, 1975); generating organisational renewal (McWhinney & Batista, 1988); transferring knowledge in the workplace when mentoring others (Swap et al., 2001); facilitating internal and external communications, developing teams and leadership skills, and engaging clients and customers (Collison & Mackenzie, 1999); and communicating complex ideas and persuading others to change (Prusak et al., 2012). According to Snowden (2003) when stories are told and retold over time, they create or reinforce themes as well as characters. A good example are the many stories told at Virgin about the founder, Richard Branson. Strategic leaders construct the shared meanings that provide the rationale for the continuity of the organisation's past, present, and future through dialogue and storytelling (Boal & Schultz, 2007).

5.5.4 Learning Agility

Learning agility is "the willingness and ability to learn new competencies in order to perform under first-time, tough, or different conditions" (Lombardo & Eichinger, 2000, p. 323). As highlighted earlier, contemporary organisations operate in an environment of constant change due to increased globalisation, turbulent economic conditions, working with temporary virtual interactions and social media, working across cultures, and rapidly adapting to technological advancements. As a result, leaders must develop agility as a core capability, as a means to be able to respond effectively to the uncertainty and ambiguity of contemporary markets. This entails flexible and adaptive leadership. This is the type of leadership that involves adapting behaviour appropriately as the situation changes by being adaptable, agile, flexible, and versatile (Yukl & Mahsud, 2010). Highly learning agile individuals continuously seek out new challenges and feedback from others to be able to grow and develop, and are reflective. Such individuals are likely to succeed when promoted, placed in to international assignments, or assigned with challenging projects.

Building on Lombardo and Eichinger's (2000) seminal work, De Meuse et al. (2011, p. 7) conceptualised learning agility comprising the following five factors:

1. Mental agility – The extent to which an individual is comfortable with complexity, examines problems carefully, is inquisitive, and can make fresh connections between different concepts.
2. People agility – The degree to which a person is open-minded toward others, interpersonally skilled, and can deal readily with a diversity of people and difficult situations.
3. Change agility – The extent to which an individual is comfortable with change, interested in continuous improvement, and in leading change efforts.
4. Results agility – The degree to which an individual can deliver results in first time and/or tough situations through sheer personal drive and by inspiring teams.
5. Self-awareness – The depth to which a person knows him or herself, recognising skills, strengths, weaknesses, blind spots, and hidden strengths.

Learning agility has been identified as an example of meta-competency in that it is an individual's attribute which is a prerequisite for the development of other competencies (De Meuse et al., 2012).

5.5.5 GLOBAL LEADERSHIP, GLOBAL MINDSET, AND CULTURAL INTELLIGENCE

Globalisation – the worldwide cultural, political, and economic interconnections resulting from the abolishment of communication and trade barriers (Giddens, 1999) – has created a single global society. Leading effectively in this global world requires three interrelated constructs: global leadership; global mindset; and cultural intelligence.

5.5.5.1 Global Leadership

Global leadership (GL) is "a process of influencing the thinking, attitudes and behaviors of a global community to work together synergistically toward a common vision and common goals" (Osland et al., 2006. p. 204). GL is exercised in a unique context, characterised by strategic and cultural complexity that crosses mental, organisational, and physical boundaries. This leadership requires dealing with paradoxes and the establishment of common ground.

The GLOBE (Global Leadership and Organizational Behavior Effectiveness) programme (Dorfman et al., 2012; House et al., 2004), for example, explored the effects of culture on leadership and organisational effectiveness. This was a 20-year project that began in 1993, and involved over 170 researchers studying the culture and leadership in 62 nations, using survey responses of 17,300 participants. National cultures were studied using the following nine dimensions: (1) Power distance – the degree to which members of a collective expect power to be distributed equally; (2) Uncertainty avoidance – the extent to which a society, organisation, or group relies on social norms, rules, and procedures to alleviate unpredictability of future events; (3) Humane orientation – the degree to which a collective encourages and rewards individuals for being fair, altruistic, generous, caring and kind to others; (4) Institutional collectivism – the degree to which organisational and societal institutional practices encourage and reward collective distribution of resources and action; (5) In-group collectivism – the degree to which individuals express pride, loyalty, and cohesiveness in their organisations or families; (6) Assertiveness – the degree to which individuals are assertive, confrontational and aggressive in their relationships with others; (7) Gender egalitarianism – the degree to which a collective minimises gender inequality; (8) Future orientation – the extent to which individuals engage in future-oriented behaviours such as delaying gratification, planning, and investing in the future; and (9) Performance orientation – the degree to which a collective encourages and rewards group members for performance improvement and excellence.

Results of comparing cultures and attributes of effective leadership yielded six global leadership styles:

1. Performance-oriented (or "charismatic/value-based") – stresses high standards, decisiveness, and innovation; seeks to inspire and motivate people around a vision; and expects high-performance outcomes from people based on firmly held core values.
2. Team-oriented – instils pride, loyalty, and collaboration among organisational members; and highly values team cohesiveness and a common purpose or goals.
3. Participative – encourages input from others in decision-making and implementation; and emphasises delegation and equality.
4. Humane-oriented – stresses compassion and generosity; it is patient, supportive, and concerned with the well-being of others.
5. Autonomous – the leader is independent, individualistic, and self-centric.
6. Self-protective (and group-protective) – emphasises procedural, status-conscious, and "face-saving" behaviours; and focuses on ensuring the safety and security of the individual and the group.

Further, there were various leadership attributes that emerged from the GLOBE study that were universally rated as examples of facilitating outstanding leadership; namely – being trustworthy, planful, dynamic, and communicative. Other attributes were universally rejected as examples of inhibiting outstanding leadership (e.g., being asocial, irritable, egocentric, and dictatorial). Despite the fact that some leadership attributes were universally endorsed or rejected, the majority of attributes were culturally contingent. In sum, national culture indirectly influences leadership behaviours through the expectations of societies.

Clearly, contemporary leaders must be knowledgeable of, and sensitive to, leading cultural differences in increasingly diverse organisations, which represent workforces of people from all over the world. In addition, corporations often deal with clients and partners from different parts the globe. The many benefits of multicultural diversity, by providing businesses with a limitless pool of talent, ideas, viewpoints and opinions, has already been acknowledged (Connerley & Pedersen, 2005; Wibbeke & McArthur, 2013), including in EM (Forbes, 2008; James, 2008; Layne, 2002; Porter, 1995; Richardson, 2005). Hence, there is a need for leaders to acquire a global mindset, intercultural sensitivity, cultural intelligence, and cross-cultural competence (Osland et al., 2006).

5.5.5.2 Global Mindset

Global mindset relates to "the ability to develop and interpret criteria for personal and business performance that are independent from the assumptions of a single country, culture, or context; and to implement those criteria appropriately in different countries, cultures, and contexts" (Maznevski & Lane, 2004, p. 172). A global mindset is "a set of attributes and skills that contribute to effective leadership in a global corporation" (Pucik, 2005, p. 86). This includes attributes such as the ability to accept and work with cultural diversity, a cosmopolitan outlook, tolerance of uncertainty, and ability to handle a high degree of cognitive complexity. Beechler and Javidan (2007), define global mindset as "the process of influencing individuals, groups, and organizations (inside and outside the boundaries of the global organization) representing diverse cultural/political/institutional systems to help achieve the global organization's goals" (p. 38). Levy et al. (2007) define it as "a highly complex cognitive structure characterized by an openness to and articulation of multiple cultural and strategic realities on both global and local levels, and the cognitive ability to mediate and integrate across this multiplicity" (p. 32). This definition reflects the need to have a global mindset while working locally by capturing the cultural diversity, even in national organisations, as alluded to in the introduction. More recently, Javidan and Walker (2012) have defined global mindset as "the capability to influence others unlike yourself – and that is the key difference between leadership and global leadership" (p. 38). More specifically, the authors identify a global mindset comprising three key dimensions: (1) an openness and attentiveness to multiple realms of action and meaning; (2) a complex representation and expression of cultural and strategic dynamics; and (3) a moderation and incorporation of ideals and actions oriented toward both global and local levels. An empirical analysis conducted from this perspective yielded the global mindset construct, comprising the following three major dimensions or types of capital: (1) Intellectual Capital; (2) Psychological Capital; and (3) Social Capital. Each capital has three components, and each component has four sub-components or building blocks, as captured in Table 5.12.

TABLE 5.12

The Structure of Global Mindset

Capital Type	Components and Sub-Components	Source
Intellectual Capital (IC)	The cognitive component of Global Mindset (three sub-components) 1. Global Business Savvy: Knowledge of the way world business works • Knowledge of global industry • Knowledge of global competitive business and marketing strategies • Knowledge of how to transact business and manage risk in other countries • Knowledge of supplier options in other parts of the world 2. Cosmopolitan Outlook: Understanding that the managers' home country is not the centre of the universe: • Knowledge of cultures in different parts of the world • Knowledge of geography, history and important persons of several countries • Knowledge of economic and political issues, concerns, hot topics, etc., of major regions of the world • Up-to-date knowledge of important world events 3. Cognitive Complexity: Global is just more complicated than domestic only • Ability to grasp complex concepts quickly • Strong analytical and problem-solving skills • Ability to understand abstract ideas • Ability to take complex issues and explain the main points simply and understandably	Javidan and Walker (2012)
Psychological Capital (PC)	The affective or emotional component of Global Mindset (three sub-components) 1. Passion for diversity: Do not just tolerate or appreciate diversity – thrive on it • Interest in exploring other parts of the world • Interest knowing people from other parts of the world • Interest in living in another country • Interest in variety 2. Quest for Adventure: The Marco Polos of the world • Interest in dealing with challenging situations • Willingness to take risk • Willingness to test one's abilities • Interest in dealing with unpredictable situations 3. Self-Assurance: The source of psychological resilience and coping • Energetic • Self-confident • Comfortable in uncomfortable situations • Witty in tough situations	

(Continued)

TABLE 5.12 (*Continued*)
The Structure of Global Mindset

Capital Type	Components and Sub-Components	Source

Social Capital (SC)	The behavioural aspect of Global Mindset (three sub-components)
	1. Intercultural Empathy: Display "global" emotional intelligence
	  • Ability to work well with people from other parts of the world
	  • Ability to understand nonverbal expressions of people from other cultures
	  • Ability to emotionally connect to people from other cultures
	  • Ability to engage people from other parts of the world to work together
	2. Interpersonal Impact: Difference maker; seldom ignored across boundaries
	  • Experience in negotiating contracts in other cultures
	  • Strong networks with people from other cultures and with influential people
	  • Reputation as a leader
	  • Credibility
	3 Diplomacy: Seeks first to understand, then to be understood
	  • Ease of starting a conversation with a stranger
	  • Ability to integrate diverse perspectives
	  • Ability to listen to what others have to say
	  • Willingness to collaborate	

Intellectual Capital (IC) carputers the cognitive aspect of Global Mindset. IC relates to the leaders' knowledge of their global surroundings, as well the ability to process and leverage the additional layer of complexity embedded in global contexts or environments.

IC consists of three components: (1) Global Business Savvy; (2) Cosmopolitan Outlook; and (3) Cognitive Complexity. Each component has four corresponding sub-components or building blocks, as highlighted in Table 5.7.

Psychological Capital (PC) captures the affective or emotional aspect of Global Mindset. PC unable leaders to leverage their IC, and comprises three components: (1) Passion for Diversity; (2) Quest for Adventure; and (3) Self-Assurance. Each component has four corresponding sub-components or building blocks, as highlighted in Table 5.7.

Social Capital (SC) is the behavioural aspect of Global Mindset. SC reflects leaders' ability to act in a way that builds trusting relationships with people from other parts of the world. It also comprises three components: (1) Intercultural Empathy; (2) Interpersonal Impact; and (3) Diplomacy. Each component also has four corresponding sub-components or building blocks, as highlighted in Table 5.7.

5.5.5.3 Cultural Intelligence

Strength lies in differences, not in similarities.

(Stephen Covey)

Cultural intelligence or cultural quotient (CQ) is "an individual's capability to function and manage effectively in culturally diverse settings" (Ang et al., 2007, p. 336). CQ has also been defined as "a person's capability to adapt effectively to new cultural contexts" (Earley & Ang, 2003, p. 59), "an individual's cultural knowledge of norms, practices, and conventions in different cultural settings" (Van Dyne et al., 2015, p. 17), and "the capability to function effectively in culturally diverse settings" (Van Dyne et al., 2015, p. 16). Related and often interchangeable constructs to CQ include cultural, intercultural, or cross-cultural competence, and cultural adaptability – the ability of an individual to effectively interact, work, and develop meaningful relationships with people of various cultural backgrounds. They all relate to desirable attributes of globally competent engineering graduates in appreciating other cultures and communicating effectively across cultures (Parkinson, 2007), and engineering managers' requirements to exercise effective cross-cultural leadership (Frost & Walker, 2007). Consequently, CQ has been recognised as important component of contemporary engineering education (Jesiek et al., 2012; Goldfinch et al., 2012; Grandin & Hedderich, 2009; Hoffmann et al., 2011). CQ or cultural competence is also relevant for managing diversity within multicultural workforces in a leader's own country. In a modern economies such Australia – one of the most ethnically diverse societies in the world – this is now the norm. Cultural diversity, in fact, is the engine of innovation and the source of the necessary competitive advantage for a 21st century global economy. This explains the increasing number of workplace initiatives aimed at managing and leveraging cultural diversity and inclusion, and at promoting innovation (Salicru, 2017).

CQ can be measured using the Cultural Intelligence Scale (CQS, Ang et al., 2007). As a multidimensional construct that addresses cross-cultural interactions arising from cultural differences, the CQS is a 20-item the following four factors or components of CQ:

Cognitive, meta-cognitive, motivational, and behavioural. Table 5.13 captures the definitions and three examples for each of these four CQ dimensions.

TABLE 5.13

The Four Dimension of the Cultural Intelligence

Component	Definition and Examples
Cognitive CQ	Person's knowledge of specific norms, practices, and conventions in new cultural settings. This dimension deals with knowledge of cultural norms and practices based on personal or learned experience *Examples* • Knowing the legal and economic systems of other cultures. • Knowing the cultural values and religious beliefs of other cultures. • Knowing the rules for expressing non-verbal behaviours in other cultures.

(Continued)

TABLE 5.13 (*Continued*)
The Four Dimension of the Cultural Intelligence

Component	Definition and Examples
Metacognitive CQ	Individual's cultural awareness during interactions with people from different cultural backgrounds. It refers to individuals' judgment of their thought process, as well as judgment of the thought processes of others.
	Examples
	• Being conscious of the cultural knowledge required when interacting with people with different cultural backgrounds.
	• Being conscious of how to adjust the cultural knowledge required when interacting with people from a culture that is unfamiliar to oneself.
	• Being conscious of the cultural knowledge required to use in cross-cultural interactions.
Motivational CQ	Individual's drive to learn more about and function effectively in culturally varied situations. It refers to the energy directed toward learning how to function effectively in an environment that is culturally different from one's own.
	Examples
	• Enjoying the interaction with people from different cultures.
	• Being confident to socialise with locals in a culture that is unfamiliar to one's own.
	• Enjoying living in cultures that are unfamiliar to one's own.
Behavioural CQ	Individual's flexibility in demonstrating the appropriate actions when interacting with people from different cultural backgrounds. Refers to an individual's capability of using appropriate observable actions during interactions with people from a different culture.
	Examples
	• Being able change one's verbal behaviour (e.g., tone or inflexion) when a cross-cultural interaction requires it.
	• Being able to adjust or vary the rate of one's speech when a cross-cultural situation requires it.
	• Being able to adjust or change one's own non-verbal behaviour when a cross-cultural interaction requires it.

5.5.6 EMOTIONAL INTELLIGENCE

Anyone can get angry — that is easy. But to do this to the right person, to the right extent, at the right time, with the right motive, and in the right way, that is not for everyone, nor is it easy. (II.1109a27)

(Aristotle, Nicomachean Ethics, c. 325 BC)

The term emotional intelligence (EI), or emotional quotient (EQ), was first coined by Salovey and Mayer (1990) and defined as "a set of skills hypothesized to contribute to the accurate appraisal and expression of emotion in oneself and in others, the effective regulation of emotion in self and others, and the use of feelings to motivate, plan, and achieve in one's life" (p. 185). Subsequently, EI was popularised by Goleman (1998), who defined it as "the capacity for recognizing our own feelings and those of others, for motivating ourselves, and for managing emotions well in ourselves and in our relationships" (p. 317). EI, then, relates to the ability to identify or recognise, understand, evaluate, regulate, manage or control, and express emotions (Cherniss et al., 2001). EI has been recognised as an important component in engineering education (Chisholm, 2010; Palethorpe, 2006; Riemer, 2003) and a "missing priority" in engineering management education (Antoniadou et al., 2021, p. 92). Hence, the importance and rationale for including this capability in the EMLCF.

EI has also been linked to leadership effectiveness and outcomes (Coetzee & Schaap, 2005; Kerr et al., 2006; Palmer et al., 2001), and team outcomes (Hur et al., 2011). This is line with the assertion that "leaders have always played a primordial emotional role" (Goleman et al., 2013, p. 5). Given its gained popularity in recent times, several models of EI exist. Arguably, the two most popular ones in the development of leaders are Goleman's (1998) competency framework, and Mayer et al.'s (2008) four-branch ability model. These are outlined next. Table 5.14 summarises Goleman's (1998) EI competence model.

The above set of 18 competencies can be measured using the Emotional and Social Competence Inventory (ESCI, Hay Group, 2011). Emotional and social intelligence makes the difference between a highly effective leader and an average one. The real benefit comes from the 360° view into the behaviours that differentiate outstanding from average performers. It helps managers and professionals create competitive advantage for their organisations by increasing performance, innovation, and teamwork, ensuring time and resources are used effectively, and building motivation and trust

TABLE 5.14
The Emotional Competence Framework (18 Competencies)

Personal Skills (How to manage oneself) Nine competencies	Self-Awareness Knowing one's emotions, strengths, weaknesses, drives, values, and goals – and their impact on others.	
	Three competencies	Hallmarks
	1. Emotional Self-Awareness	• Reading one's own emotions and recognising their impact and using them to guide decisions
	2. Accurate Self-Assessment	• Realistic self-assessment
		• Knowing one's strengths and limits
		• Self-deprecating sense of humour
	3. Self-Confidence	• Desire for constructive criticism
		• A sound sense of one's self-worth and capabilities
	Self-Management or Self-regulation Controlling or redirecting disruptive emotions, and impulses.	
	Six competencies	Hallmarks
	1. Emotional Self-Control	• Keeping disruptive emotions and impulses under control
	2. Transparency	• Displaying honesty, integrity, and trustworthiness
	3. Adaptability	• Flexibility in adapting to changing situations or overcoming obstacles
	4. Achievement Orientation	• The drive to improve performance to meet inner standards of excellence
	5. Initiative	• Readiness to act and seize opportunities
	6. Optimism	• Seeing the upside in events

(Continued)

TABLE 5.14 (*Continued*)
The Emotional Competence Framework (18 Competencies)

Social Skills	Social Awareness
(How to manage relationships) Nine competencies	The ability to take the perspective of and empathise with others, including those from diverse backgrounds and cultures to understand social and ethical norms for behaviour.

Three competencies	Hallmarks
1. Empathy 2. Organisational Awareness 3. Service Orientation	• Sensing others' emotions, understanding their perspective, and taking active interest in their concerns • Reading the currents, decision networks, and politics at the organisational level • Recognising and meeting follower, client, or customer needs and expectations • Expertise in attracting and retaining talent • Ability to develop others • Sensitivity to culture differences

Relationship Management
Managing relationship to influence, guide, or move people in desired directions.

Six competencies	Hallmarks
1. Developing Others 2. Inspirational Leadership 3. Change Catalyst 4. Influence 5. Conflict Management 6. Teamwork and Collaboration	• Bolstering others' abilities through feedback and guidance • Guiding and motivating with a compelling vision • Effectiveness in leading change by initiating, managing, and leading in new directions • Wielding a range of tactics for persuasion • Negotiating and resolving agreements and disputes • Extensive networking by cultivating and maintaining relationship webs • Expertise in building and leading teams • Cooperating and collaborating with others

Mayer et al. (2008) define EI as the ability to perceive and express emotions, to use emotions to facilitate thinking, to understand and reason with emotions, and to effectively manage emotions within oneself and in relationships with others This set of skills contribute to the accurate appraisal and expression of emotion in oneself and in others, the effective regulation of emotion in self and others, and the use of feelings to motivate, plan, and achieve in one's life (Mayer & Salovey, 1997). In their model, the authors present four branches that are arranged from more basic to higher or more integrated psychological processes. For example, the lowest level branch describes the relatively simple abilities of perceiving and expressing emotion, while the highest-level branch represents the conscious and reflective regulation of emotion. Table 5.15 captures Mayer and Salovey's (1997) four-branch ability EI model.

TABLE 5.15
The Four-Branch Ability EI Model

Branch 1: Perceiving Emotions	Perceiving Emotions The ability to perceive emotions in oneself and others, as well as in objects, art, stories, music, and other stimuli.
	Perceiving emotions is about identifying or recognising emotions. Emotions are data and contain information about ourselves, other people, and the world around us. Paying attention to emotions is important to be accurate in identifying how we, and others, feel. This includes perceiving, identifying, or recognising the nonverbal and facial expressions such as happiness, sadness, anger, and fear, which are universally recognisable in human beings. The capacity to accurately perceive emotions in the face or voice of others provides a crucial starting point for more advanced understanding of emotions.
Branch 2: Facilitating Thought	Facilitating Thought The ability to generate, use, and feel emotion as necessary to communicate feelings or employ them in other cognitive processes.
	This ability is concerned with using emotions to facilitate thought. That is, the capacity to use motions to guide the cognitive system and promote thinking, and help direct thinking toward matters that are truly important. This is important for certain kind of creativity to emerge. This ability also includes how to generate an emotion, and then reason with this emotion. Emotions enter the cognitive system as notice signals and as influenced of cognition. Our emotions influence both what we think about, and how we think. For example, if you are in a positive mood, you will see things differently than if you were in a more negative mood.
Branch 3: Understanding Emotions	Understanding Emotions The ability to understand emotional information, how emotions combine and progress through relationship transitions, and to appreciate such emotional meanings.
	This ability relates to the capacity understand and reason about emotions and their meanings. This includes understand complex emotions and emotional chains – how emotions shift from one stage to another. Emotions convey its own pattern of possible messages, and actions associated with those messages. Understanding emotions is important to figure out why we feel a certain way, and how these feelings will change over time. By understand our emotions, for example, we can predict how an idea will be received, and how others might react to it.

(Continued)

TABLE 5.15 (*Continued*)
The Four-Branch Ability EI Model

Branch 4:	Managing Emotions
Managing Emotions	The ability to be open to emotions, and to modulate them in oneself and others, to promote personal understanding and growth.
	Since emotions contain data or information, we need to stay open to this information, and use it to help us makegood decisions.
	Emotions often can be managed. To the extent that it is under self-control, a person may want to remain open to emotional signals, as long as they are not too painful, and block out those that are overwhelming.
	We can't always go with the current feeling, but we can return to that feeling later. If we permanently suppress feelings we will be ignoring critical information.
	In between, within the person's emotional comfort zone, it becomes possible to regulate and manage one's own and others' emotions to promote one's own and others' personal and social goals. An emotionally intelligent leader, for example, can guide his team members in a better way.

Finally, the above set of abilities can be measured using The Mayer-Salovey-Caruso Emotional Intelligence Test (MSCEIT, Mayer et al., 2003). The MSCEIT is an ability-based test developed from an intelligence-testing tradition formed by the emerging scientific understanding of emotions and their function and from the first published ability measure specifically intended to assess emotional intelligence – the Multifactor Emotional Intelligence Scale (MEIS). The MSCEIT comprises 141 items and takes 30–45 minutes to complete. Results include 15 main scores: Total EI score, two Area scores, four Branch scores, and eight Task scores.

5.5.7 CREATIVE THINKING AND INNOVATIVE BEHAVIOUR

To raise new questions, new possibilities, to regard old problems from a new angle, requires creative imagination and marks real advance in science.

(Albert Einstein)

Creativity and innovation have become essential capabilities for graduate engineers, as well as major drivers of sustainability, economic growth, and competitive advantage in the engineering world (Badran, 2007). Creativity is "essential to human progress" (Hennessey & Amabile, 2010, p. 569), was projected to become the third most important skill needed in 2020 (World Economic Forum, 2016), and is a critical driver for innovation in engineering design (Charyton, 2015). Consequently, the teaching of creativity has been recognised as a modern practice in the digital era in engineering (Lunevich, 2022; Lunevich & Wadaani, 2023). This section defines creativity, innovation, innovative workplace behaviour (IWB), and explains their link to leadership.

Creativity relates to the production of novel and useful ideas, or socially valued products or services (Mumford & Gustafson, 1988). Innovation is related to the production or adoption of useful ideas and idea implementation (Van de Ven, 1986), and is central to the long-term survival of organisations (Ancona & Caldwell, 1987). Leadership is a chief predictor of creativity – the precursor of all innovation (Amabile et al., 1996).

Strategically, leaders establish work environments that are conducive to creative thinking and innovation, which in turn leads to a competitive advantage. In doing so, they drive and manage innovation goals (Salicru, 2017). Leader behaviour also shapes company culture (Whitehurst, 2016), and is an important predictor and innovative workplace behaviour (IWB) – the behaviour that guides the initiation and intentional introduction of new and useful ideas, processes, products, services, or procedures (Afsar et al., 2014; Scott & Bruce, 1994), which has been recognised as paramount in today's uncertain global economy (Janssen, 2001). To elicit this type of behaviour leader also needs to create a climate for innovation (support for innovation and supply of resources) via transformational, and participative or collaborative leadership (Scott & Bruce, 1994). These forms of leadership generate employees' intrinsic motivation, psychological empowerment, and ultimately creative thinking and IWB (Hennessey & Amabile, 2010).

As useful model for leaders to use to achieve creative thinking and IWB from their teams is that of the leadership psychological contract (LPC, Salicru, 2017; Salicru & Chelliah, 2014). The LPC is a relational leadership model that ingrates LMX theory, transactional leadership, and other contemporary leadership approaches (e.g., positive, ethical, and authentic leadership), and provides an analytical framework for studying relationships within organisations. The LPC links self-leadership, leader credibility, and leadership impact in three functional dimensions:

1. Cognitive or rational – thought (head). Relates to the credibility of the leader, and comprises three indicators: fulfilment of leaders' expectations, trust, and fairness);
2. Emotional – feeling (heart). Comprises two indicators: team members' levels of affective commitment and satisfaction.
3. Behavioural – action (hands). Comprises two indicators: discretionary effort and innovation (innovative behaviour).

REFLECTIVE QUESTIONS

- Does your team search out new technologies, processes, techniques and/or concepts/ideas?
- Does your team generate new and creative ideas?
- As a leader, do you promote and champion new ideas to your team?
- Do you encourage your team to implement new and useful ideas?
- Do you and your team develop adequate plans and schedules for the implementation of new and useful ideas?

5.5.8 Ethical Behaviour and Ethical Leadership

Organisational misconduct, cheating, deception, and many other forms of unethical behaviour are some of the greatest challenges in our society (Gino, 2015). The importance of ethical behaviour and the ethical dimension of leadership has been widely recognised as critical in the contemporary business world (Brown & Trevino, 2006; Brown & Mitchell, 2010; Ciulla, 2003, 2005; Knights & O'Leary, 2005; Lawton

& Páez, 2015; Treviño et al., 2000). In fact, "ethics is central to science and engineering" (Wang & Thompson, 2013, p. 287), and ethics is at the heart of leadership (Ciulla, 2014). 'Having high ethical and moral standards' was rated the highest, and most important leadership competency – among 74 leaders' attributes, by leaders around the world in results published in Harvard Business Review (Giles, 2016).

In recent times, the need for broadening the scope of teaching ethics to engineers to promote sustainability principles and sustainable development, as well as encouraging corporate social responsibility (CSR), has been strongly acknowledged (Bucciarelli, 2008; Byrne, 2012; Conlon & Zandvoort, 2011; Haws, 2001; Smith et al., 2017, 2021). Engineering managers are required to embed sustainability in their engineering practices by formulating and disseminating the relevant engineering codes, and develop their education by making sustainability a core component of engineering curriculum (Jones et al., (2015). According to Smith et al. (2021), engineering ethics need to make more explicit reference to CSR. The mining and energy industries, for example, present unique challenges to engineers. As a result, they must navigate often competing responsibilities and codes of conduct. This includes their personal sense of right and wrong, professional codes of ethics, and CSR policies (Smith et al., 2017).

Ethical behaviour refers to actions judged by, and consistent with, one's personal principles and the commonly held values of the group, organisation, or society (Salicru, 2017). (Un)ethical behaviour focuses on behaviour that is consistent or inconsistent with societal or organisational norms (Treviño et al., 2014). Prior to the 1990s, professional schools rarely taught business ethics and ethical leadership. However, following harrowing ethical cultural failures, which resulted in the commercial dissolution of high-profile cooperate global giants such Arthur Andersen and Enron in 2001, the existing landscape at that time changed for ever (Den Hartog, 2015). Despite of the efforts made at the time to rectify such disturbing situation, ethical cultural failures persisted. Some examples include Volkswagen's CEO Martin Winterkorn lies about his ignorance of his company efforts to manipulate their vehicles' emissions data, and Facebook's CEO Mark Zuckerberg admission of not taking full responsibility for hi company's mismanagement of around 87 million user profiles that were used for political purposes (De Cremer & Moore, 2020). Not surprisingly, much greater attention has been placed recently on ethical leadership in organisations by both researchers and education institutions (e.g., business schools), and new research perspectives and practises have merged. The main realisations of this new research, new perspectives, and practise related to ethical behaviour and the ethical leadership include the following.

First, there is now a clearer business case for ethical leadership and ethical behaviour. Thomas et al. (2004), for example, document the real and hefty costs of ethical failures that seldom are reflected on annual reports, balance sheets or income statements – and which in extreme cases can literally destroy a firm, at the following three levels of an escalating continuum:

1. Level one costs – Government penalties and fines;
2. Level two costs – Administrative and audit, legal and investigative, remedial education, corrective cations, and government oversight;
3. Customer desertions, loss of reputation, employee cynicism, loss of employee morale, employee turnover, and government cynicism and regulation.

According to the authors, the following paradox happens in relation to ethics and ethical behaviour in organisations. The cots at level one, which are less demanding and least understated, are the ones that capture greater executives'/leaders' attention. This is because they are the easiest quantify and calculate. Contrastingly, level three costs, because they are the most difficult to quantify, tend to be underappreciated and chronically undervalued by executives/leaders in their decision-making, or even they go completely unnoticed. However, level three costs that are the costliest. In fact, according to the Cone-Roper poll – National Survey Finds Americans Intend to Punish Corporate "Bad Guys," and Reward Good Ones (2002, as cited in Thomas et al., 2004), found that our days the public is prepared to raise level three costs by punishing unethical organisations in the following ways: 91% of respondents reported they would consider switching to another company's products or services; 85% stated they would speak out against that company among family and friends; 83% reported they would refuse to invest in that company's stock; 80% declared they would refuse to work at that company; 76% said would boycott that company's products or services; and 68% affirmed they would be less loyal to a job at that company. In short, it is critical for leaders to realise that the bottom line of business success always includes an ethics component.

The second finding related to ethical behaviour and the ethical leadership include the following. Traditionally, researchers adopted a normative or prescriptive approach to business ethics and how to resolve ethical dilemmas by using insights from a philosophical and morality perspective (Bazerman & Gino, 2012; Treviño et al. 2003). These approaches to ethics, however, were proven to be somewhat ineffective due to the following five main reasons:

1. The normative perspective focused on the moral understanding between right and wrong, and then prescribing what people should do via compliance. Regrettably, such approach does not explain why individuals deviate from ethical standards (e.g., the impact of psychological factors that force them to blindly engage in self-deception).

2. Typical ethical training programs assume that if people know what's the right thing to do, they will do it. The reality is that often people don't have the confidence and courage (voice) to take action;

3. 'Compliance' approaches to ethics require employees to stay out of trouble, but do not assist them to do so, neither lessen legal violations. In fact, the opposite is true. Compliance approaches to ethics may serve as window dressing to deflect attention or culpability from illegal activities. They also tend to suppress ethical reflection, as people have less need to form their own opinions and take personal responsibility for their decisions. Thus, replacing 'accountability' for 'responsibility';

4. Corporate documenting such as mission statements and codes of conduct, or appointing ethics officials, are also arguably ineffective. They tend to generate cynicism, as employees witness the gap between rhetoric and practice; and

5. Organisations fail to integrate ethics and ethical leadership, along with more contemporary forms of leadership, as strategic imperatives and drivers of performance and competitive advantage.

Finally, and as a result of the above, the third main shift investigating ethics relates to the emergence of the field of behavioural ethics (De Cremer et al., 2010; De Cremer & Moore, 2020; Gino, 2015; Mitchell et al., 2017; Treviño et al., 2006). Behavioural ethics refers to "individual behavior that is subject to or judged according to generally accepted moral norms of behavior" (Treviño et al., 2006, p. 952). Drawing from psychological research, behavioural ethics can help to comprehend "why it is the case that apparently good people sometimes do bad things" (De Cremer et al., 2010, p. 2). Behavioural ethics has shed light on the fact that many individuals who engage in unethical behaviour may not necessarily be doing so consciously or willingly (De Cremer & Moore, 2020). The main difference then between normative (or moral) versus behavioural (or descriptive) approaches to ethics is that the goal of the first is constructing arguments related to what people should, ought, or must do; while the former is about studying what people actually do drawing on research from behavioural psychology, cognitive science, neuroscience, and evolutionary biology. Hence, the core principle of behavioural ethics is that most ethical misconduct is not enacted by dishonest people (e.g., the stereotypical 'bad apples'), but rather by normal people who, while valuing morality and considering themselves ethical, make errors of judgement, fail to resist temptation, social pressures, expectations, or fail to recognise decisions that have moral, ethical or legal implications. Therefore, behavioural ethics add a valuable contribution to the field of business ethics by helping to better understand why good people can still do bad things (e.g., damaging to the reputation of the firm) via the activation of unconscious bias/tendencies, which act as primary drivers of unethical behaviour. A major finding from behavioural ethics research is that people simultaneously think of themselves as good people, yet frequently lie and cheat – generally in a minor way (University of Texas at Austin, 2023). In fact, according to research (Mazar et al., 2008), "people behave dishonestly enough to profit but honestly enough to delude themselves of their own integrity" (p. 633). That is, they cheat to the degree that allows them to retain their self-image as reasonably honest people. Further, Gino et al.'s (2009) research found that when a team member behaves unethically, and the behaviour is visible to others, they follow suit and behave unethically themselves. Hence, the importance of modelling positive ethical behaviour, not only as leaders but also as a team and an organisational member.

Engineering managers are required to exercise the ethical leadership needed to achieve sustainability. This transformation is particularly challenging due to the deeply ingrained daily habits of the profession. Namely, challenges inherent to the engineered system, and challenges deeply-rooted to the engineering professional culture and practice (Jones et al., 2015). From this perspective, engineering managers are critical for transformative change towards sustainable systems across the multiple stages of any engineering process. In addition to supporting this cultural change, ethical leadership also improves employee attitudes, job satisfaction, affective commitment, and work engagement, as well as reducing employee turnover intentions (Tanner et al., 2010).

In summary, ethics and leadership are two inseparable research disciplines and critical business practices. Ethical leadership entails exercising influence in ways that are ethical in both means and in ends. Behavioural ethics go beyond traditional approaches to ethics which have focused on the moral understanding between right and wrong and suggesting what people should do. Instead, behavioural ethics explains how good people can do bad things, and provides more effective countermeasures. To this end, behavioural ethics studies why people make ethical and unethical decisions with a view to improve ethical decision-making and promote ethical cultures. Behavioural ethics reveals the unavoidable limitations and inherent biases in the psychological processes that drive behaviour in organisations, cause flawed reasoning, and what needs to be recognised for effective ethical decision-making. Ethical, responsible, and distributed leadership empowers employees to act by building their confidence to perform at their best, and the courage to speak up. Hence, they should be part of the strategic agenda of contemporary organisations wishing to follow high ethical standards, and aspiring to outperform their competitors.

REFLECTIVE QUESTIONS

- Do you have character and integrity that will assist you when faced with difficult moral choices?
- Have you known good people to do bad things? (either personally, or who you've heard or read about in the media)
- If so, how would you explain their behaviour?

5.6   CONCLUSION

This chapter has presented the Engineering Managers' Leadership Capability Framework (EMLCF) – a contemporary, global, and futuristic framework of leadership education and development for engineering managers. The EMLCF is, arguably, the first ever holistic Leadership Capability Framework developed for engineering managers, and comprises eight high-level capabilities or meta-competencies. This framework will be valuable for students of engineering management, and emerging and established engineering managers wishing to further develop their leadership knowledge, capability, and impact to lead the current digital disruption and associated future challenges. The EMLCF will also be a valuable resource for those responsible for the design, delivery, and evaluation of leadership development programs, and other learning and development initiatives in organisations (e.g., human resources personnel, organisational learning and development professionals, trainers, organisational psychologists, internal and external coaches, and consultants).

REFERENCES

Abbott, G., Gilbert, K., & Rosinski, P. (2013). Cross-cultural working in coaching and mentoring. In J. Passmore, D. B. Peterson, & T. Freire (Eds.), *The Wiley-Blackwell handbook of the psychology of coaching and mentoring* (pp. 483–500). Wiley Blackwell.
Adams, B. (2019). *Green development: Environment and sustainability in a developing world.* Routledge.

Afsar, B., Badir, Y. F., & Saeed, B. B. (2014). Transformational leadership and innovative work behavior. *Industrial Management & Data Systems, 114*(8), 1270–1300. https://doi.org/10.1108/IMDS-05-2014-0152

Allison, S. T., Goethals, G. R., & Kramer, R. M. (Eds.). (2016). *Handbook of heroism and heroic leadership.* Taylor & Francis.

AlSagheer, A., & Al-Sagheer, A. (2011). Faculty's perceptions of teaching ethics and leadership in engineering education. *Journal of International Education Research, 7*(2), 55–66. https://doi.org/10.19030/jier.v7i2.4250

Amabile, T. M., & Khaire, M. (2008). Creativity and the role of the leader. *Harvard Business Review, 86*(10), 100–109. https://hbr.org/2008/10/creativity-and-the-role-of-the-leader

Amabile, T. M., Conti, R., Coon, H., Lazenby, J., & Herron, M. (1996). Assessing the work environment for creativity. *Academy of Management Journal, 39*(5), 1154–1184. https://doi.org/10.2307/256995

Amirianzadeh, M., Jaafari, P., Ghourchian, N., & Jowkar, B. (2011). Role of student associations in leadership development of engineering students. *Procedia-Social and Behavioral Sciences, 30*, 382–385. https://doi.org/10.1016/j.sbspro.2011.10.075

Ancona, D. (2005). Leadership in an age of uncertainty. *Center for Business Research Brief, 6*(1), 1–3. Retrieved from: https://web.mit.edu/curhan/www/docs/Articles/15341_Readings/Leadership/Ancona_LeadershipinanAgeofUncertainty-researchbrief.pdf

Ancona, D. (2012). Framing and acting in the unknown. In S. Snook, N. Nohria, & R. Khurana (Eds.), *The handbook for teaching leadership*: Knowing, doing, and being (pp. 3–19). Sage Publications.

Ancona, D., & Caldwell, D. (1987). Management issues facing new product teams in high technology companies. In D. Lewin, D. Lipsky, & D. Sokel (Eds.), *Advances in industrial and labor relations* (pp. 191–221). JAI Press.

Andrasik, F. and Heimberg, J.S. (1982). Self-management procedures. In L. W. Frederikson, (Ed.), *Handbook of organizational behavior management* (pp. 219–47). Wiley.

Ang, S., & Van Dyne, L. (2015). *Handbook of cultural intelligence.* Routledge.

Ang, S., Van Dyne, L., Koh, C., Ng, K. Y., Templer, K. J., Tay, C., & Chandrasekar, N. A. (2007). Cultural intelligence: Its measurement and effects on cultural judgment and decision making, cultural adaptation and task performance. *Management and Organization Review, 3*(3), 335–371. https://doi.org/10.1111/j.1740-8784.2007.00082.x

Antonakis, J., Avolio, B. J., & Sivasubramaniam, N. (2003). Context and leadership: An examination of the nine-factor full-range leadership theory using the Multifactor Leadership Questionnaire. *The Leadership Quarterly, 14*(3), 261–295. https://doi.org/10.1016/S1048-9843(03)00030-4

Antoniadou, M., Crowder, M., & Andreakos, G. (2021). Emotional intelligence in engineering management education: The missing priority. In *Cases on engineering management education in practice* (pp. 92–104). IGI Global.

Aron, D. C., & Leykum, L. (2022). Sensemaking: Appreciating patterns and coherence in complexity. In D. C. Aron, & L. Leykum (Eds.), *Implementation science* (pp. 96–98). Routledge.

Arthur, W. B. (1996). Increasing returns and the two worlds of business. *Harvard Business Review, 74*(4), 100–109. https://hbr.org/1996/07/increasing-returns-and-the-new-world-of-business

Arvey, R. D., Rotundo, M., Johnson, W., Zhang, Z., & McGue, M. (2006). The determinants of leadership role occupancy: Genetic and personality factors. *The Leadership Quarterly, 17*(1), 1–20. https://doi.org/10.1016/j.leaqua.2005.10.009

ASCE. (2008, February). Civil engineering body of knowledge for the 21st century: Preparing the civil engineer for the future. *American Society of Civil Engineers, 3*, 23–28. https://doi.org/10.1061/9780784409657

Ashforth, B. E. (1989). The experience of powerlessness in organizations. *Organizational Behavior and Human Decision Processes, 43*(2), 207–242. https://doi.org/10.1016/0749-5978(89)90051-4

Ashforth, B. E., Schinoff, B. S., & Rogers, K. M. (2016). "I identify with her," "I identify with him": Unpacking the dynamics of personal identification in organizations. *The Academy of Management Review, 41*(1), 28–60. https://doi.org/10.5465/amr.2014.0033

Ashkanasy, N. M., Härtel, C. E. J., & Daus, C. S. (2002). Diversity and emotion: The new frontiers in organizational behavior research. *Journal of Management, 28*(3), 307–338. https://doi.org/10.1177/014920630202800304

Auyang, S. Y. (2006). *Engineering-an endless frontier.* Harvard University Press.

Avolio, B. J., Gardner, W. L., Walumbwa, F. O., Luthans, F., & May, D. R. (2004). Unlocking the mask: A look at the process by which authentic leaders impact follower attitudes and behaviors. *The Leadership Quarterly, 15*(6), 801–823. https://doi.org/10.1016/j.leaqua.2004.09.003

Badran, I. (2007). Enhancing creativity and innovation in engineering education. *European Journal of Engineering Education, 32*(5), 573–585. https://doi.org/10.1080/03043790701433061

Bandura, A. (1977). Self-efficacy: Toward a unifying theory of behavioral change. *Psychological Review, 84*(2), 191–215. https://doi.org/10.1037/0033-295X.84.2.191

Bandura, A. (1982). Self-efficacy mechanism in human agency. *American Psychologist, 37*(2), 122–147. https://doi.org/10.1037/0003-066X.37.2.122

Bailey, K. D. (2005). Beyond system internals: Expanding the scope of living systems theory. *Systems Research and Behavioral Science, 22*(6), 497–508. https://doi.org/10.1002/sres.659

Baldwin, T. T., & Padgett, M. Y. (1994). Management development: A review and commentary. In C. L. Cooper, I. T. Robertson, & Associates (Eds.), *Key reviews in managerial psychology: Concepts and research for practice* (pp. 270–320). John Wiley & Sons.

Barrow, J. C. (1977). The variables of leadership: A review and conceptual framework. *Academy of Management Review, 2*(2), 231–251. https://doi.org/10.5465/amr.1977.4409046

Basadur, M. (2004). Leading others to think innovatively together: Creative leadership. *The Leadership Quarterly, 15*(1), 103–121. https://doi.org/10.1016/j.leaqua.2003.12.007

Bass, B. M. (1985). *Leadership and performance beyond expectations.* Free Press.

Bass, B. M. (1990). *Bass & Stogdil's handbook of leadership: Theory, research, and managerial applications* (3rd Ed.). Free Press.

Bass, B. M. (2009). *The Bass handbook of leadership: Theory, research, & managerial applications* (4th Ed.). Free Press.

Bass, B. M., Avolio, B. J., Jung, D. I., & Berson, Y. (2003). Predicting unit performance by assessing transformational and transactional leadership. *Journal of Applied Psychology, 88*(2), 207–218. https://doi.org/10.1037/0021-9010.88.2.207

Bayless, D. J., & Robe, T. R. (2010, October). Leadership education for engineering students. In 2010 *IEEE Frontiers in Education Conference (FIE)* (pp. S2J–1). IEEE. https://doi.org/10.1109/FIE.2010.5673554

Bazerman, M. H., & Gino, F. (2012). Behavioral ethics: Toward a deeper understanding of moral judgment and dishonesty. *Annual Review of Law and Social Science, 8,* 85–104. https://doi.org/10.1146/annurev-lawsocsci-102811-173815

Becker, G. S. (2002). The age of human capital. *Education in the Twenty-First Century,* 3–8.

Beechler, S., & Javidan, M. (2007). Leading with a global mindset. In M. Javidan, R. M. Steers, & M. A. Hitt (Eds.), *The global mindset* (pp. 131–169), Emerald Group Publishing. https://doi.org/10.1016/S1571-5027(07)19006-9

Benefiel, M. (2005). The second half of the journey: Spiritual leadership for organizational transformation. *The Leadership Quarterly, 16*(5), 723–747. https://doi.org/10.1016/j.leaqua.2005.07.005

Bennett, N., & Lemoine, J. (2014). What VUCA really means for you. *Harvard Business Review, 92*(1/2). https://ssrn.com/abstract=2389563

Bennett, R. J., & Millam, E. (2013). *Leadership for engineers: The magic of mindset.* McGraw-Hill.

Bennis, W., & Nanus, B. (1985). *Leaders: The strategies for taking charge.* Harper Row.

Blake, R., & Mouton, J. (1964). *The managerial grid: The key to leadership excellence.* Gulf Publishing.

Boal, K. B., & Hooijberg, R. (2000). Strategic leadership research: Moving on. *The Leadership Quarterly, 11*(4), 515–549. https://doi.org/10.1016/S1048-9843(00)00057-6

Boal, K. B., & Schultz, P. L. (2007). Storytelling, time, and evolution: The role of strategic leadership in complex adaptive systems. *The Leadership Quarterly, 18*(4), 411–428. https://doi.org/10.1016/j.leaqua.2007.04.008

Bogner, W. C., & Barr, P. S. (2000). Making sense in hypercompetitive environments: A cognitive explanation for the persistence of high velocity competition. *Organization Science, 11*(2), 212–226. https://doi.org/10.1287/orsc.11.2.212.12511

Boisot, M., & McKelvey, B. (2010). Integrating modernist and postmodernist perspectives on organizations: A complexity science bridge. *Academy of Management Review, 35*(3), 415–433. https://doi.org/10.5465/amr.35.3.zok415

Bolden, R. (2011). Distributed leadership in organizations: A review of theory and research. *International Journal of Management Reviews, 13*(3), 251–269. https://doi.org/10.1111/j.1468-2370.2011.00306.x

Bolden, R., & Gosling, J. (2006). Leadership competencies: Time to change the tune? *Leadership, 2*(2), 147–163. https://doi.org/10.1177/1742715006062932

Bourdieu, P. (1986). The forms of capital. In J. G. Richardson (Ed.), *Handbook of theory and research for the sociology of education* (pp. 241–258). Greenwood.

Bouty, I. (2000). Interpersonal and interaction influences on informal resource exchanges between R&D researchers across organizational boundaries. *Academy of Management Journal, 43*(1), 50–65. https://doi.org/10.5465/1556385

Boyatzis, R. E. (1993). Beyond competence: The choice to be a leader. *Human Resource Management Review, 3*(1), 1–14. https://doi.org/10.1016/1053-4822(93)90007-Q

Boyatzis, R. E. (2009). Competencies as a behavioral approach to emotional intelligence [Editorial]. *Journal of Management Development, 28*(9), 749–770. https://doi.org/10.1108/02621710910987647

Brass, D. J., & Krackhardt, D. (1999). The social capital of twenty-first century leaders. In J. G. Hunt, G. E. Dodge, & L. Wong (Eds.), *Out-of-the-box leadership: Transforming the twenty-first-century army and other top-performing organizations* (pp. 179–194). Jai Press.

Brown, M. E., & Mitchell, M. S. (2010). Ethical and unethical leadership: Exploring new avenues for future research. *Business Ethics Quarterly, 20*(4), 583–616. https://doi.org/10.5840/beq201020439

Brown, M. E., & Treviño, L. K. (2003). Is values-based leadership ethical leadership? In S. W. Gilliland, D. D. Steiner, & D. P. Skarlicki, D. P. (Eds.), *Emerging perspectives on values in organizations* (151–173). Information Age Publishing.

Brown, M. E., & Treviño, L. K. (2006). Ethical leadership: A review and future directions. *The Leadership Quarterly, 17*(6), 595–616. https://doi.org/10.1016/j.leaqua.2006.10.004

Brown, M. E., Treviño, L. K., & Harrison, D. A. (2005). Ethical leadership: A social learning perspective for construct development and testing. *Organizational Behavior and Human Decision Processes, 97*(2), 117–134. https://doi.org/10.1016/j.obhdp.2005.03.002

Brower, H. H., Schoorman, F. D., & Tan, H. H. (2000). A model of relational leadership: The integration of trust and leader-member exchange. *The Leadership Quarterly, 11*(2), 227–250. https://doi.org/10.1016/S1048-9843(00)00040-0

Bucciarelli, L. L. (2008). Ethics and engineering education. *European Journal of Engineering Education, 33*(2), 141–149. https://doi.org/10.1080/03043790801979856

Burke, W. (1986). Leadership as empowering others. In S. Srivastra (Ed.), *Executive power* (pp. 51–77). Jossey-Bass.

Burke, C. S., Stagl, K. C., Klein, C., Goodwin, G. F., Salas, E., & Halpin, S. M. (2006). What type of leadership behaviors are functional in teams? A meta-analysis. *The Leadership Quarterly, 17*(3), 288–307. https://doi.org/10.1016/j.leaqua.2006.02.007

Burns, J. M. (1978). *Leadership.* Harper & Row.

Byrne, E. P. (2012). Teaching engineering ethics with sustainability as context. *International Journal of Sustainability in Higher Education, 13*(3), 232–248. https://doi.org/10.1108/1467637121124255

Cambridge Dictionary. (2023). *Engineering.* https://dictionary.cambridge.org/dictionary/english/engineering

Cameron, K. S., & Caza, A. (2004). Introduction: Contributions to the discipline of positive organizational scholarship. *American Behavioral Scientist, 47*(6), 731–739. https://doi.org/10.1177/0002764203260207

Cameron, K. & Spreitzer G. M. (Ed.). (2012). *Oxford handbook of positive organizational scholarship.* Oxford University Press.

Carlyle, T. (1907). *On heroes, hero-worship, and the heroic in history.* Houghton Mifflin.

Carroll, B., Levy, L., & Richmond, D. (2008). Leadership as practice: Challenging the competency paradigm. *Leadership, 4*(4), 363–379. https://doi.org/10.1177/1742715008095186

Carsten, M. K., Uhl-Bien, M., West, B. J., Patera, J. L., & McGregor, R. (2010). Exploring social constructions of followership: A qualitative study. *The Leadership Quarterly, 21*(3), 543–562. https://doi.org/10.1016/j.leaqua.2010.03.015

Carver, C. S., & Scheier, M. F. (1981). *Attention and self-regulation: A control theory approach to human behavior.* Springer-Verlag.

Cerri, S. T. (2016). *The fully integrated engineer: Combining technical ability and leadership prowess.* John Wiley & Sons.

Chaleff, (1995). *The courageous follower: Standing up to and for our leaders.* Berrett-Koehler Publishers.

Charyton, C. (2015). Creative engineering design: The meaning of creativity and innovation in engineering. *Creativity and innovation among science and art: A discussion of the two cultures*, 135–152.

Chemers, M. M. (2014). *An integrative theory of leadership.* Psychology Press.

Chemers, M. M. (2001). Efficacy and effectiveness: Integrating models of leadership and intelligence. In R. E. Riggio & S. E. Murphy (Eds.), *Multiple intelligences and leadership* (pp. 139–160). Erlbaum.

Cherniss, C., & Goleman, D. (2001). *The emotionally intelligent workplace: How to select for, measure, and improve emotional intelligence in individuals, groups, and organizations.* Jossey-Bass.

Chisholm, C. U. (2010). The formation of engineers through the development of emotional intelligence and emotional competence for global practice. *Global Journal of Engineering Education, 12*(1), 6–11.

Churchman, C. W. (1967). Wicked problems. *Management Science. 14*(4), 141–142.

Ciulla, J. B. (2003). *[Introduction to] The ethics of leadership.* Wadsworth/Thomson.

Ciulla, J. B. (2014). *Ethics, the heart of leadership.* ABC-CLIO.

Ciulla, J. B. (2005). The state of leadership ethics and the work that lies before us. *Business Ethics: A European Review, 14*(4), 323–335. https://doi.org/10.1111/j.1467-8608.2005.00414.x

Clapp-Smith, R., & Lester, G. V. (2014). Defining the "mindset" in global mindset: Modeling the dualities of global leadership. In *Advances in global leadership*, 8, (pp. 205–228). Emerald Group Publishing. https://doi.org/10.1108/S1535-120320140000008017

Clerkin, C., & Ruderman, M. N. (2016). Holistic leader development: A tool for enhancing leader well-being. In W. A. Gentry & C. Clerkin (Eds.), *The role of leadership in occupational stress* (pp. 161–186). Emerald Group Publishing. https://doi.org/10.1108/S1479-355520160000014007

Clutterbuck, D., Whitaker, C., & Lucas, M. (2016). *Coaching supervision: A practical guide for supervisees*. Routledge.

Coetzee, C., & Schaap, P. (2005). The relationship between leadership behaviour, outcomes of leadership and emotional intelligence. *SA Journal of Industrial Psychology, 31*(3), 31–38. https://hdl.handle.net/10520/EJC89058

Colbert, A. E., Judge, T. A., Choi, D., & Wang, G. (2012). Assessing the trait theory of leadership using self and observer ratings of personality: The mediating role of contributions to group success. *The Leadership Quarterly, 23*(4), 670–685. https://doi.org/10.1016/j.leaqua.2012.03.004

Collison, C., & Mackenzie, A. (1999). The power of story in organisations. *Journal of Workplace Learning, 11*(1), 38–40.

Connerley, M. L., & Pedersen, P. B. (2005). *Leadership in a diverse and multicultural environment: Developing awareness, knowledge, and skills*. Sage Publications.

Conger, J. A. (1999). Charismatic and transformational leadership in organizations: An insider's perspective on these developing streams of research. *The Leadership Quarterly, 10*(2), 145–179. https://doi.org/10.1016/S1048-9843(99)00012-0

Conger, J. A., & Ready, D. A. (2004). Rethinking leadership competencies. *Leader to Leader, 2004*(32), 41–47. https://doi.org/10.1002/ltl.75

Conlon, E., & Zandvoort, H. (2011). Broadening ethics teaching in engineering: Beyond the individualistic approach. *Science and Engineering Ethics, 17*, 217–232. https://doi.org/10.1007/s11948-010-9205-7

Copeland, M. K. (2014). The emerging significance of values-based leadership: A literature review. *International Journal of Leadership Studies, 8*(2), 105–135.

Cornelissen, J. P., & Werner, M. D. (2014). Putting framing in perspective: A review of framing and frame analysis across the management and organizational literature. *Academy of Management Annals, 8*(1), 181–235. https://doi.org/10.5465/19416520.2014.875669

Crawley, E. F., Malmqvist, J., Östlund, S., Brodeur, D. R., & Edström, K. (2014). Historical accounts of engineering education. In *Rethinking engineering education* (pp. 231–255). Springer.

Crevani, L., Lindgren, M., & Packendorff, J. (2007). Shared leadership: A post-heroic perspective on leadership as a collective construction. *International Journal of Leadership Studies, 3*(1), 40–67. https://www.diva-portal.org/smash/record.jsf?dswid=1537&pid=diva2%3A455741

Crowne, K. A. (2008). What leads to cultural intelligence?. *Business Horizons, 51*(5), 391–399. https://doi.org/10.1016/j.bushor.2008.03.010

Crumpton-Young, L., McCauley-Bush, P., Rabelo, L., Meza, K., Ferreras, A., Rodriguez, B., ... & Kelarestani, M. (2010). Engineering leadership development programs: A look at what is needed and what is being done. *Journal of STEM Education: Innovations and Research, 11*(3), 10–21. https://www.jstem.org/jstem/index.php/JSTEM/article/view/1604

Cullen, K. L., Palus, C. J., Chrobot-Mason, D., & Appaneal, C. (2012). Getting to "we": Collective leadership development. *Industrial and Organizational Psychology, 5*(4), 428–432. https://doi.org/10.1111/j.1754-9434.2012.01475.x

Cullen, K., Willburn, P., Chrobot-Mason, D., & Palus, C. (2014). *Networks: How collective leadership really works*. Center for Creative Leadership.

Da Bormida, M. (2021). The big data world: Benefits, threats and ethical challenges. *In Ethical issues in covert, security and surveillance research* (Vol. 8, pp. 71–91). Emerald Publishing Limited.

Daniels, C. B. (2009). Improving leadership in a technical environment: A case example of the ConITS Leadership Institute. *Engineering Management Journal, 21*(1), 47–52. https://doi.org/10.1080/10429247.2009.11431798

Dawkins, S., Martin, A., Scott, J., & Sanderson, K. (2013). Building on the positives: A psychometric review and critical analysis of the construct of Psychological Capital. *Journal of Occupational and Organizational Psychology, 86*(3), 348–370. https://doi.org/10.1111/joop.12007

Day, D. V. (2000). Leadership development: A review in context. *The Leadership Quarterly, 11*(4), 581–613. https://doi.org/10.1016/S1048-9843(00)00061-8

Day, D. V., & Dragoni, L. (2015). Leadership development: An outcome-oriented review based on time and levels of analyses. *Annual Review of Organizational Psychology and Organizational Behavior, 2*(1), 133–156. https://doi.org/10.1146/annurev-orgpsych-032414-111328

Day, D. V., & Lord, R. G. (1988). Executive leadership and organizational performance: Suggestions for a new theory and methodology. *Journal of Management, 14*(3), 453–464. https://doi.org/10.1177/014920638801400308

Day, D. V., & Zaccaro, S. J. (2007). Leadership: A critical historical analysis of the influence of leader traits. In L. L. Koppes (Ed.), *Historical perspectives in industrial and organizational psychology* (pp. 383–405). Erlbaum.

Deci, E. L., Connell, J. P., & Ryan, R. M. (1989). Self-determination in a work organization. *Journal of Applied Psychology, 74*(4), 580–590. https://doi.org/10.1037/0021-9010.74.4.580

De Cremer, D., & Moore, C. (2020). Toward a better understanding of behavioral ethics in the workplace. *Annual Review of Organizational Psychology and Organizational Behavior, 7*, 369–393. https://doi.org/10.1146/annurev-orgpsych-012218-015151

De Cremer, D., Mayer, D. M., & Schminke, M. (2010). Guest editors' introduction: On understanding ethical behavior and decision making: A behavioral ethics approach. *Business Ethics Quarterly, 20*(1), 1–6. https://doi.org/10.5840/beq20102012

De Meuse, K. P., Dai, G., Swisher, V. V., Eichinger, R. W., & Lombardo, M. M. (2012). Leadership development: Exploring, clarifying, and expanding our understanding of learning agility. *Industrial and Organizational Psychology: Perspectives on Science and Practice, 5*(3), 280–286. https://doi.org/10.1111/j.1754-9434.2012.01445.x

De Meuse, K. P., Dai, G., Eichinger, R. W., Page, R. C., Clark, L. P., & Zewdie, S. (2011, January). The development and validation of a self-assessment of learning agility. In *Society for Industrial and Organizational Psychology Conference*, Chicago, Illinois.

Den Hartog, D. N. (2015). Ethical leadership. *Annual Review of Organizational Psychology and Organizational Behavior, 2*, 409–434. https://doi.org/10.1146/annurev-orgpsych-032414-111237

Denning, S. (2005). *The leader's guide to storytelling: Mastering the art and discipline of business narrative*. Jossey-Bass.

Denis, J. L., Langley, A., & Sergi, V. (2012). Leadership in the plural. *The Academy of Management Annals, 6*(1), 211–283. https://doi.org/10.1080/19416520.2012.667612

DeRue, D. S. (2011). Adaptive leadership theory: Leading and following as a complex adaptive process. *Research in Organizational Behavior, 31*, 125–150. https://doi.org/10.1016/j.riob.2011.09.007

DeRue, D. S., Nahrgang, J. D., Wellman, N. E., & Humphrey, S. E. (2011). Trait and behavioral theories of leadership: An integration and meta-analytic test of their relative validity. *Personnel Psychology, 64*(1), 7–52. https://doi.org/10.1111/j.1744-6570.2010.01201.x

Dent, E. B., Higgins, M. E., & Wharff, D. M. (2005). Spirituality and leadership: An empirical review of definitions, distinctions, and embedded assumptions. *The Leadership Quarterly, 16*(5), 625–653. https://doi.org/10.1016/j.leaqua.2005.07.002

Dhiman, S. (2017). *Holistic leadership: A new paradigm for today's leaders*. Springer.

Dijk, C. F. V., & Freedman, J. (2007). Differentiating emotional intelligence in leadership. *Journal of Leadership Studies, 1*(2), 8–20. https://doi.org/10.1002/jls.20012

Dolbier, C. L., Soderstrom, M., & Steinhardt, M. A. (2001). The relationships between self-leadership and enhanced psychological, health, and work outcomes. *The Journal of Psychology: Interdisciplinary and Applied, 135*(5), 469–485. https://doi.org/10.1080/00223980109603713

Dorfman, P., Javidan, M., Hanges, P., Dastmalchian, A., & House, R. (2012). GLOBE: A twenty year journey into the intriguing world of culture and leadership. *Journal of World Business, 47*(4), 504–518. https://doi.org/10.1016/j.jwb.2012.01.004

Doyle, A. (2017). Adaptive challenges require adaptive leaders. *Performance Improvement, 56*(9), 18–26. https://doi.org/10.1002/pfi.21735

Drath, W. H. (1998). Approaching the future of leadership development. In C. D. McCauley, R. S. Moxley, & E. Van Velsor (Eds.), *The center for creative leadership handbook of leadership development* (pp. 403–432). Jossey-Bass.

Drath, W. H., & Palus, C. J. (1994). *Making common sense: Leadership as meaning-making in a community of practice.* Center for Creative Leadership.

Dunwoody, A. B., Teslenko, T. N., Cramond, P. J., Paterson, C. S., Nesbit, S. E. & Reilly, J., (2017). *Fundamental competencies for the 21st-century engineer* (2nd Ed.). Oxford University Press.

Dweck, C. S. (2012). Implicit theories. In P. A. M. Van Lange, A. W. Kruglanski, & E. T. Higgins (Eds.), *Handbook of theories of social psychology* (pp. 43–61). Sage Publications. https://doi.org/10.4135/9781446249222.n28

Earley, P. C., & Ang, S. (2003). *Cultural intelligence: Individual interactions across cultures.* Stanford University Press.

Earley, P. C., & Mosakowski, E. (2004). Cultural intelligence. *Harvard Business Review, 82*(10), 139–146. https://hbr.org/2004/10/cultural-intelligence

Easterly, W., & Pennings, S. (2020). *Leader value added: Assessing the growth contribution of individual national leaders* (No. w27153). National Bureau of Economic Research.

Edmondson, A. (1999). Psychological safety and learning behavior in work teams. *Administrative Science Quarterly, 44*(2): 350–383. https://doi.org/10.2307/2666999

Edmondson, A.C. (2012). *Teaming: How organizations learn, innovate, and compete in the knowledge economy.* John Wiley & Sons.

Edwards, G., & Bolden, R. (2022). Why is collective leadership so elusive?. *Leadership,* 17427150221128357. https://doi.org/10.1177/17427150221128357

Ekvall, G. (1996). Organizational climate for creativity and innovation. *European Journal of Work and Organizational Psychology, 5*(1), 105–123. https://doi.org/10.1080/13594329608414845

Elder, L., & Paul, R. (1997). Critical thinking: Crucial distinctions for questioning. *Journal of Developmental Education, 21*(2), 34–35. https://www.proquest.com/openview/2cd6acb1e860476b8c162248ad8976ac/1?pq-origsite=gscholar&cbl=2030483

Elias, S. (2008). Fifty years of influence in the workplace: The evolution of the French and Raven power taxonomy. *Journal of Management History, 14*(3), 267–283. https://doi.org/10.1108/17511340810880634

Elliott, C.J. (2023). Reflecting on leadership, leading, and leaders. In: J. C., Collins & J. L. Callahan (Eds), *The Palgrave handbook of critical human resource development* (pp. 243–255). Palgrave Macmillan. https://doi.org/10.1007/978-3-031-10453-4_14

Epitropaki, O., & Martin, R. (2004). Implicit leadership theories in applied settings: Factor structure, generalizability, and stability over time. *Journal of Applied Psychology, 89*(2), 293–310. https://doi.org/10.1037/0021-9010.89.2.293

Facione, P. A. (2011). Critical thinking: What it is and why it counts. *Insight Assessment, 2007*(1), 1–23.

Fairhurst, G. T. (2005). Reframing the art of framing: Problems and prospects for leadership. *Leadership, 1*(2), 165–185. https://doi.org/10.1177/1742715005051857

Fairhurst, G. T. (2009). Considering context in discursive leadership research. *Human Relations, 62*(11), 1607–1633. https://doi.org/10.1177/0018726709346379

Fairhurst, G. T. (2010). *The power of framing: Creating the language of leadership.* John Wiley & Sons.

Fairhurst, G., & Sarr, R. (1996). *The art of framing.* Jossey-Bass.

Fairhurst, G. T., Jackson, B., Foldy, E. G., & Ospina, S. M. (2020). Studying collective leadership: The road ahead. *Human Relations, 73*(4), 598–614. https://doi.org/10.1177/00187267198987

Farr, J. V., & Brazil, D. M. (2009). Leadership skills development for engineers. *Engineering Management Journal, 21*(1), 3–8. https://doi.org/10.1080/10429247.2009.11431792

Farr, J., & Brazil, D. (2012). Leadership skills development for engineers. *IEEE Engineering Management Review, 3*(40), 13–22. https://doi.org/10.1109/EMR.2012.6291576

Farr, J. V., Walesh, S. G., & Forsythe, G. B. (1997). Leadership development for engineering managers. *Journal of Management in Engineering, 13*(4), 38–41. https://doi.org/10.1061/(ASCE)0742-597X(1997)13:4(38)

Fasano, A. (2011). From a design engineer to a leader: Not an easy road. *Leadership and Management in Engineering, 11*(4), 342–343. https://doi.org/10.1061/(ASCE)LM.1943-5630.0000148

Fiedler, F. E. (1967). *A theory of leadership effectiveness.* McGraw-Hill.

Fiedler, F. E. (1971). Validation and extension of the contingency model of leadership effectiveness: A review of empirical findings. *Psychological Bulletin, 76*(2), 128–148. https://doi.org/10.1037/h0031454

Fiedler, F. E. (1996). Research on leadership selection and training: One view of the future. *Administrative Science Quarterly, 41*(2), 241–250. https://doi.org/10.2307/2393716

Fiedler, F. E. (2002). The curious role of cognitive resources in leadership. In R. E. Riggio, S. E. Murphy, & F. J. Pirozzolo (Eds.), *Multiple intelligences and leadership* (pp. 91–104). Erlbaum.

Fleishman, E. A. (1953). The description of supervisory behavior. *Personnel Psychology, 37,* 1–6.

Fleishman, E. A., & Harris, E. F. (1962). Patterns of leadership behavior related to employee grievances and turnover. *Personnel Psychology, 15*(2), 43–56. https://doi.org/10.1111/j.1744-6570.1962.tb01845.x

Fleishman, E. A., Mumford, M. D., Zaccaro, S. J., Levin, K. Y., Korotkin, A. L., & Hein, M. B. (1991). Taxonomic efforts in the description of leader behavior: A synthesis and functional interpretation. *The Leadership Quarterly, 2*(4), 245–287. https://doi.org/10.1016/1048-9843(91)90016-U

Flowers, R. B. (2002). Leadership as a responsibility. *Leadership and Management in Engineering, 2*(3), 15–19. https://doi.org/10.1061/(ASCE)1532-6748(2002)2:3(15)

Flyverbom, M., Deibert, R., & Matten, D. (2019). The governance of digital technology, big data, and the internet: New roles and responsibilities for business. *Business & Society, 58*(1), 3–19. https://doi.org/10.1177/0007650317727540

Forbes, L. H. (2008). Diversity is key to a world-class organization. *Leadership and Management in Engineering, 8*(1), 11–15. https://doi.org/10.1061/(ASCE)1532-6748(2008)8:1(11)

Forster, N., Cebis, M., Majteles, S., Mathur, A., Morgan, R., Preuss, J., ... & Wilkinson, D. (1999). The role of story-telling in organizational leadership. *Leadership & Organization Development Journal, 20*(1), 11–17. https://doi.org/10.1108/01437739910251134

Freeman, L. (2004). *The development of social network analysis: A study in the sociology of science.* Empirical Press

French, J., & Raven, B. H. (1959). The bases of social power. In D. Cartwright (Ed.), *Studies of social power* (pp. 150–167). Institute for Social Research. University of Michigan Press.

Friedrich, T. L., Griffith, J. A., & Mumford, M. D. (2016). Collective leadership behaviors: Evaluating the leader, team network, and problem situation characteristics that influence their use. *The Leadership Quarterly, 27*(2), 312–333. https://doi.org/10.1016/j.leaqua.2016.02.004

Frost, J., & Walker, M. (2007). Cross cultural leadership. *Engineering Management, 17*(3), 27–29. https://doi.org/10.1049/em:20070303

Fry, L. W. (2003). Toward a theory of spiritual leadership. *The Leadership Quarterly, 14*(6), 693–727. https://doi.org/10.1016/j.leaqua.2003.09.001

Fry, L. W. (2005). Editorial: Introduction to *The Leadership Quarterly* special issue: Toward a paradigm of spiritual leadership. *The Leadership Quarterly, 16*(5), 619–622. https://doi. org/10.1016/j.leaqua.2005.07.001

Fry, L. W., Vitucci, S., & Cedillo, M. (2005). Spiritual leadership and army transformation: Theory, measurement, and establishing a baseline. *The Leadership Quarterly, 16*(5), 835–862. https://doi.org/10.1016/j.leaqua.2005.07.012

Fullan, M. (2001). *Leading in a culture of change.* Jossey-Bass.

Gardner, W. L., Avolio, B. J., Luthans, F., May, D. R., & Walumbwa, F. (2005). "Can you see the real me?" A self-based model of authentic leader and follower development. *The Leadership Quarterly, 16*(3), 343–372. https://doi.org/10.1016/j.leaqua.2005.03.003

Gardner, W. L., Cogliser, C. C., Davis, K. M., & Dickens, M. P. (2011). Authentic leadership: A review of the literature and research agenda. *The Leadership Quarterly, 22*(6), 1120–1145. https://doi.org/10.1016/j.leaqua.2011.09.007

Gates, B. (2015). Empower others to lead. In N. Craig, N., B. George, & S. Snook, (Eds.), *The discover your true north fieldbook: A personal guide to finding your authentic leadership* (pp. 167–186). John Wiley & Sons.

Gharajedaghi, J. (1999). *Systems thinking: Managing chaos and complexity.* Butterworth Heinemann.

Gibb, C. A. (1954). Leadership. In Lindzey, G. (Ed.), *Handbook of social psychology, Vol. 2* (pp. 877–917). Addison-Wesley.

Gibson, J. L, Ivancevich, J. M., Donnelly, J. H., & Konopaske, R. (2012). *Organizations: Behavior, structure, processes* (14th Ed.). McGraw-Hill.

Giddens, A. (1999). *Runaway world: How globalization is reshaping our lives.* Profile Books.

Giegold, W. C. (1981). Leadership-The essential of engineering management. *Engineering Management International, 1*(1), 49–56. https://doi.org/10.1016/0167-5419(81)90008-9

Giles, S. (2016). The most important leadership competencies, according to leaders around the world. *Harvard Business Review, 15*(3), 2–6.

Gino, F. (2015). Understanding ordinary unethical behavior: Why people who value morality act immorally. *Current Opinion in Behavioral Sciences, 3*, 107–111. https://doi. org/10.1016/j.cobeha.2015.03.001

Gino, F., Ayal, S., & Ariely, D. (2009). Contagion and differentiation in unethical behavior: The effect of one bad apple on the barrel. *Psychological Science, 20*(3), 393–398. https:// doi.org/10.1111/j.1467-9280.2009.02306

Gluckman, P. (2018). The digital economy and society: A preliminary commentary. *Policy Quarterly, 14*(1). https://doi.org/10.26686/pq.v14i1.4763

Goethals, G. R., & Allison, S. T. (2014). Kings and charisma, Lincoln and leadership: An evolutionary perspective. In Goethals, G. R., Allison, S. T., Kramer, R., & Messick, D. (Eds.), *Conceptions of leadership: Enduring ideas and emerging* insights (pp. 111–124). Palgrave Macmillan. https://doi.org/10.1057/9781137472038_7

Goethals, G. R., & Allison, S. T. (2019). *The romance of heroism and heroic leadership.* Emerald Group Publishing.

Goleman, D. (1995). *Emotional intelligence: Why it can matter more than IQ.* Bantam Books.

Goleman, D. (1998). *Working with emotional intelligence.* Bantam Books.

Goleman, D. (2011). *Leadership: The power of emotional intelligence.* More Than Sound LLC.

Goleman, D., & Boyatzis, R. (2017). Emotional intelligence has 12 elements. Which do you need to work on. *Harvard Business Review, 84*(2), 1–5. https://hbr.org/2017/02/ emotional-intelligence-has-12-elements-which-do-you-need-to-work-on

Goleman, D., Boyatzis, R. E., & McKee, A. (2013). *Primal leadership: Unleashing the power of emotional intelligence*. Harvard Business Press.

Goldfinch, T., Abuodha, P., Hampton, G., Hill, F., Dawes, L., & Thomas, G. (2012, January). *Intercultural competence in engineering education: Who are we teaching*. In Australasian Association of Engineering Education Annual Conference.

Gordon, B. M. (2021). *Capabilities of effective engineering leaders*. MIT Engineering Leadership Program. Retrieved from: https://gelp.mit.edu/capabilitiesofeffectiveengineeringleaders

Graen, G. B., & Uhl-Bien, M. (1995). Relationship-based approach to leadership: Development of leader-member exchange (LMX) theory of leadership over 25 years: Applying a multi-level multi-domain perspective. *The Leadership Quarterly, 6*(2), 219–247. https://doi.org/10.1016/1048-9843(95)90036-5

Graham, J. W. (1991). Servant-leadership in organizations: Inspirational and moral. *The Leadership Quarterly, 2*(2), 105–119. https://doi.org/10.1016/1048-9843(91)90025-W

Grandin, J. M., & Hedderich, N. (2009). Global competence for engineers. In D K. Deardorff, *The Sage handbook of intercultural competence* (pp. 362–373). Sage Publications. https://doi.org/10.4135/9781071872987

Grayson, C., & Baldwin, D. (2011). *Leadership networking: Connect, collaborate, create*. CCL Press.

Graves, N. (Ed.). (2013). *Learner managed learning: Practice, theory and policy*. Routledge.

Greenleaf, R. (1977). *Servant leadership*. Jossey-Bass.

Grint, K. (2005). Problems, problems, problems: The social construction of 'leadership'. *Human Relations, 58*(11), 1467–1494. https://doi.org/10.1177/001872670506131

Gronn, P. (2000). Distributed properties: A new architecture for leadership. *Educational Management Administration & Leadership, 28*(3), 317–338. https://doi.org/10.1177/0263211X000283006

Gronn, P. (2002). Distributed leadership as a unit of analysis. *The Leadership Quarterly, 13*(4), 423–451. https://doi.org/10.1016/S1048-9843(02)00120-0

Hackman, J. R., & Oldham, G. R. (1980). *Work redesign*. Addison-Wesley.

Hackman, J. R., & Wageman, R. (2005). A theory of team coaching. *Academy of Management Review, 30*(2), 269–287. https://doi.org/10.5465/amr.2005.16387885

Hackman, J. R., & Walton, R. E. (1986). Leading groups in organizations. In: P. S. Goodman, et al. (Eds.), *Designing effective work groups* (pp. 72–119). Jossey-Bass.

Halpin, A. W., & Winer, B. J. (1957). A factorial study of the leader behavior descriptions. In R. Stogdill & A. Coons (Eds.), *Leader behavior: Its description and measurement* (pp. 39–51). Bureau of Business Research, Ohio State University.

Harari, M. B., Williams, E. A., Castro, S. L., & Brant, K. K. (2021). Self-leadership: A meta-analysis of over two decades of research. *Journal of Occupational and Organizational Psychology, 94*(4), 890–923. https://doi.org/10.1111/joop.12365

Hartmann, B. L. & Jahren, C. T. (2015). Leadership: Industry needs for entry-level engineering positions. *Journal of STEM Education: Innovations and Research 16*(3): 13–19. https://www.learntechlib.org/p/151966/

Hase, S., & Davis, L. (1999). From competence to capability: The implications for human resource development and management. *Association of International Management, 17th Annual Conference*, San Diego, August.

Hausmann, R. (2022). *Green growth opportunities, IMF: International Monetary Fund*. United States of America. Available at https://policycommons.net/artifacts/3217276/green-growth-opportunities/4016347/

Hawkins, P. (2004). A centennial tribute to Gregory Bateson 1904-1980 and his influence on the fields of organizational development and action research. *Action Research, 2*(4), 409–423. https://doi.org/10.1177/1476750304047984

Hawkins, P. (2021) (4th Ed.). *Leadership team coaching: Developing collective transformational leadership*. Kogan Page.

Haws, D. R. (2001). Ethics instruction in engineering education: A (mini) meta-analysis. *Journal of Engineering Education, 90*(2), 223–229. https://doi.org/10.1002/j.2168-9830.2001.tb00596.x

Hay Group. (2011). *Emotional and social competency inventory (ESCI): A user guide for accredited practitioners.* L&T direct and the McClelland Center for Research and Innovation. Hay Group. Available at: https://www.eiconsortium.org/measures/eci_360.html

Haynes, C. J. (2009). Holistic human development. *Journal of Adult Development, 16*(1), 53–60. https://doi.org/10.1007/s10804-009-9052-4

Heifetz, R. A. (1998). *Leadership without easy answers.* Harvard University Press.

Heifetz, R. A., & Laurie, D. L. (2001). The work of leadership. *Harvard Business Review, 79*(11). https://hbsp.harvard.edu/product/R0111K-PDF-ENG

Heifetz, R. A., Heifetz, R., Grashow, A., & Linsky, M. (2009). *The practice of adaptive leadership: Tools and tactics for changing your organization and the world.* Harvard Business Press.

Hensey, M. (1999). The why and how of facilitative leadership. *Journal of Management in Engineering, 15*(3), 43–46. https://doi.org/10.1061/(ASCE)0742-597X(1999)15:3(43)

Hennessey, B. A., & Amabile, T. M. (2010). Creativity. *Annual Review of Psychology, 61*, 569–598. https://doi.org/10.1146/annurev.psych.093008.100416

Hersey, P., & Blanchard, K.H. (1969). Life cycle theory of leadership. *Training & Development Journal, 23*(5), 26–34.

Hersey, P., & Blanchard, K. H. (1993). *Management of organizational behavior: Utilizing human resources* (6th Ed.). Prentice-Hall, Inc.

Hersey, P., Blanchard, K.H., & Natemeyer, W.E. (1979). Situational leadership, perception and the impact of power. *Group & Organization Management, 4*(4), 418–428. https://doi.org/10.1177/105960117900400404

Hess, D. W. (2018). *Leadership by engineers and scientists: Professional skills needed to succeed in a changing world.* John Wiley & Sons.

Hester, J. (2012). Values-based leadership: A shift in attitude. *The Journal of Values-Based Leadership, 5*(1), 5. https://scholar.valpo.edu/jvbl/vol5/iss1/5

Hickel, J., & Kallis, G. (2020). Is green growth possible?. *New Political Economy, 25*(4), 469–486. https://doi.org/10.1080/13563467.2019.1598964

Hinkle, G. C. (2007). All engineers need leadership skills. *IEEE USA Today's Engineer Online, 4*(7), 34–41.

Hirt, M. (2020, January 31). Competency out, capability in. *TD Magazine, 74*(2), 28–33. https://www.td.org/magazines/td-magazine/competency-out-capability-in

Hitt, M. A., & Ireland, R. D. (2002). The essence of strategic leadership: Managing human and social capital. *Journal of Leadership & Organizational Studies, 9*(1), 3–14. https://doi.org/10.1177/107179190200900010

Hoffmann, B., Jørgensen, U., & Christensen, H. P. (2011, June). *Culture in engineering education.* CDIO Framing Intercultural Competences. In Proceedings of the 7th International CDIO Conference (pp. 20–23). https://orbit.dtu.dk/en/publications/culture-in-engineering-education-cdio-framing-intercultural-compe

Hogg, M. A., Martin, R., Epitropaki, O., Mankad, A., Svensson, A., & Weeden, K. (2005). Effective leadership in salient groups: Revisiting leader-member exchange theory from the perspective of the social identity theory of leadership. *Personality and Social Psychology Bulletin, 31*(7), 991–1004. https://doi.org/10.1177/0146167204273309

Hollander, E. P., & Offermann, L. R. (1990). Power and leadership in organizations: Relationships in transition. *American Psychologist, 45*(2), 179–189. https://doi.org/10.1037/0003-066X.45.2.179

Horney, N., Pasmore, B., & O'Shea, T. (2010). Leadership agility: A business imperative for a VUCA world. *People & Strategy, 33*(4), 32–38. https://luxorgroup.fr/coaching/wp-content/uploads/Leadership-agility-model.pdf

Horth, D., & Buchner, D. (2014). *Innovation Leadership: How to use innovation to lead effectively, work collaboratively and drive results*. Center for Creative Leadership. Retrieved from: https://imamhamzatcoed.edu.ng/library/ebooks/resources/Innovation_Leadership_by_david_horth.pdf

Houghton, J. D., & DiLiello, T. C. (2010). Leadership development: The key to unlocking individual creativity in organizations. *Leadership & Organization Development Journal, 31*(3), 230–245. https://doi.org/10.1108/01437731011039343

Houghton, J. D., & Neck, C. P. (2002). The revised self-leadership questionnaire: Testing a hierarchical factor structure for self-leadership. *Journal of Managerial Psychology, 17*(8), 672–691. https://doi.org/10.1108/02683940210450484

House, R. J. (1971). A path goal theory of leader effectiveness. *Administrative Science Quarterly, 16*(3), 321–339. https://doi.org/10.2307/2391905

House, R. J. (1996). Path-goal theory of leadership: Lessons, legacy and a reformulated theory. *The Leadership Quarterly, 7*(3), 323–352. https://doi.org/10.1016/S1048-9843(96)90024-7

House, R. J., Hanges, P. J., Javidan, M., Dorfman, P. W., & Gupta, V. (Eds.). (2004). *Culture, leadership, and organizations: The GLOBE study of 62 societies*. Sage Publications.

Hoyt, C. L., Murphy, S. E., Halverson, S. K., & Watson, C. B. (2003). Group Leadership: Efficacy and Effectiveness. *Group Dynamics: Theory, Research, and Practice, 7*(4), 259–274. https://doi.org/10.1037/1089-2699.7.4.259

Hughes, D. J., Lee, A., Tian, A. W., Newman, A., & Legood, A. (2018). Leadership, creativity, and innovation: A critical review and practical recommendations. *The Leadership Quarterly, 29*(5), 549–569. https://doi.org/10.1016/j.leaqua.2018.03.001

Hunter, J. C. (2004). *The world's most powerful leadership principle: How to become a servant leader*. Crown Publishing.

Hunter, S. T., Cushenbery, L., Fairchild, J., & Boatman, J. (2012). Partnerships in leading for innovation: A dyadic model of collective leadership. *Industrial and Organizational Psychology: Perspectives on Science and Practice, 5*(4), 424–428. https://doi.org/10.1111/j.1754-9434.2012.01474.x

Hur, Y., van den Berg, P. T., & Wilderom, C. P. M. (2011). Transformational leadership as a mediator between emotional intelligence and team outcomes. *The Leadership Quarterly, 22*(4), 591–603. https://doi.org/10.1016/j.leaqua.2011.05.002

Ireland, R. D., & Hitt, M. A. (2005). Achieving and maintaining strategic competitiveness in the 21st century: The role of strategic leadership. *Academy of Management Perspectives, 19*(4), 63–77. https://doi.org/10.5465/ame.2005.19417908

Ivey, J. M. (2002). Five critical components of leadership. *Leadership and Management in Engineering, 2*(2), 26–28. https://doi.org/10.1061/(ASCE)1532-6748(2002)2:2(26)

James, D. T. (2008). Importance of diversity in a successful firm. *Leadership and Management in Engineering, 8*(1), 16–18. https://doi.org/10.1061/(ASCE)1532-6748(2008)8:1(16)

Janssen, O. (2001). Fairness perceptions as a moderator in the curvilinear relationships between job demands, and job performance and job satisfaction. *Academy of Management Journal, 44*(5), 1039–1050. https://doi.org/10.5465/3069447

Javidan, M., & Walker, J. L. (2012). A whole new global mindset for leadership. *People & Strategy, 35*(2), 36–41.

Jaworski, J., & Scharmer, C. O. (2000). *Leadership in the new economy: Sensing and actualizing emerging futures*. Generon Consulting.

Jenkins, W. O. (1947). A review of leadership studies with particular reference to military problems. *Psychological Bulletin, 44*(1), 54–79. https://doi.org/10.1037/h0062329

Jesiek, B. K., Shen, Y., & Haller, Y. (2012). Cross-cultural competence: A comparative assessment of engineering students. *International Journal of Engineering Education, 28(1)*, 144–155.

Jones, B. F., & Olken, B. A. (2005). Do leaders matter? National leadership and growth since World War II. *The Quarterly Journal of Economics, 120*(3), 835–864. https://doi.org/10.1093/qje/120.3.835

Jones, S. A., Michelfelder, D., & Nair, I. (2015). Engineering managers and sustainable systems: The need for and challenges of using an ethical framework for transformative leadership. *Journal of Cleaner Production, 140*, 205–212. https://doi.org/10.1016/j.jclepro.2015.02.009

Judge, T. A., & Bono, J. E. (2000). Five-factor model of personality and transformational leadership. Journal of Applied Psychology, 85(5), 751–765. https://doi.org/10.1037/0021-9010.85.5.751

Judge T. A, Bono, J. E., Ilies R., Gerhardt, M. W. (2002). Personality and leadership: A qualitative and quantitative review. *Journal of Applied Psychology, 87*, 765–780. https://psycnet.apa.org/buy/2002-15406-013

Judge, T. A., Colbert, A. E., & Ilies, R. (2004). Intelligence and Leadership: A Quantitative Review and Test of Theoretical Propositions. *Journal of Applied Psychology, 89*(3), 542–552. https://doi.org/10.1037/0021-9010.89.3.542

Juma, C. (February 28, 2013). *"Engineering green growth"*. Global grand challenges summit blog. https://www.belfercenter.org/publication/engineering-green-growth

Junker, N. M., & van Dick, R. (2014). Implicit theories in organizational settings: A systematic review and research agenda of implicit leadership and followership theories. *The Leadership Quarterly, 25*(6), 1154–1173. https://doi.org/10.1016/j.leaqua.2014.09.002

Kaiser, R. B., & Overfield, D. V. (2010). Assessing flexible leadership as a mastery of opposites. *Consulting Psychology Journal: Practice and Research, 62*(2), 105–118. https://doi.org/10.1037/a0019987

Kane, G. (2019). The technology fallacy: People are the real key to digital transformation. *Research-Technology Management, 62*(6), 44–49. https://doi.org/10.1080/08956308.2019.1661079

Karlen, Y., & Hertel, S. (2021). The power of implicit theories for learning in different educational contexts. *Frontiers in Education, 6*:788759. https://doi.org/10.3389/feduc.2021.788759

Kaufman, S., Elliott, M., & Shmueli, D. (2003). Frames, framing and reframing. *Beyond Intractability, 1*, 1–8. https://www.beyondintractability.org/

Katz, R. L. (1955). Skills of an effective administrator. *Harvard Business Review, 33*(1), 33–42.

Kegan, R. (1980). Making meaning: The constructive-developmental approach to persons and practice. *The Personnel and Guidance Journal, 58*(5), 373–380. https://doi.org/10.1002/j.2164-4918.1980.tb00416.x

Kegan, R. (1982). *The evolving self: Problem and process in human development*. Harvard University Press.

Kegan, R. (1994). *In over our heads: The mental demands of modern life*. Harvard University Press.

Kegan, R., & Lahey, L. L. (1984). Adult leadership and adult development: A constructionist view. In B. Kellerman (Ed.), *Leadership: Multidisciplinary perspectives* (pp. 199–230). Prentice-Hall.

Keller, T. (1999). Images of the familiar: Individual differences and implicit leadership theories. *The Leadership Quarterly, 10*(4), 589–607. https://doi.org/10.1016/S1048-9843(99)00033-8

Kellerman, B. (2008). *Followership: How followers are creating change and changing leaders*. Harvard Business Press

Kerr, R., Garvin, J., Heaton, N., & Boyle, E. (2006). Emotional intelligence and leadership effectiveness. *Leadership & Organization Development Journal, 27*(4), 265–279. https://doi.org/10.1108/01437730610666028

Khanna, T. (2014). Contextual intelligence. *Harvard Business Review, 92*(9), 58–68.

Khanna, T. (2015). A case for contextual intelligence. *Management International Review, 55*(2), 181–190. https://doi.org/10.1007/s11575-015-0241-z

Killgore, M. W. (2014, June), *Visions of the Future of Engineering Education: Sharpening the Focus*. Paper presented at 2014 ASEE Annual Conference & Exposition, Indianapolis, Indiana.

King, A. S. (1990). Evolution of leadership theory. *Vikalpa, 15*(2), 43–54.

Kipley, D., Lewis, A. & Jewe, R. (2012). Entropy-disrupting Ansoff's five levels of environmental turbulence. *Business Strategy Series, 13*(6), 251–262. https://doi.org/10.1108/17515631211286083

Kipnis, D., Schmidt, S. M., & Wilkinson, I. (1980). Intraorganizational influence tactics: Explorations in getting one's way. *Journal of Applied Psychology, 65*(4), 440–452. https://doi.org/10.1037/0021-9010.65.4.440

Kjellström, S., Stålne, K., & Törnblom, O. (2020). Six ways of understanding leadership development: An exploration of increasing complexity. *Leadership, 16*(4), 434–460. https://doi.org/10.1177/1742715020926731

Klassen, M., Reeve, D., Evans, G. J., Rottmann, C., Sheridan, P. K., & Simpson, A. (2020). Engineering: Moving leadership from the periphery to the core of an intensely technical curriculum. *New Directions for Student Leadership, 2020*(165), 113–124. https://doi.org/10.1002/yd.20373

Knies, E.; Jacobsen, C. & Tummers, L.G. (2016). Leadership and organizational performance: State of the art and research agenda. In: Storey, J., Denis, J. L., Hartley, J. & 't Hart, P. (Eds.). *Routledge companion to leadership* (pp. 404–418). Routledge.

Knight, J. (2005). All leaders manage but not all managers lead. *Engineering Management, 15*(1), 36–37. https://doi.org/10.1049/em:20050109

Knights, D. & O'Leary, M. (2005). Reflecting on corporate scandals: The failure of ethical leadership. *Business Ethics: A European Review, 14*(4), 359–366. https://doi.org/10.1111/j.1467-8608.2005.00417.x

Kocaoglu, D. F. (1991, October). Education for leadership in management of engineering and technology. In *Technology management: The new international language* (pp. 78–83). IEEE. https://doi.org/10.1109/PICMET.1991.183567

Kohnke, O. (2017). It's not just about technology: The people side of digitization. In *Shaping the digital enterprise* (pp. 69–91). Springer. https://doi.org/10.1007/978-3-319-40967-2_3

Kong, M., Xu, H., Zhou, A., & Yuan, Y. (2019). Implicit followership theory to employee creativity: The roles of leader-member exchange, self-efficacy and intrinsic motivation. *Journal of Management & Organization, 25*(1), 81–95. https://doi.org/10.1017/jmo.2017.18

Kouzes, J. M., & Posner, B. Z. (2017). *The leadership challenge: How to get extraordinary things done in organizations* (6th Ed.). Jossey-Bass.

Kotter, J. P. (2013). Management is (still) not leadership. *Harvard Business Review, 9*(1).

Kotterman, J. (2006). Leadership versus management: What's the difference?. *The Journal for Quality and Participation, 29*(2), 13–17.

Kozlowski, S. W. J., Mak, S., & Chao, G. T. (2016). Team-centric leadership: An integrative review. *Annual Review of Organizational Psychology and Organizational Behavior, 3*, 21–54. https://doi.org/10.1146/annurev-orgpsych-041015-062429

Krasikova, D. V., Green, S. G., & LeBreton, J. M. (2013). Destructive leadership: A theoretical review, integration, and future research agenda. *Journal of Management, 39*(5), 1308–1338. https://doi.org/10.1177/0149206312471388

Kutz, M. (2008). Toward a conceptual model of contextual intelligence: A transferable leadership construct. *Leadership Review, 8*(2), 18–31.

Kutz, M. (2017). What is contextual intelligence?. In *Contextual intelligence*. Palgrave Macmillan. https://doi.org/10.1007/978-3-319-44998-2_2

Kutz, M. R., & Bamford-Wade, A. (2014). Contextual intelligence: A critical competency for leading. In N. D. Erbe (Ed.), *Approaches to managing organizational diversity and innovation* (pp. 42–59). IGI Global.

Ladkin, D. (2010). *Rethinking leadership: A new look at old leadership questions.* Edward Elgar Publishing.

Lancer, N., Clutterbuck, D., & Megginson, D. (2016). *Techniques for coaching and mentoring* (2nd Ed.). Routledge. https://doi.org/10.4324/9781315691251

Lapsley, D. K., & Narvaez, D. (2004). A social- cognitive approach to moral personality. In D. K. Lapsley & D. Narvaez (Eds.), *Moral development, self, and identity* (pp. 189–212). Erlbaum.

Latham, G. P., & Seijts, G. H. (1998). Management development. In P. J. D. Drenth, H. Thierry, & Associates (Eds.), *Handbook of work and organizational psychology* (vol. 3; 2nd ed.; pp. 257–272). Psychology Press.

Lawton, A., & Páez, I. (2015). Developing a framework for ethical leadership. *Journal of Business Ethics, 130*(3), 639–649. https://doi.org/10.1007/s10551-014-2244-2

Layne, P. (2002). Best practices in managing diversity. *Leadership and Management in Engineering, 2*(4), 28–30. https://doi.org/10.1061/(ASCE)1532-6748(2002)2:4(28)

Lee, A., Legood, A., Hughes, D., Tian, A. W., Newman, A., & Knight, C. (2020). Leadership, creativity and innovation: A meta-analytic review. *European Journal of Work and Organizational Psychology, 29*(1), 1–35. https://doi.org/10.1080/13594 32X.2019.1661837

Lepak, D. P., & Snell, S. A. (1999). The human resource architecture: Toward a theory of human capital allocation and development. *Academy of Management Review, 24,* 31–48. https://doi.org/10.5465/amr.1999.1580439

Levy, O., S. Taylor, N. A. Boyacigiller, & S. Beechler, (2007). Global mindset: A review and proposed extensions. In M. Javidan, R. M. Steers & M. Hitt (Eds.), *The global mindset. advances in international management, 19* (pp. 11–47). JAI Press. https://doi. org/10.1016/S1571-5027(07)19002-1

Lichtenthaler, U. (2009). Absorptive capacity, environmental turbulence, and the complementarity of organizational learning processes. *Academy of Management Journal, 52*(4), 822–846. https://doi.org/10.5465/amj.2009.43670902

Liden, R. C., & Antonakis, J. (2009). Considering context in psychological leadership research. *Human Relations, 62*(11), 1587–1605. https://doi.org/10.1177/0018726709346374

Likert, R. (1961). An emerging theory of organizations, leadership and management. In L. Petrullo & B. M. Bass (Eds.), *Leadership and interpersonal behavior* (pp. 201–215). New York: Holt, Rinehart & Winston.

Lombardo, M. M., & Eichinger, R. W. (2000). High potentials as high learners. *Human Resource Management, 39*(4), 321–329. https://doi.org/10.1002/1099-050X(200024) 39:4<321::AID-HRM4>3.0.CO;2–1

Lord, R. G., & Maher, K. J. (1993). *Leadership and information processing: Linking perceptions and performance.* Routledge. https://doi.org/10.4324/9780203423950

Lord, R. G., de Vader, C. L., & Alliger, G. M. (1986). A meta-analysis of the relation between personality traits and leadership perceptions: An application of validity generalization procedures. *Journal of Applied Psychology, 71*(3), 402–410. https://doi. org/10.1037/0021-9010.71.3.402

Lord, R. G., Epitropaki, O., Foti, R. J., & Hansbrough, T. K. (2020). Implicit leadership theories, implicit followership theories, and dynamic processing of leadership information. *Annual Review of Organizational Psychology and Organizational Behavior, 7,* 49–74. https://doi.org/10.1146/annurev-orgpsych-012119-045434

Ludwig, D. (2001). The era of management is over. *Ecosystems, 4*(8), 758–764. https://doi. org/10.1007/s10021-001-0044-x

Lunenburg, F. C. (2011). Leadership versus management: A key distinction-At least in theory. *International Journal of Management, Business, and Administration, 14*(1), 1–4.

Lunevich, L. (2022). Critical digital pedagogy: Alternative ways of being and educating, connected knowledge and connective learning. *Creative Education, 13*(6), 1884–1896. https://doi.org/10.4236/ce.2022.136118

Lunevich, L., & Wadaani, M. R. (2023). *Creativity in Teaching and Teaching for Creativity: Modern Practices in the Digital Era.* CRC Press.

Luthans, F. (2002). Positive organizational behavior: Developing and managing psychological strengths. *Academy of Management Perspectives, 16*(1), 57–72. https://doi.org/10.5465/ame.2002.6640181

Luthans, F., & Youssef, C. M. (2004). Human, social, and now positive psychological capital management: Investing in people for competitive advantage. *Organizational Dynamics, 33*(2), 143–160. https://doi.org/10.1016/j.orgdyn.2004.01.003

Luthans, F., Luthans, K. W., & Luthans, B. C. (2004). Positive psychological capital: Beyond human and social capital. *Business Horizons, 47*(1), 45–45. https://doi.org/10.1016/j.bushor.2003.11.007

Luthans, F., Youssef, C. M., & Avolio, B. J. (2007). *Psychological capital: Developing the human competitive edge.* Oxford: Oxford University Press.

Luthans, F., & Youssef-Morgan, C. M. (2017). Psychological capital: An evidence-based positive approach. *Annual Review of Organizational Psychology and Organizational Behavior, 4*, 339–366. https://doi.org/10.1146/annurev-orgpsych-032516-113324

Maccoby, M. (2000). The human side: Understanding the difference between management and leadership. *Research-Technology Management, 43*(1), 57–59. https://doi.org/10.1080/08956308.2000.11671333

MacGregor, D. (1960). *The human side of enterprise.* McGraw-Hill.

Mainemelis, C., Kark, R., & Epitropaki, O. (2015). Creative leadership: A multi-context conceptualization. *Academy of Management Annals, 9*(1), 393–482. https://doi.org/10.5465/19416520.2015.1024502

Maitlis, S., & Christianson, M. (2014). Sensemaking in organizations: Taking stock and moving forward. *The Academy of Management Annals, 8*(1), 57–125. https://doi.org/10.5465/19416520.2014.873177

Manz, C. C. (2015). Taking the self-leadership high road: Smooth surface or potholes ahead? *Academy of Management Perspectives, 29*(1), 132–151. https://doi.org/10.5465/amp.2013.0060

Mawson, T. C. (2001). Can we really train leadership?. *Leadership and Management in Engineering, 1*(3), 44–45. Retrieved from: https://ascelibrary.org/doi/pdf/10.1061/(ASCE)1532-6748(2001)1%3A3(44)

Mayer, J. D., & Salovey, P. (1997). What is emotional intelligence? In P. Salovey & D. J. Sluyter (Eds.), *Emotional development and emotional intelligence: Educational implications* (pp. 3–34). Basic Books.

Mayer, J. D., Salovey, P., & Caruso, D. R. (2008). Emotional intelligence: New ability or eclectic traits? *American Psychologist, 63*(6), 503–517. https://doi.org/10.1037/0003-066X.63.6.503

Mayer, J. D., Salovey, P., Caruso, D. R., & Sitarenios, G. (2003). Measuring emotional intelligence with the MSCEIT V2.0. *Emotion, 3*(1), 97–105. https://doi.org/10.1037/1528-3542.3.1.97

Mazar, N., Amir, O., & Ariely, D. (2008). The dishonesty of honest people: A theory of self-concept maintenance. *Journal of Marketing Research, 45*(6), 633–644. https://doi.org/10.1509/jmkr.45.6.63

Maznevski, M. L. & Lane, H. W. (2004). Shaping the global mindset: Designing educational experiences for effective global thinking and action. In N. Boyacigiller, R. M. Goodman, & M. Phillips (Eds), *Crossing cultures: Insights from master teachers* (pp. 171–184). Routledge.

McCauley, C. D., Moxley, R. S., & Van Velsor, E. (Eds.). (1998). *The Center for Creative Leadership handbook of leadership development*. Jossey-Bass.

McCauley, C. D., Van Velsor, E., & Ruderman, M. N. (2010). Introduction: Our view of leadership development. In E. Van Velsor, C. D. McCauley, & M. N. Ruderman, (Eds.), *The center for creative leadership handbook of leadership development* (pp. 1-26). Jossey-Bass.

McCauley, C. D., Drath, W. H., Palus, C. J., O'Connor, P. M., & Baker, B. A. (2006). The use of constructive-developmental theory to advance the understanding of leadership. *The Leadership Quarterly, 17*(6), 634–653. https://doi.org/10.1016/j.leaqua.2006.10.006

McClelland, D. C. (1995). Power is the great motivator. *Harvard Business Review, 73*(1), 126–134.

McCleskey, J. A. (2014). Situational, transformational, and transactional leadership and leadership development. *Journal of Business Studies Quarterly, 5*(4), 117–130. https://www.proquest.com/docview/1548766781?pq-origsite=gscholar&fromopenview=true

McCuen, R. H. (1999). A course on engineering leadership. *Journal of Professional Issues in Engineering Education and Practice, 125*(3), 79–82. https://doi.org/10.1061/(ASCE)1052-3928(1999)125:3(79)

McWhinney, W., & Batista, J. (1988). How remythologizing can revitalize organizations. *Organizational Dynamics, 17*(2), 46–58. https://doi.org/10.1016/0090-2616(88)90018-6

Meindl, J. R., & Ehrlich, S. B. (1987). The romance of leadership and the evaluation of organizational performance. *Academy of Management Journal, 30*(1), 91–109. https://doi.org/10.2307/255897

Meindl, J. R., Ehrlich, S. B., & Dukerich, J. M. (1985). The romance of leadership. *Administrative Science Quarterly, 30*(1), 78–102. https://doi.org/10.2307/2392813

Mhatre, K. H., & Riggio, R. E. (2014). Charismatic and transformational leadership: Past, present, and future. In D. Day (Ed.), *The Oxford handbook of leadership and organizations* (221–240). Oxford University Press.

Mintzberg, G. (1983). *Power in and around organisations*. Sage Publications.

Mitchell, M. S., Reynolds, S. J., & Treviño, L. K. (2017). The study of behavioral ethics within organizations. *Personnel Psychology, 70*(2), 313–314. https://doi.org/10.1111/peps.12227

Mitroff, I.I., & Kilmann, R.H. (1975). Stories managers tell: A new tool for organizational problem solving. *Management Review, 64*(7), 18–28.

Monat, J., Amissah, M., & Gannon, T. (2020). Practical applications of systems thinking to business. *Systems, 8*(2), 14. https://doi.org/10.3390/systems8020014

Monnier, J. (2021, April 20). What was the Deepwater Horizon disaster? *LiveScience*. https://www.livescience.com/deepwater-horizon-oil-spill-disaster.html

Morgan, G. (1997). *Images of organization*. Sage Publications.

Morgeson, F. P., DeRue, D. S., & Karam, E. P. (2010). Leadership in teams: A functional approach to understanding leadership structures and processes. *Journal of Management, 36*(1), 5–39. https://doi.org/10.1177/0149206309347376

Mumford, M. D., & Gustafson, S. B. (1988). Creativity syndrome: Integration, application, and innovation. *Psychological Bulletin, 103*(1), 27–43. https://doi.org/10.1037/0033-2909.103.1.27

Mumford, M. D., Zaccaro, S. J., Connelly, M. S., & Marks, M. A. (2000). Leadership skills: Conclusions and future directions. *The Leadership Quarterly, 11*(1), 155–170. https://doi.org/10.1016/S1048-9843(99)00047-8

Mumford, M. D., Friedrich, T. L., Vessey, W. B., & Ruark, G. A. (2012). Collective leadership: Thinking about issues vis-à-vis others. *Industrial and Organizational Psychology: Perspectives on Science and Practice, 5*(4), 408–411. https://doi.org/10.1111/j.1754-9434.2012.01469.x

Munduate, L., & Medina, F. (2004). Power, authority, and leadership. *Encyclopedia of Applied Psychology, 10*, 91–99.

Murgai, A. (2018). Transforming digital marketing with artificial intelligence. *International Journal of Latest Technology in Engineering, Management & Applied Science, 7*(4), 259–262.

Nadkarni, S., & Narayanan, V.K. (2007). Strategic schemas, strategic flexibility, and firm performance: The moderating role of industry clockspeed. *Strategic Management Journal, 28*(3), 243–270. https://doi.org/10.1002/smj.576

Nagarajan, R., & Prabhu, R. (2015). Competence and capability: A new look. *International Journal of Management, 6*(6), 7–11.

Nahapiet, J., & Ghoshal, S. (1998). Social capital, intellectual capital, and the organizational advantage. *Academy of Management Review, 23*(2), 242–266. https://doi.org/10.5465/amr.1998.533225

National Academy of Engineering, U. S. (2004). *The engineer of 2020: Visions of engineering in the new century.* National Academies Press. https://hdl.voced.edu.au/10707/320922.

Nayar, V. (2013). Three differences between managers and leaders. *Harvard Business Review, 8*(1). https://hbr.org/2013/08/tests-of-a-leadership-transiti

Neck, C. P., & Houghton, J. D. (2006). Two decades of self-leadership theory and research. *Journal of Managerial Psychology, 21*(4), 270–295. https://doi.org/10.1108/02683940610663097

Neck, C. P., & Manz, C. C. (1996). Thought self-leadership: The impact of mental strategies training on employee cognition, behavior, and affect. *Journal of Organizational Behavior, 17*(5), 445–467. https://doi.org/10.1002/(SICI)1099-1379(199609)17:5<445::AID-JOB770>3.0.CO;2-N

Neck, C. P., & Manz, C. C. (2012). *Mastering self-leadership: Empowering yourself for personal excellence.* Pearson.

Nelson, J. K., Zaccaro, S. J., & Herman, J. L. (2010). Strategic information provision and experiential variety as tools for developing adaptive leadership skills. *Consulting Psychology Journal: Practice and Research, 62*(2), 131–142. https://doi.org/10.1037/a0019989

Northouse, P. G. (2019). *Leadership: Theory and practice* (8th Ed.). Sage Publications.

Oc, B. (2018). Contextual leadership: A systematic review of how contextual factors shape leadership and its outcomes. *The Leadership Quarterly, 29*(1), 218–235. https://doi.org/10.1016/j.leaqua.2017.12.004

Offermann, L. R., Kennedy, J. K., & Wirtz, P. W. (1994). Implicit leadership theories: Content, structure, and generalizability. *The Leadership Quarterly, 5*(1), 43–58. https://doi.org/10.1016/1048-9843(94)90005-1

Organ, D. W. (1996). Leadership: The great man theory revisited. *Business Horizons, 39*(3), 1–4. https://doi.org/10.1016/S0007-6813(96)90001-4

Osland, J. S., Bird, A., Mendenhall, M., & Osland, A. (2006). Developing global leadership capabilities and global mindset: A review. In G. K. Stahl & I. Björkman (Eds.), *Handbook of research in international human resource management* (pp. 197–222). Edward Elgar Publishing. https://doi.org/10.4337/9781845428235.00017

Osland, J., Bird, A., & Oddou, G.R. (2012). The context of expert global leadership. In W. H. Mobley, Y. Wang, & M. Li (Eds.), *Advances in Global Leadership, vol.7* (pp. 107–124). Elsevier.

Osborn, R. N., & Marion, R. (2009). Contextual leadership, transformational leadership and the performance of international innovation seeking alliances. *The Leadership Quarterly, 20*(2), 191–206. https://doi.org/10.1016/j.leaqua.2009.01.010

Osborn, R. N., Hunt, J. G., & Jauch, L. R. (2002). Toward a contextual theory of leadership. *The Leadership Quarterly, 6*(13), 797–837. https://doi.org/10.1016/S1048-9843(02)00154-6

Osborn, R. N., Uhl-Bien, M., & Milosevic, I. (2014). The context and leadership. In D. V. Day (Ed.), *The Oxford handbook of leadership and organizations* (pp. 589–612). Oxford University Press.

Ospina, S. M., Foldy, E. G., Fairhurst, G. T., & Jackson, B. (2020). Collective dimensions of leadership: Connecting theory and method. *Human Relations, 73*(4), 441–463. https://doi.org/10.1177/0018726719899714

Palethorpe, M. (2006). Are you emotional but intelligent - or are you emotionally intelligent? [emotional intelligence]. *Engineering Management, 16*(1), 11–13. https://doi.org/10.1049/em:20060101

Palmer, B., Walls, M., Burgess, Z., & Stough, C. (2001). Emotional intelligence and effective leadership. *Leadership and Organization Development Journal, 22*(1), 5–10. https://doi.org/10.1108/01437730110380174

Parkin, J. (1997). Choosing to lead. *Journal of Management in Engineering, 13*(1), 62–66. https://doi.org/10.1061/(ASCE)0742-597X(1997)13:1(62)

Parkinson, A. (2007). Engineering study abroad programs: Formats, challenges, best practices. *Online Journal for Global Engineering Education, 2*(2), 1–15. https://digitalcommons.uri.edu/ojgee/vol2/iss2/2/

Parolini, J., Patterson, K., & Winston, B. (2009). Distinguishing between transformational and servant leadership. *Leadership & Organization Development Journal, 30*(3), 274–291. https://doi.org/10.1108/01437730910949544

Paul, R., & Falls, L. G. C. (2015, June). Engineering leadership education: A review of best practices. *In 2015 ASEE Annual Conference & Exposition* (pp. 26–634).

Paul, R., Sen, A., & Wyatt, E. (2018, June). What is engineering leadership? A proposed definition. In *ASEE Annual Conference and Exposition, Conference Proceedings*. American Society for Engineering Education.

Pearce, C. L. (2007). The future of leadership development: The importance of identity, multi-level approaches, self-leadership, physical fitness, shared leadership, networking, creativity, emotions, spirituality and on-boarding processes. *Human Resource Management Review, 17*(4), 355–359. https://doi.org/10.1016/j.hrmr.2007.08.006

Pearce, C. L., & Conger, J. A. (2002). *Shared leadership: Reframing the hows and whys of leadership*. Sage Publications.

Pearce, C. L. and Conger, J. A. (2003). All those years ago: The historical underpinnings of shared leadership. In Pearce, C. L. and Conger, J. A. (Eds), *Shared leadership: Reframing the hows and whys of leadership* (pp. 1–18). Sage Publications.

Pearce, C. L., & Manz, C. C. (2005). The new silver bullets of leadership: The Importance of self-and shared leadership in knowledge work. *Organizational Dynamics, 34*(2), 130–140. https://doi.org/10.1016/j.orgdyn.2005.03.003

Pearce, C. L., Conger, J. A., & Locke, E. A. (2008). Shared leadership theory. *The Leadership Quarterly, 19*(5), 622–628. https://doi.org/10.1016/j.leaqua.2008.07.005

Perrot, B. E. (2011). Strategic Issues Management as a change catalyst. *Strategy & Leadership, 39*(5), 20–29. https://doi.org/10.1108/10878571111161499

Petrie, N. (2014). *Vertical leadership development-part 1 developing leaders for a complex world* [White paper]. Center for Creative Leadership, 1–13. https://kairosconsulting.com/wp-content/uploads/2016/06/CCL-VerticalLeadersPart2.pdf

Petrie, N. (2015). *The how-to of vertical leadership development-Part 2. 30 experts, 3 conditions and 15 approaches* [White paper]. Center for Creative Leadership, 1–25. https://kairosconsulting.com/wp-content/uploads/2016/06/CCL-VerticalLeadersPart2.pdf

Pfeffer, J. (1992). *Managing with power: Politics and influence in organizations*. Harvard Business Press.

Piaget, J. (1954). *The construction of reality in a child*. Basic Books.

Pink, D. (2005). *A whole new mind*. Riverhead Books.

Pitichat, T., Reichard, R. J., Kea-Edwards, A., Middleton, E., & Norman, S. M. (2018). Psychological capital for leader development. *Journal of Leadership & Organizational Studies, 25*(1), 47–62. https://doi.org/10.1177/1548051817719232

Porter, J. C. (1995). Facilitating cultural diversity. *Journal of Management in Engineering, 11*(6), 39–43. https://doi.org/10.1061/(ASCE)0742-597X(1995)11:6(39)

Porter, L. W., & McLaughlin, G. B. (2006). Leadership and the organizational context: Like the weather? *The Leadership Quarterly, 17*(6), 559–576. https://doi.org/10.1016/j.leaqua.2006.10.002

Prieto, B. (2013). Establishing and building leadership skills. *Leadership and Management in Engineering, 13*(3), 209–211. https://doi.org/10.1061/(ASCE)LM.1943-5630.0000235

Prusak, L., Groh, K., Denning, S. & Brown, J.S. (2012). *Storytelling in organizations.* Routledge.

Puccio, G. J., Mance, M., & Murdock, M. C. (2010). *Creative leadership: Skills that drive change.* Sage Publications.

Pucik, V. (2005). Reframing global mindset: From thinking to acting. In W. H. Mobley & E. Weldon (Ed.), *Advances in global leadership, 4* (pp. 83–100). Emerald Group Publishing. https://doi.org/10.1016/S1535-1203(06)04007-X

Quatro, S. A., Waldman, D. A., & Galvin, B. M. (2007). Developing holistic leaders: Four domains for leadership development and practice. *Human Resource Management Review, 17*(4), 427–441. https://doi.org/10.1016/j.hrmr.2007.08.003

Raelin, J. A. (2016). *Introduction to leadership-as-practice: Theory and application.* Routledge.

Raelin, J. A. (2018). What are you afraid of: Collective leadership and its learning implications. *Management Learning, 49*(1), 59–66. https://doi.org/10.1177/1350507617729974

Raelin, J. (2011). From leadership-as-practice to leaderful practice. *Leadership, 7*(2), 195–211. https://doi.org/10.1177/1742715010394808

Raelin, J. A. (2020). Toward a methodology for studying leadership-as-practice. *Leadership, 16*(4), 480–508. https://doi.org/10.1177/17427150198828

Rahim, M. A. (1989). Relationships of leader power to compliance and satisfaction with supervision: Evidence from a national sample of managers. *Journal of Management, 15*(4), 545–556. https://doi.org/10.1177/014920638901500404

Rao, M. S. (2017). Values-based leadership. *The Journal of Values-Based Leadership, 10*(2), 5. https://doi.org/10.22543/0733.102.1185

Raven, B. H. (1965). Social influence and power. In Steiner, D. & Fishbein, M. (Eds), *Current studies in social psychology* (pp. 371–382). Holt, Rinehart and Winston.

Rego, A., Sousa, F., Marques, C., & Cunha, M. P. e. (2012). Authentic leadership promoting employees' psychological capital and creativity. *Journal of Business Research, 65*(3), 429–437. https://doi.org/10.1016/j.jbusres.2011.10.003

Reiter-Palmon, R., & Illies, J. J. (2004). Leadership and creativity: Understanding leadership from a creative problem-solving perspective. *The Leadership Quarterly, 15*(1), 55–77. https://doi.org/10.1016/j.leaqua.2003.12.005

Reynolds, M., & Holwell, S. (Eds.). (2010). *Systems approaches to managing change: A practical guide.* Springer.

Richardson, P. (2005). Managing cultural diversity. *Engineering Management, 15*(2), 24–27. https://doi.org/10.1049/em:20050207

Riemer, M. J. (2003). Integrating emotional intelligence into engineering education. *World Transactions on Engineering & Technology Education, 2*(2), 189–194.

Riggio, R. E. (Ed.). (2019). *What's wrong with leadership? Improving leadership research and practice.* Routledge.

Rittel, H. W., & Webber, M. M. (1973). Dilemmas in a general theory of planning. *Policy Sciences, 4*(2), 155–169. https://doi.org/10.1007/BF01405730

Robbins, T. L., Crino, M. D., & Fredendall, L. D. (2002). An integrative model of the empowerment process. *Human Resource Management Review, 12*(3), 419–443. https://doi.org/10.1016/S1053-4822(02)00068-2

Rogers, G., Mentkowski, M., & Hart, J. R. (2006). Adult holistic development and multidimensional performance. In C. H. Hoare (Ed.), *Handbook of adult development and learning* (pp. 498–534). Oxford University Press.

Rost, J. C. (1991). *Leadership for the twenty-first century*. Praeger.

Rottmann, C., Sacks, R., & Reeve, D. (2015). Engineering leadership: Grounding leadership theory in engineers' professional identities. *Leadership, 11*(3), 351–373. https://doi.org/10.1177/1742715014543581

Ruderman, M.N., Clerkin, C., & Connolly, C. (2014). *Leadership development beyond competencies: Moving to a holistic approach*. Center for Creative Leadership. Retrieved from: https://mobiusleadership.co.uk/wp-content/uploads/2014/01/LeadershipDevelopmentCompetencies.pdf

Runco, M. A., Pritzker, S. R. (1999). Implicit theories. In *Encyclopedia of creativity, 2*, 27–30. Academic Press.

Sabatini, D. A., & Knox, R. C. (1999). Results of a student discussion group on leadership concepts. *Journal of Engineering Education, 88*(2), 185–188. https://doi.org/10.1002/j.2168-9830.1999.tb00433.x

Sage, A. P., & Rouse, W. B. (2014). *Handbook of systems engineering and management*. John Wiley & Sons.

Salancik, G. R., & Pfeffer, J. (1977). Who gets power – And how they hold on to it: A strategic-contingency model of power. *Organizational Dynamics, 5*(3), 2–21. https://doi.org/10.1016/0090-2616(77)90028-6

Salicru, S. (2017). *Leadership results: How to create adaptive leaders and high-performing organisations for an uncertain world*. John Wiley & Sons.

Salicru, S. (2020). A new model of leadership-as-practice development for consulting psychologists. *Consulting Psychology Journal: Practice and Research, 72*(2), 79–99. https://doi.org/10.1037/cpb0000142

Salicru, S. (2023). Leadership, group leadership, and functional leadership, In L. Lunevich (Ed.), *Handbook of engineering management: The digital economy*. CRC Press Taylor & Francis Group. https://www.routledge.com/Handbook-of-Engineering-Management-The-Digital-Economy/Lunevich/p/book/9781032448107

Salicru, S., & Chelliah, J. (2014). Messing with corporate heads? Psychological contracts and leadership integrity. *Journal of Business Strategy, 35*(3), 38–46. https://doi.org/10.1108/JBS-10-2013-0096

Salovey, P., & Mayer, J. D. (1990). Emotional intelligence. *Imagination, Cognition and Personality, 9*(3), 185–211. https://doi.org/10.2190/DUGG-P24E-52WK-6CDG

Samimi, M., Cortes, A. F., Anderson, M. H., & Herrmann, P. (2022). What is strategic leadership? Developing a framework for future research. *The Leadership Quarterly, 33*(3), 101353. https://doi.org/10.1016/j.leaqua.2019.101353

Santos, J. P., Caetano, A., & Tavares, S. M. (2015). Is training leaders in functional leadership a useful tool for improving the performance of leadership functions and team effectiveness? *The Leadership Quarterly, 26*(3), 470–484. https://doi.org/10.1016/j.leaqua.2015.02.010

Sarros, J. C. (1992). What leaders say they do: An Australian example. *Leadership & Organization Development Journal, 13*(5), 21–27. https://doi.org/10.1108/01437739210016204

Schell, W., Hughes, B. E., Tallman, B., Kwapisz, M., Sybesma, T., Annand, E., ... & Krejci, C. C. (2022). Understanding the joint development of engineering and leadership identities. *Engineering Management Journal, 34*(3), 497–507. https://doi.org/10.1080/10429247.2021.1952021

Schuhmann, R. J. (2010). Engineering leadership education-The search for definition and a curricular approach. *Journal of STEM education: Innovations and Research, 11*(3/4). 61–69.

Schyns, B., & Meindl, J. R. (2005). An overview of implicit leadership theories and their application in organization practice. In B. Schyns, & J. R. Meindl (Eds.), *Implicit leadership theories: Essays and explorations* (pp. 15–36). Information Age Publishing.

Schyns, B. & Riggio, R. E. (2016). *Implicit leadership theories*. In A. Farazmand (Eds), *Global encyclopedia of public administration, public policy, and governance* (pp. 1–7). Springer. https://doi.org/10.1007/978-3-319-31816-5_2186-1

Scott, S. G., & Bruce, R. A. (1994). Determinants of innovative behavior: A path model of individual innovation in the workplace. *Academy of Management Journal, 37*(3), 580–607. https://doi.org/10.5465/256701

Sergi, V., Denis, J.- L., & Langley, A. (2012). Opening up perspectives on plural leadership. *Industrial and Organizational Psychology: Perspectives on Science and Practice, 5*(4), 403–407. https://doi.org/10.1111/j.1754-9434.2012.01468.x

Shamir, B., & Howell, J. M. (1999). Organizational and contextual influences on the emergence and effectiveness of charismatic leadership. *The Leadership Quarterly, 10*(2), 257–283. https://doi.org/10.1016/S1048-9843(99)00014-4

Shaw, W. H. (2002, August). Engineering management in our modern age. In *IEEE International Engineering Management Conference* (Vol. 2, pp. 504–509). IEEE. https://doi.org/10.1109/IEMC.2002.1038486

Shawn Burke, C., Hess, K. P., & Salas, E. (2006). Building the adaptive capacity to lead multi-cultural teams. In C. Shawn Burke, L. G. Pierce, & E. Salas (Eds), *Understanding adaptability: A prerequisite for effective performance within complex environments* (pp. 175–211). Emerald Group Publishing. https://doi.org/10.1016/S1479-3601(05)06006-6

Shondrick, S. J., & Lord, R. G. (2010). Implicit leadership and followership theories: Dynamic structures for leadership perceptions, memory, and leader-follower processes. In G. P. Hodgkinson & J. K. Ford (Eds.), *International review of industrial and organizational psychology 2010* (pp. 1–33). Wiley Blackwell. https://doi.org/10.1002/9780470661628.ch1

Singh, A. (2009). Organizational power in perspective. *Leadership and Management in Engineering, 9*(4), 165–176. https://doi.org/10.1061/(ASCE)LM.1943-5630.0000018

Singh, A., Lim, W. M., Jha, S., Kumar, S., & Ciasullo, M. V. (2023). The state of the art of strategic leadership. *Journal of Business Research, 158*, 113676. https://doi.org/10.1016/j.jbusres.2023.113676

Sirmon, D.G., Hitt, M.A. & Ireland, R.D. (2007). Managing firm resources in dynamic environments to create value: Looking inside the black box. *Academy of Management Review, 32*(1), 273–292. https://doi.org/10.5465/amr.2007.23466005

Škerlavaj, M. (2022). Post-heroic leadership of creativity and innovation: From idea to excel. In M. Skerlavaj (Ed.), *Post-heroic leadership: Context, process and outcomes* (pp. 157–171). Palgrave Macmillan. https://doi.org/10.1007/978-3-030-90820-1_8

Smith, J. E., Carson, K. P., & Alexander, R. A. (1984). Leadership: It can make a difference. *Academy of Management Journal, 27*(4), 765–776. https://doi.org/10.5465/255877

Smith, J. M., McClelland, C. J., & Smith, N. M. (2017). Engineering students' views of corporate social responsibility: A case study from petroleum engineering. *Science and Engineering Ethics, 23*, 1775–1790. https://doi.org/10.1007/s11948-016-9859-x

Smith, N. M., Zhu, Q., Smith, J. M., & Mitcham, C. (2021). Enhancing engineering ethics: Role ethics and corporate social responsibility. *Science and Engineering Ethics, 27*, 1–21. https://doi.org/10.1007/s11948-021-00289-7

Snowden, D. J. (2000). The art and science of story or 'are you sitting uncomfortably?'. *Business Information Review, 17*(4), 215–226. https://doi.org/10.1177/026638200423778

Spreitzer, G. M. (1995). Psychological empowerment in the workplace: Dimensions, measurement, and validation. *Academy of Management Journal, 38*(5), 1442–1465. https://doi.org/10.2307/256865

Sternberg, R. J., Kaufman, J. C., & Pretz, J. E. (2003). A propulsion model of creative leadership. *The Leadership Quarterly, 14*(4-5), 455–473. https://doi.org/10.1016/S1048-9843(03)00047-X

Stephenson, J. (1994). Capability and competence: Are they the same and does it matter. *Capability, 1*(1), 3–4.

Stephenson, J. & Weil, S. (1992) *Quality in learning: A capability approach in higher education*. Kogan Page.

Stewart, G. L., Carson, K. P., & Cardy, R. L. (1996). The joint effects of conscientiousness and self-leadership training on employee self-directed behavior in a service setting. *Personnel Psychology, 49*(1), 143–164. https://doi.org/10.1111/j.1744-6570.1996.tb01795.x

Stogdill, R. M. (1948). Personal factors associated with leadership; a survey of the literature. *The Journal of Psychology: Interdisciplinary and Applied, 25*, 35–71. https://doi.org/10.1080/00223980.1948.9917362

Stogdill, R. M. (1974). *Handbook of leadership: A survey of theory and research*. Free Press.

Swap, W., Leonard, D., Shields, M., & Abrams, L. (2001). Using mentoring and storytelling to transfer knowledge in the workplace. *Journal of Management Information Systems, 18*(1), 95–114. https://doi.org/10.1080/07421222.2001.11045668

Sy, T. (2010). What do you think of followers? Examining the content, structure, and consequences of implicit followership theories. *Organizational Behavior and Human Decision Processes, 113*(2), 73–84. https://doi.org/10.1016/j.obhdp.2010.06.001

Tannenbaum, R. and Schmidt, W. (1958). How to choose a leadership pattern. *Harvard Business Review 36*(2), 95–101.

Tanner, C., Brügger, A., van Schie, S., & Lebherz, C. (2010). Actions speak louder than words: The benefits of ethical behaviors of leaders. *Journal of Psychology, 218*(4), 225–233. https://doi.org/10.1027/0044-3409/a000032

Thomas, A. B. (1988). Does leadership make a difference to organizational performance? *Administrative Science Quarterly, 33*(3), 388–400. https://doi.org/10.2307/2392715

Thomas, K. W., & Velthouse, B. A. (1990). Cognitive elements of empowerment: An "interpretive" model of intrinsic task motivation. *The Academy of Management Review, 15*(4), 666–681. https://doi.org/10.2307/258687

Thomas, T., Schermerhorn Jr, J. R., & Dienhart, J. W. (2004). Strategic leadership of ethical behavior in business. *Academy of Management Perspectives, 18*(2), 56–66. https://doi.org/10.5465/ame.2004.13837425

Thoresen, C. E. & Mahoney, M. J. (1974). *Behavioral Self-control*. Holt, Rinehart & Winston.

Thorndike, E. L. (1920). A constant error in psychological ratings. *Journal of Applied Psychology, 4*(1), 25–29. https://doi.org/10.1037/h0071663

Toor, S. U. R. (2011). Differentiating leadership from management: An empirical investigation of leaders and managers. *Leadership and Management in Engineering, 11*(4), 310–320. https://doi.org/10.1061/(ASCE)LM.1943-5630.0000138

Toor, S. U. R., & Ofori, G. (2008). Leadership versus management: How they are different, and why. *Leadership and Management in Engineering, 8*(2), 61–71. https://doi.org/10.1061/(ASCE)1532-6748(2008)8:2(61)

Tracy, D. (1992). *10 steps to empowerment: A common-sense guide to managing people*. William Morrow.

Travis Maynard, M., Resick, C. J., Cunningham, Q. W., & DiRenzo, M. S. (2017). Ch-Ch-Ch-changes: How action phase functional leadership, team human capital, and interim vs. permanent leader status impact post-transition team performance. *Journal of Business and Psychology, 32*(5), 575–593. https://doi.org/10.1007/s10869-016-9482-5

Treviño, L. K., Hartman, L. P., & Brown, M. (2000). Moral person and moral manager: How executives develop a reputation for ethical leadership. *California Management Review, 42*(4), 128–142. https://doi.org/10.2307/411660

Treviño, L. K., Brown, M., & Hartman, L. P. (2003). A qualitative investigation of perceived executive ethical leadership: Perceptions from inside and outside the executive suite. *Human Relations, 56*(1), 5–37. https://doi.org/10.1177/0018726703056001448

Treviño, L. K., Weaver, G. R., & Reynolds, S. J. (2006). Behavioral Ethics in Organizations: A Review. *Journal of Management, 32*(6), 951–990. https://doi. org/10.1177/0149206306294258

Treviño, L. K., den Nieuwenboer, N. A., & Kish-Gephart, J. J. (2014). (Un)ethical behavior in organizations. *Annual Review of Psychology, 65*, 635–660. https://doi.org/10.1146/ annurev-psych-113011-143745

Triandis, H. C. (1998). Vertical and horizontal individualism and collectivism: Theory and research implications for international comparative management. In J. L. Cheng & R. B. Peterson (Eds.), *Advances in international and comparative management* (pp. 7–36). JAI Press.

Tsai, W., & Ghoshal, S. (1998). Social capital and value creation: The role of intrafirm networks. *Academy of Management Journal, 41*(4), 464–476. https://doi.org/10.2307/257085

Tubbs, S. L., & Schulz, E. (2006). Exploring a taxonomy of global leadership competencies and meta-competencies. *Journal of American Academy of Business, 8*(2), 29–34. https:// commons.emich.edu/mgmt_facsch/62/

University of Texas at Austin (2023). *Intro to behavioral ethics. Ethics unwrapped.* McCombs School of Business. https://ethicsunwrapped.utexas.edu/video/intro-to-behavioral-ethics

van Dierendonck, D. (2011). Servant leadership: A review and synthesis. *Journal of Management, 37*(4), 1228–1261. https://doi.org/10.1177/0149206310380462

Van Dyne, L., Ang, S., & Koh, C. (2015). Development and validation of the CQS: The cultural intelligence scale. *In Handbook of cultural intelligence* (pp. 16–38). Routledge.

Van de Ven, A. H. (1986). Central problems in the management of innovation. *Management Science, 32*(5), 590–607. https://doi.org/10.1287/mnsc.32.5.590

Van der Merwe, L., & Verwey, A. (2007). Leadership meta-competencies for the future world of work. *SA Journal of Human Resource Management, 5*(2), 33–41. https://hdl.handle. net/10520/EJC95858

Vera, D., & Crossan, M. (2004). Strategic leadership and organizational learning. *Academy of Management Review, 29*(2), 222–240. https://doi.org/10.5465/amr.2004.12736080

Vincent, N., Ward, L., & Denson, L. (2015). Promoting post-conventional consciousness in leaders: Australian community leadership programs. *The Leadership Quarterly, 26*(2), 238–253. https://doi.org/10.1016/j.leaqua.2014.11.007

Vogus, T.J., & Welbourne, T.M. (2003). Structuring for high reliability: HR practices and mindful processes in reliability-seeking organizations. *Journal of Organizational Behavior, 24*(7), 877–903. https://doi.org/10.1002/job.221

Volckmann, R. (2012). Fresh perspective: Barbara Kellerman and the leadership industry. Articles from *Integral Leadership Review, 20*(12), 6–8. https://integralleadershipreview. com/7064-barbara-kellerman-and-the-leadership-industry/

Vroom, V. H., & Jago, A. G. (2007). The role of the situation in leadership. *American Psychologist, 62*(1), 17.

Vroom, V., & Yetton, P. (1973). Leadership and decision-making. University of Pittsburgh Press. https://doi.org/10.2307/j.ctt6wrc8r

Walesh, S. G. (2012). The leader within you: Let it come out!. *Leadership and Management in Engineering, 12*(1), 37–38. https://doi.org/10.1061/(ASCE)LM.1943-5630.0000154

Walumbwa, F. O., Avolio, B. J., Gardner, W. L., Wernsing, T. S., & Peterson, S. J. (2008). Authentic leadership: Development and validation of a theory-based measure. *Journal of Management, 34*(1), 89–126. https://doi.org/10.1177/01492063073089

Wang, G., & Thompson, R. G. (2013). Incorporating global components into ethics education. *Science and Engineering Ethics, 19*, 287–298. https://doi.org/10.1007/ s11948-011-9295-x

Wang, Y., Chen, Y., & Zhu, Y. (2021). Promoting innovative behavior in employees: The mechanism of leader psychological capital. *Frontiers in Psychology, 11*, 598090. https://doi. org/10.3389/fpsyg.2020.598090

Waterman, R. H. (1990). *Adhocracy: The power to change.* Whittle Direct Books.

Wellington, P. (2009). *Effective leadership for engineers*. The Institution of Engineering and Technology.

Weick, K.E. (1995). *Sensemaking in organizations*. Sage Publications.

Weick, K. E. (1993). The collapse of sensemaking in organizations: The Mann Gulch disaster. *Administrative Science Quarterly, 38*, 628–652. https://doi.org/10.2307/2393339

Weick, K. E. (2001). Leadership as the legitimization of doubt. In W. Bennis, G. M. Spreitzer, & T. G. Cummings (Eds.), *The future of leadership: Today's top leadership thinkers speak to tomorrow's leaders* (pp. 91–102). Jossey-Bass.

Weick, K. E. (2009*). Making sense of the organization, Vol. 2: The impermanent organization*. Chichester: John Wiley & Sons.

Weick, K.E. (2010). Reflections on enacted sensemaking in the Bhopal disaster. *Journal of Management Studies, 47*(3), 537–550. https://doi.org/10.1111/j.1467-6486.2010.00900.x

Weick, K., Sutcliffe, K.M., & Obstfeld, D. (2005). Organizing and the process of sensemaking. *Organization Science, 16*(4), 409–421. https://doi.org/10.1287/orsc.1050.0133

Weingardt, R. G. (2000). Leaving a legacy. *Journal of Management in Engineering, 16*(2), 42–47. https://doi.org/10.1061/(ASCE)0742-597X(2000)16:2(42)

Whitehurst, J. (2016). Leaders can shape company culture through their behaviors. *Harvard Business Review*, 1–5.

Wibbeke, E. S., & McArthur, S. (2013). *Global business leadership*. Routledge.

World Economic Forum. (2016). The future of jobs: Employment, skills and workforce strategy for the fourth industrial revolution. *Global Challenge Insight Report*. https://hdl.voced.edu.au/10707/393272

Xenikou, A., & Simosi, M. (2006). Organizational culture and transformational leadership as predictors of business unit performance. *Journal of Managerial Psychology, 21*(6), 566–579. https://doi.org/10.1108/02683940610684409

Yammarino, F. J., Salas, E., Serban, A., Shirreffs, K., & Shuffler, M. L. (2012). Collectivistic leadership approaches: Putting the "we" in leadership science and practice. *Industrial and Organizational Psychology: Perspectives on Science and Practice, 5*(4), 382–402. https://doi.org/10.1111/j.1754-9434.2012.01467.x

Yeager, D. S., & Dweck, C. S. (2020). What can be learned from growth mindset controversies? *American Psychologist, 75*(9), 1269–1284. https://doi.org/10.1037/amp0000794

Yukl, G. (1989). Managerial leadership: A review of theory and research. *Journal of Management, 15*(2), 251–289. https://doi.org/10.1177/014920638901500207

Yukl, G. (2006). *Leadership in organizations* (6th Ed.). Pearson-Prentice Hall.

Yukl, G. A., & Becker, W. S. (2006). Effective empowerment in organizations. *Organization Management Journal, 3*(3), 210–231. https://doi.org/10.1057/omj.2006.20

Yukl, G., & Mahsud, R. (2010). Why flexible and adaptive leadership is essential. *Consulting Psychology Journal: Practice and Research, 62*(2), 81–93. https://doi.org/10.1037/a0019835

Yukl, G., Mahsud, R., Hassan, S., & Prussia, G. E. (2013). An improved measure of ethical leadership. *Journal of Leadership & Organizational Studies, 20*(1), 38–48. https://doi.org/10.1177/1548051811429352

Zaccaro, S. J., Rittman, A. L., & Marks, M. A. (2001). Team leadership. *The Leadership Quarterly, 12*(4), 451–483. https://doi.org/10.1016/S1048-9843(01)00093-5

Zaccaro, S. J., & Klimoski, R. J. (Eds.). (2002). *The nature of organizational leadership: Understanding the performance imperatives confronting today's leaders*. John Wiley & Sons.

Zaleznik, A. (1997). Managers and leaders: Are they different? *Harvard Business Review, 55*, 67–78.

Zheng, M. (2015). Intercultural competence in intercultural business communication. *Open Journal of Social Sciences, 3*(03), 197–200. https://doi.org/10.4236/jss.2015.33029

6 Decision Analysis Driven by Big Data for Engineering Managers

John V. Farr and David Farr
United States Army

6.1 INTRODUCTION

So, what is big data and how big is big? The definitions of big data vary but most talk about extremely large data sets that have extremely large volume, value, variety, velocity, or the fast rate at which data are received and processed with low latency and veracity or the accuracy of the data. According to Bulao (2023), in 2021 people created 2.5 quintillion bytes (1 quintillion bytes is equal to 1 million terabytes) of data every day. For example, there are over 333.2 billion emails sent every day in 2022. Companies are spending large amounts of money on big data such as:

- GE spent $1 billion in 2022 alone to analyze data from sensors on gas turbines, jet engineers, oil pipelines, and other machines and aims to triple sales of software products by 2023 to roughly $15 billion (Winig, 2016);
- Uber, with more than 8 million users and 160,000 people driving cars, the secret to this $51 billion startup has been the use of big data for surge pricing, better cars, detecting fake rides, fake cards, fake ratings, estimating fares, and driver ratings (Project Pro, 2023);
- Netflix with over 65 million members (as of 2020) uses big data in every aspect of business to include predicting what customers will enjoy watching and making suggestions on what to watch, all with the goal of trying to predict what customers will enjoy watching (Marr, 2020).

These are just some of the examples of how big data shapes our daily life.

The statistics are amazing with the amount of data doubling every 3.4 months. Entrepreneurs, governments, big businesses, and even small businesses are mining these data with the hope of uncovering patterns/trends, behavior, and other valuable information. Some big data will be converted into high-quality assets, or commercial products and/or productive capital.

DOI: 10.1201/9781003374879-6

The bigger question is how engineering managers can make decisions using these large amounts at every increasing velocity and volume data? Can big data improve our decision-making ability for complex problems?

Decisions are important events due to their irrevocable commitment of resources, such as time, money, or personnel commitments. Engineering managers make many decisions every day some intuitively and some after a comprehensive analysis. Decisions are made or they are not, and often the failure to make a decision is one in itself. According to Howard and Abbas (2016), a decision is defined as "A choice between two or more alternatives that involves an irrevocable allocation of resources." In our world of complexity, most decisions worthy of analysis involve many alternatives, competing stakeholders and value propositions, and scenarios with different risks and rewards? Some decisions are straightforward, while others are complex with unforeseen second- and third-order effects that make them difficult to navigate.

Complex decisions are difficult due to the various potential outcomes and corresponding uncertainties that are involved with multiple choices. Diverse stakeholders with conflicting goals are often involved in the decision-making process and outcomes. However, the explosion (i.e., acquisition, organizing, processing, and analyzing) of data has allowed us to make complex decisions literally unimaginable a decade ago.

6.2   ROLE OF THE DATA SCIENTIST IN ENGINEERING MANAGEMENT

In many ways, a data scientist is a complimentary profession to an engineering manager. Yes, in some instances, a data scientist can be the decision-maker. However, in most cases, they are a valuable member of an interdisciplinary and/or multidisciplinary team.

Figure 6.1 shows how a data scientist might interact with other engineers and support traditional decision sciences techniques to support the decision-making process.

A data scientist is responsible to:

- Manipulate data using advanced techniques such as artificial intelligence (i.e., mainly machine learning) to find patterns, trends, and insights in data;
- Develop algorithms/models to make forecasts and visualize results;
- Communicate recommendations to other teams and senior staff; and
- Utilize the appropriate analytical tools such as Python, JAVA, R, and SQL in data analysis.

Most engineering managers need to understand the methods, processes, and tools used by a data scientist to be able to ask intelligent questions to a data science, to consider their applicability, strengths, and weaknesses. The introduction of data scientists into the decision-making process allows for more timely decisions for many

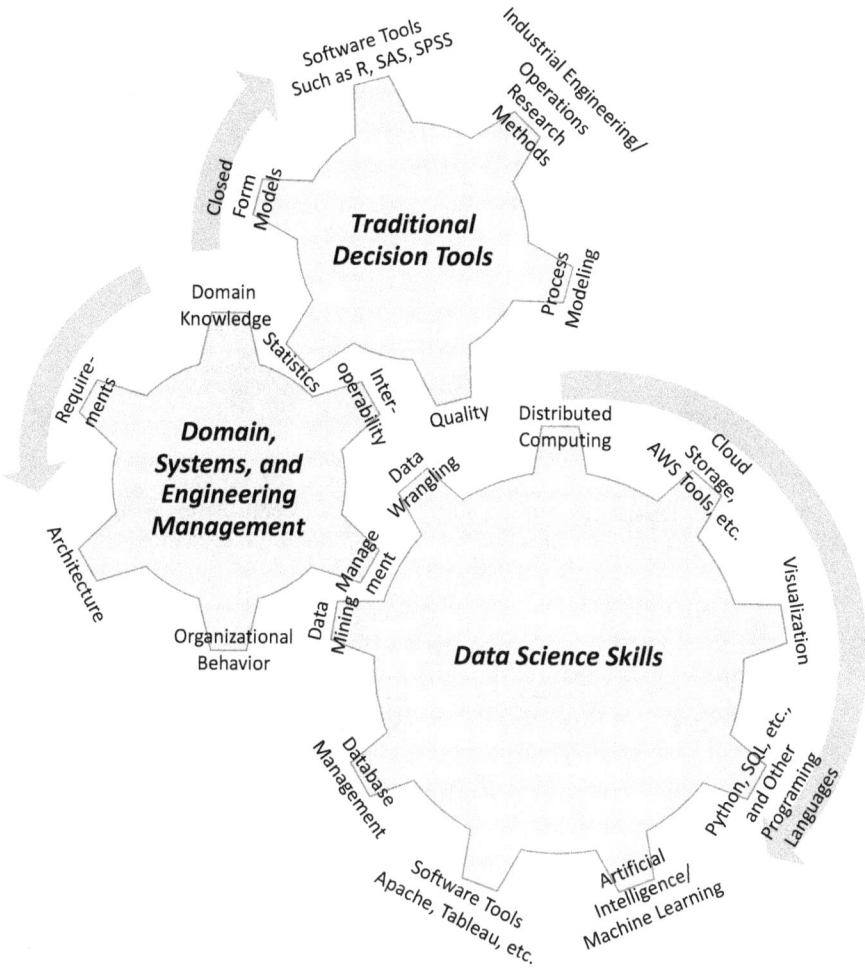

FIGURE 6.1 Interdisciplinary decision-making skills utilizing data scientists.

complex problems. Their skills contribute to more effective decisions because their toolset is suited for environments characterized by growing levels of complexity and high dynamism. Figure 6.2 presents the elements of making decision using big data.

6.3 TRADITIONAL DECISION ANALYSIS TOOLS

From a math modeling perspective, uncertainty and probability are the most difficult aspects of decision-making. This also applies to big data. If one could know, in advance, the exact outcomes and their effect on stakeholders the decision becomes simple. Any responsible decision should aim for the optimal outcome, whether that

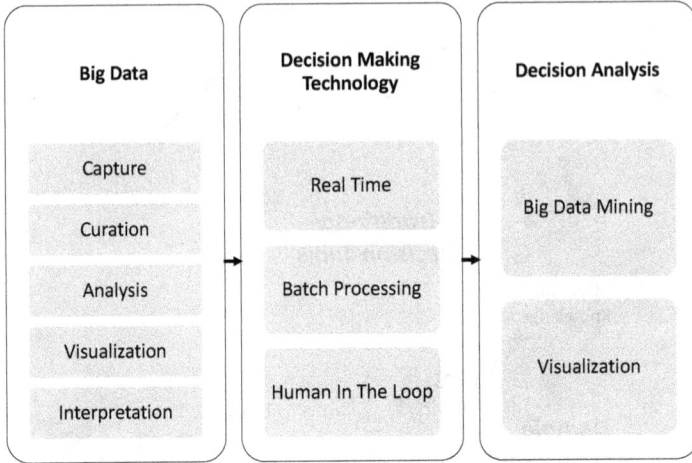

FIGURE 6.2 Decision-making elements utilizing big data. (Modified from Wang, H., Xu, Z., Fujita, H., and Liu, S., *Information Sciences, 367*, 747–765, 2016. With permission.)

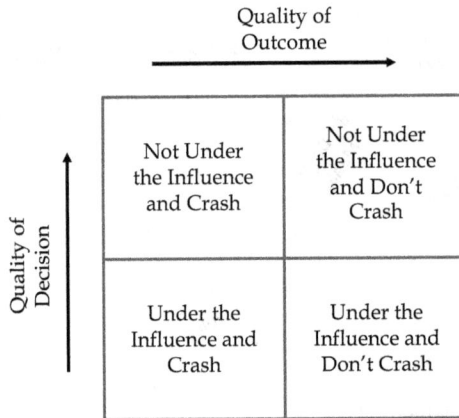

FIGURE 6.3 Decisions vs. outcomes.

be happiness, profit, time to project completion, or some blend of multiple important values. Uncertainty is easily understood yet hard to approximate and assign real values. Big data helps bound uncertainty, but as shown repeatedly, past performance and data do not necessarily predict future events.

Consider the comparison in Figure 6.3 which contrasts decision quality with decision outcome. The common logical fallacies in the judgment of decisions are confusing these two concepts. For example, the figure shows two outcomes for a person who drives themselves home (perhaps from a social event). The person can arrive home safely, or they can crash their vehicle along the way. This is a good outcome, arrive safely, and a bad outcome, crash. However, many factors that determine the outcome are out of control or uncertain for the individual. Who is in control of the decision at a social event to consume alcohol which is known to impair driving ability?

The decision to consume alcohol and drive is a bad decision (or low quality) versus abstaining, knowing that they must drive afterward. The four scenarios are the possible results of:

- Make a good decision, have a good outcome;
- Make a good decision, have a bad outcome;
- Make a bad decision, have a good outcome; or
- Make a bad decision, have a bad outcome.

Separating decision quality for outcomes can only be achieved when considering uncertainty as the key feature of a decision.

According to Barclay et al. (1997), decision analysis builds upon four basic elements as a way to conceptualize and resolve complex decisions and include:

1. Initial courses of action. You have a decision to make only if you face a choice among alternative possible acts. Each of the choices you want to consider should be made explicit.
2. The possible consequences of each initial act. What are the important things that can happen that will make one act more valuable or worth more than another act? Relevant sequences of subsequent events and follow-up acts must be identified for each initial act.
3. How attractive or unattractive each possible consequence of each act is to you. How undesirable is one outcome compared to others which might result from the same or another decision? This value could be measured in terms of money, utility, or some other carefully defined index.
4. How likely is it that a particular act will result in each of the consequences? This uncertainty may be measured either by a numerical probability from 0 to 1 or in the form of odds.

The quality of a decision is determined by the consideration of available information. Specifically, the reduction of uncertainty about the decision is critical. The introduction of big data analytics has allowed for the analysis of many more variables with a quicker turnaround thus reducing risk. The greater the possible consequences of a decision the greater the consideration a decision should receive. Anyone who has struggled with making complex and important decisions understand that assessing and conducting trade-offs of risk, value, and cost are needed to make an informed finding. To make sound decisions you must understand the stakeholder requirements and have a sound and defensible process supported by appropriate analytical results. In this chapter, we will present a brief overview of how to address decisions based not only on costs but also value and risk. We will not address the challenges of making decisions other than a brief discussion of bias.

Decision analysis should be used by decision-makers and stakeholders in a structured way to think about decisions and allows for quantitatively making trade space studies. Table 6.1 shows some of the many of the traditional decisions analysis techniques used. The list is by no means all encompassing. Also, any of the techniques could be used for any step in the decision process. However, they are often meant to be used when resource trade-offs are required as demonstrated by Example 6.1.

TABLE 6.1

Process for Performing a Decision Analysis Study

Technique	Decision Framing	Identify Objectives and Alternatives	Determine the Value of Each Alternative	Select the Best Alternative	Conduct Sensitivity Analysis	Execute
Probabilistic						
• Simulation			✔		✔	
• Game theory						
• Bayesian analysis						
Multi criteria						
• Value modeling						
• Kepner-Tregoe					✔	
• Linear programming (LP) and non-LP		✔	✔	✔		
• Analytic hierarchy process						
Network						
• Queuing			✔			
• Bayesian						
Tabular/Graphical						
• Pareto analysis						
• Pugh matrix						
• Strengths, weaknesses, opportunities, and threats						
• Affinity diagram	✔	✔	✔	✔		
• Fishbone						
• Influence diagrams						
• Fault tree						
• Decision tree						
Economic						
• Net present value			✔	✔	✔	✔
• Return on investment						
Informal						
• Delphi	✔	✔	✔	✔		✔
• Brainstorming						

Steps in the Decision Analysis Process

Example 6.1

Consider the value hierarchy shown in Figures 6.4 and 6.5. This value hierarchy from the overall objective down to the evaluation measures for a data set from the National Reconnaissance Office (NRO) of R&D projects (see Farr and Parnell,

Fundamental Objective

Provide technology innovations to revolutionize global reconnaissance

Objectives

| Provide information superiority to enable NRO customers to 50 revolutionize future capabilities | Reduce life cycle costs by an order of magnitude 20 | Rapidly design and deploy innovative technologies 30 solutions |

Functions

Visualize operational space: anywhere, anything/anyone, anytime 40	NRO Systems 80	Introduce new sources and methods to solve intractable 60 intelligence problems
Plan in real-time 20	User Systems 20	Integrate with other systems 10
Communicate any information anywhere, anytime 20		Rapidly transition system concepts 30
Resolve political, economic, social and military conflicts with no loss of life or resources 20		Manage technical risk 0

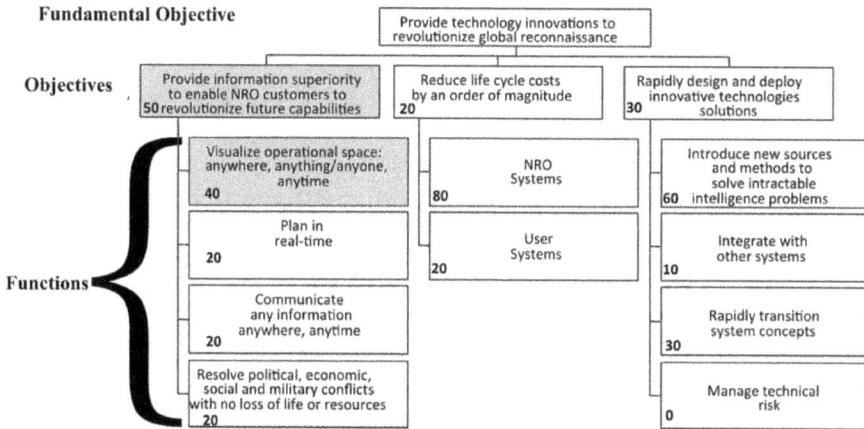

FIGURE 6.4   Value analysis high-level model for the NRO data set.

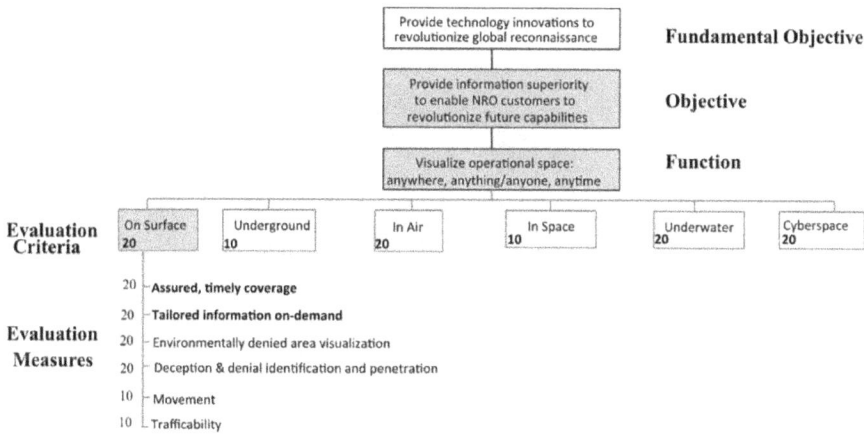

Provide technology innovations to revolutionize global reconnaissance   **Fundamental Objective**

Provide information superiority to enable NRO customers to revolutionize future capabilities   **Objective**

Visualize operational space: anywhere, anything/anyone, anytime   **Function**

Evaluation Criteria

| On Surface 20 | Underground 10 | In Air 20 | In Space 10 | Underwater 20 | Cyberspace 20 |

Evaluation Measures

20 ⊢ Assured, timely coverage
20 ⊢ Tailored information on-demand
20 ⊢ Environmentally denied area visualization
20 ⊢ Deception & denial identification and penetration
10 ⊢ Movement
10 ⌐ Trafficability

FIGURE 6.5   Evaluation criteria and evaluation measures for a function for the NRO data set.

2000). Figure 6.6 contains two of the scoring functions used to evaluate project value using a weighted scoring methodology. Figure 6.7 is a plot of the cumulative value when the projects are sorted by budget. The lowest budget projects are put in the portfolio before the higher budget projects.

This is an example of how a traditional cost/benefit analysis study can be used to look across a portfolio of projects.

Notice that the volume, velocity, and, to some respect, the variety (i.e., the characteristics of big data) of the data are lacking for this methodology and data set. Though certainly a defensible decision, this is more of a static decision-making problem.

Assured Timely Coverage

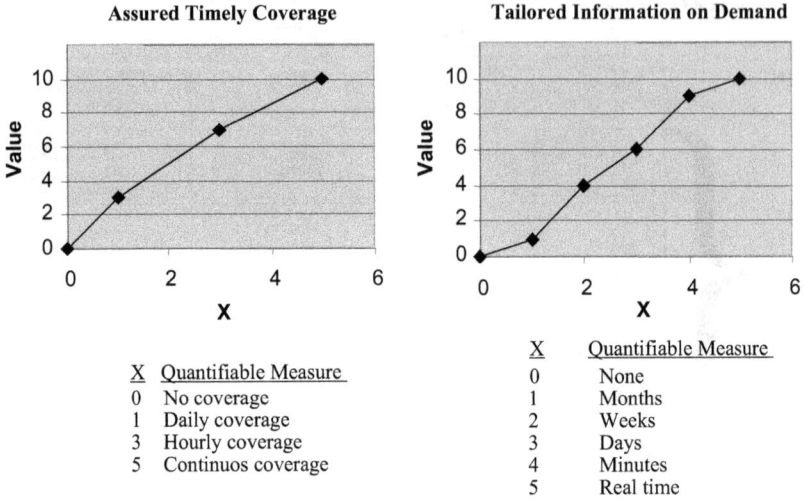

X	Quantifiable Measure
0	No coverage
1	Daily coverage
3	Hourly coverage
5	Continuos coverage

Tailored Information on Demand

X	Quantifiable Measure
0	None
1	Months
2	Weeks
3	Days
4	Minutes
5	Real time

FIGURE 6.6   Scoring functions for two evaluation measures.

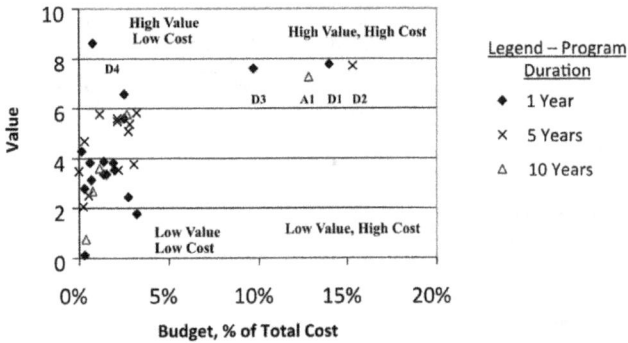

FIGURE 6.7   Plot of value versus of percentage of the total portfolio cost.

6.4   MAKING DECISIONS INVOLVING COMPLEXITY AND BIG DATA

Complex decisions using traditional decision-making involve:

- asking the right questions with input from multiple stakeholders;
- relies more on intuition, judgment, and experience;
- have many different answers; and
- often little insight into the interdependent components/systems that can each affect the behavior of the total systems often in an unforeseen manner.

Data mining offers new levels of near real-time ability to analyze large amounts of data. Data mining is the process of finding patterns, including causality and relative importance, of large data sets and develops predictive behaviors and characteristics. Big data is one of the ways to unravel interdependencies of a complex system. However, data mining will not work for many problems that include many physical systems. Some of the data mining techniques used for big data are shown in Table 6.2.

All of these techniques require significant storage and computational power. As discussed in Example 6.2, the costs and complexity can be significant.

TABLE 6.2
Data Mining Techniques

Technique	Characteristics
Data cleaning and preparation	• Data is cleansed and formatted which can provide insights into interdependencies, quality, trends, aggregation, etc. • First step in any technique
Tracking patterns	• Identifies/monitors trend or patterns • Can be accomplished either visually or algorithmically
Classification	• Analyzing the attributes/grouping of the data • Can provide insight into grouping, interdependencies, causality, etc.
Association	• Used to link data or data-driven events
Outlier detection	• Detects any aberrations in the data • Can be used in real time event detection • Can be accomplished either visually or algorithmically
Clustering	• Uses graphical techniques to look patterns, distributions, behaviors, metrics, etc.
Sequential patterns	• Used to uncover a series of events that take place in sequence • Can be accomplished graphical or numerically
Statistical techniques	• Can utilize both static and dynamic techniques to look at outliers, predictive models, etc.
Visualization	• Can utilized both statically and dynamically
Regression	• Helps identify the causality between variables, i.e., how one variable depends on others • Generally used for forecasting
Neural networks/machine learning	• This the most common means for machine learning • Very complex and offers little insight into first principle understanding of the system behavior
Data warehousing	• Modern cloud data warehouses including semi and unstructured data allow for real-time in-depth analysis
Association rule learning	• Identify variables within data and the behavior of different variables that appear very frequently (hidden patterns) in the dataset

Example 6.2

AI/ML has been lauded as a way to significantly improve productivity. For example, large complex data sets can now be used to make a whole host of decisions into consumer behavior and optimization. According to Bughin and van Zeebroeck (2018):

> it could add some $13 trillion to total output by 2030 and boost global GDP by about 1.2%/year. This is comparable to – or even larger than – the economic impact of past general-purpose technologies, such as steam power during the 1800s, industrial manufacturing in the 1900s, and information technology during the 2000s.

One such company hoping to help realize this improved productivity is OpenAI. With investments of many billions of dollars by such companies as Microsoft, OpenAI has been able to develop a family of AI-based models. One of these is a natural language processing model called GPT-3. GPT-3 model contains over 175 billion parameters (Wiggers, 2020). This model uses unsupervised ML. Although the costs are really unknown, most estimates place just the cost of training the model between $5 million and $12 million. This is just for computational time alone. Note that with funding from Microsoft, OpenAI has built a supercomputer to specifically train the company's AI models. According to Techxplore (2020), this would be the world's fifth most powerful computer with the sole purpose of training AI models. The amount of computing power needed for training AI models has been increasing exponentially since 2012 with a 3.4-month doubling time (OpenAI, 2022). Utilizing large-scale ML models offers many promises to solve many complex problems with multiple inputs; however, the costs are significant. Note that CPT-3 used a machine learning technique requiring training data to develop the algorithm. Open AI utilized existing websites to train the model. Thus, the cost of the training sets was very small compared to many AI models in which training data had to be developed in order to develop predictive models.

6.4.1 Big Challenges and Big Opportunities

Big data brings new opportunities to modern society and challenges to data scientists. According to Fan et al. (2014), the massive sample size and high dimensionality of big data introduce unique computational and statistical challenges, including scalability and storage bottleneck, noise accumulation, spurious correlation, incidental endogeneity, and measurement errors. The case study with OpenAI demonstrates these challenges. The engineering manager must learn to utilize big data as one of the inputs when making a decision. Big data cannot often assess the human, ethical, emotional, and economic aspects of decisions (Ulvila and Brown, 1982).

The numbers being espoused in the open literature about the economic value of big are tremendous and will reshape our way of doing business. Articles talking about how big data will transform how we work, live, and think are everywhere. This much we do know that big data, combined with social media, digitization, etc., has already significantly affected our lives in many ways.

6.4.2 DATA ETHICS

According to Cote (2021), "data ethics encompasses the moral obligations of gathering, protecting, and using personally identifiable information and how it affects individuals." Cote (2021) presents five principles of data ethics to include:

1. Ownership – an individual has ownership over their personal information.
2. Transparency – an individual has the right to know how you plan to collect, store, and use the data.
3. Privacy – companies have an obligation to ensure an individual's privacy.
4. Intention – the intentions of company matter.
5. Outcomes – companies must protect individuals from unintended outcomes.

Currently, there are no laws protecting personal information in many nations such as the United States. Whether having your DNA analyzed, cell phone records/locations, facial recognition software, tracking IP addresses, loyalty programs, etc., most believe that privacy is a thing of the past. So how does the engineering manager play a role in data ethics? These issues are much bigger than the EM profession. However, engineering managers must try to adhere to the five principles previously presented.

6.5 SUMMARY AND FUTURE NEEDS

Engineering managers are living in a world of complexity and big data. With over 95% of business needing to manage some kind of unstructured data (Perez, 2020), engineering managers need to understand the limitations of the various tools and how to make decisions using the results. They must move beyond traditional decision analysis and use big data analytics to make more informed decisions on complex problems.

REFERENCES

Barclay, S., Brown, R., Kelly, C., Peterson, C., Phillips, L., and Selvidge, J., "Handbook of Decision Analysis," *Advanced Decision Technology Program, Office of Naval Research*, Technical Report TR-77-6-30, September, 1977.

Bughin, J. and van Zeebroeck, N., "The Promise and Pitfalls of AI," *McKinsey Global Institute*, 6 September 2018, accessed 21 December 2022 at https://www.mckinsey.com/mgi/overview/in-the-news/the-promise-and-pitfalls-of-ai

Bulao, J., "How Much Data Is Created Every Day in 2022?" *techjury*, 5 January 2023, accessed 5 January 2023 at https://techjury.net/blog/how-much-data-is-created-every-day/#gref

Cote, C., "5 Principle of Data Ethics for Business," *Harvard Business School*, 16 March 2021, accessed 16 January 2023 at https://online.hbs.edu/blog/post/data-ethics

Farr, J. V., and Parnell, G. S., "A Comparison of Portfolio Analysis Techniques for Research and Development Program," *21st Annual American Society of Engineering Management Conference*, Washington, DC, pp. 293–304, October, 2000.

Fan, J, Han, F, and Liu, H., "Challenges of Big Data Analysis," *National Science Review*, Volume 1, Issue 2, June 2014, pp. 293–314, accessed 17 January 2023 at https://doi.org/10.1093/nsr/nwt032

Howard, R. E., and Abbas, A. E., *Foundation of Decision Analysis*, Pearson Education, Inc., Pearson, NY, 2016.

Keeny, R. L. R., and Raiffa, H. "Decisions with multiple objectives." In *Cambridge Books*. Cambridge University Press, 1993. https://EconPapers.repec.org/RePEc:cup:cbooks:9780521438834

Manyika, J., Chui, M., Farrell, D., Van Kuiken, S., Groves, P, and Doshi, E. A., "Open Data: Unlocking Innovation and Performance with Liquid Information," 1 October 2013, accessed 17 January 2023 at https://www.mckinsey.com/capabilities/mckinsey-digital/our-insights/open-data-unlocking-innovation-and-performance-with-liquid-information.

Marr, B., "Netflix and Uber: Getting Big Data Right", *Technology*, 17 May 2022 accessed 16 January 2023 at https://technologymagazine.com/data-and-data-analytics/netflix-and-uber-getting-big-data-right

Open AI, "AI and Compute," accessed December 19, 2022 at https://openai.com/blog/ai-and-compute/

Perez, E., "How to Manage Complexity and Realize the Value of Big Data," *IBM*, 28 May, 2020 accessed 11 February, 2023 at https://www.ibm.com/blogs/services/2020/05/28/how-to-manage-complexity-and-realize-the-value-of-big-data/

ProjectPro, "How Uber Uses Data Science to Reinvent Transportation," accessed 16 January 2023 at https://www.projectpro.io/article/how-uber-uses-data-science-to-reinvent-transportation/290

Techxplore, "Microsoft OpenAI Computer Is World's 5th Most Powerful," 20 May 2020, accessed 19 December 2023 at https://techxplore.com/news/2020-05-microsoft-openai-world-5th-powerful.html

Ulvila, J. W., and Brown, R. V., "Decision Analysis Comes of Age," *Harvard Business Review*, September 1982, accessed 23 August 2021, at https://hbr.org/1982/09/decision-analysis-comes-of-age

Wang, H., Xu, Z., Fujita, H., and Liu, S., "Towards Felicitous Decision Making: An Overview on Challenges and Trends of Big Data," *Information Sciences*, *367*, 747–765, 2016.

Wiggers, K, "OpenAI Launches and API to Commercialize Its Research," 11 June 2020, accessed 19 December, 2022 at https://venturebeat.com/ai/openai-launches-an-api-to-commercialize-its-research/

Winig, L, "GE'S Big Bet on Data and Analytics," *MIT Sloan Management Review*, 18 February, 2016, accessed 16 January 2023 at https://sloanreview.mit.edu/case-study/ge-big-bet-on-data-and-analytics/

7 Forming Alliances Strategically

John Mo
RMIT University

Matthew C. Cook
British Engineering Council

7.1 INTRODUCTION

In the contemporary world, many highly complex engineering projects such as building aircraft, ships, buildings, and infrastructure have enormous financial and technical requirements to achieve success. These requirements generate risks that essentially require management and mitigation actions throughout the entire life cycle of the project. In many cases, this is beyond the capacity of a single organisation and results in technical failings, schedule delays, and serious cost blowouts. One option that is becoming ubiquitous within large and/or complex projects is to "share" the risks by forming an alliance between several organisations.

The formation of an alliance is generally thought of as a risk reduction strategy for sharing the technical challenges, tapping into appropriate resources, developing new capability and know-how, ensuring a competitive edge, sharing the financial burden, and relieving the schedule pressure of large challenging projects. Project risks are essentially spread across two or more organisations, the theory being that each organisation should have the attributes essential to meet key project requirements and thus mitigate risks associated with these requirements. There are many examples of such alliances, and their value has been much publicised in areas such as aerospace and defence (Keller, 2016). These industries tend to undertake extremely technically complex projects that require massive financial investment and commitment over significant periods of time.

In large defence projects, the technical, schedule, and cost challenges can be significant, and both governments and organisations will look to both distribute and spread these risks where possible. Alliances offer the opportunity to involve partners with specific skill sets, capability, and access to specific markets or finance. This could be a positive way of mitigating technical risks and reducing costs. Alliance-partnered organisations can also work concurrently on various related aspects of a major project, resulting in potential schedule pressure reductions. It is important to note that in some cases, especially with government projects, alliance partners may not always have the opportunity or luxury of choosing/selecting who they enter an alliance with, and this in itself can constitute a significant risk.

DOI: 10.1201/9781003374879-7

Unfortunately, forming an alliance between several potentially competing organisations can also bring challenges that did not exist before. Forming alliances strategically is key to competitive advantage, but involves bigger risks.

The operating conditions of a typical business environment are often characterised by frequent changes in products, services, processes, organisations, resource, markets, supply, and distribution networks. In an alliance environment, the partner organisations then form a temporary alliance (in some cases lasting many years) to deliver a project or product and dissolve the relationship when the job is completed. The partner teams should work together as an entity for a specific goal but the relationships between the organisations are often disrupted by differences in established practices, cultures, opinions, and motivations of the individual companies (Mo et al., 2006).

The formation of an alliance brings added complexity to the structure of the project and actually leads to a significant increase in the overall risk level of the project. There is growing evidence to suggest that the failure rate of alliance projects is as high as 50% (Goa and Zhang, 2008). Many factors contribute to these Figures including the complexity of controlling partnership risks, the process of how individual partners will work, emerging behaviours of alliance partners, etc. Unfortunately, there is no well-established method or system available to assess the risks in such alliances satisfactorily. Therefore, achieving project success in alliances depends on luck and perhaps relying on the persistence of some companies, more than a predictable outcome.

This chapter explores how alliances are formed and how they should be managed, in particular, focusing on analysis leading to strategic decisions for managing risks due to potential opacity existing between the alliance partners. This chapter then presents a model that can be used by managers, governance teams, and engineers alike to identify and assess how risks can multiply as internal organisations, and external project risks are generated and combined. These risks include technical, process, behavioural, and cultural issues that can exist and/or develop between organisations, thus increasing the challenge of achieving success. This novel method of capturing, assessing, and modelling alliance risks for major engineering projects is then demonstrated in a case study.

7.2   ENTERPRISE MODELS

Modern enterprises are highly agile and adaptable. The situation is more complex when these enterprises become partners in an alliance. To develop viable strategies, knowledge on enterprises can be grouped at two levels. At the enterprise level, to understand and maximise the efficiency of the enterprise system, people need a way to reduce the complexity of the enterprise system into a manageable number of entities and to understand how these entities relate to each other (Bernus and Nemes, 1996). When given an integrating framework using common sense language, people can simplify and understand the complexity around them. Such a framework enables people to think clearly about difficult issues, to build shared views, to develop/implement a roadmap with others, and to work collaboratively. It can enable communication to be done effectively in new enterprises or improved extant enterprises. This can promote a sense of predictability in an unpredictable environment.

Unfortunately, changes in enterprises often face resistance from people who believe they will lose out in the chain of actions. Conflicts are often the result of differences in opinion, culture, and many other factors (Barmeyer and Mayrhofer, 2008). Oberg et al. (2007) presented the concept of "network pictures" as the modelling framework to illustrate and analyse changes in managerial sensemaking and networking activities. Gregor et al. (2007) argued that by drawing on established alignment to architectural theory, an organisation's enterprise architecture can enable the alignment of business strategy and information systems and technology. The use of enterprise architecture can provide a good foundation for managing risks in information system development projects (Janssen and Klievink, 2012). When these enterprises are examined closely, micro architectural views are critical to exhibit different forms of the enterprise outfits. Five models portray the micro architectural views in the following sections.

7.2.1 People-Centric Model

The people-centric model (PCM) (Chattopadhyay and Mo, 2010) was developed from research conducted in a global engineering, procurement, and construction management (EPCM) company. People play a pivotal role in any organisation and their presence pervades through all layers of the organisation from the shop floor to the board room. The essence of this model is threefold: People will generate outputs utilising their skills and resources over time. The organisational structure is driven by the strategy of the organisation. A customer-focused strategy drives a continually evolving organisation structure due to ever-changing customer expectations. The PCM model is, therefore, supported by three pillars, namely, people, strategy, and customer, connected by a feedback loop.

7.2.2 Molecular Model

The molecular model (MM) perceives humans as an independent, standalone, intelligent, and effective information and communication system by themselves. In the work environment, people use both tangible and intangible resources and information to produce tangible and intangible "outputs" while utilising their skills. This model represents people as the metaphor of atoms with skills and resources represented by the orbits of the electrons in the atom (Chattopadhyay et al., 2011). The key operational functions such as planning, scheduling, shop floor management, and control are performed by the human atoms in the energy bands of the molecules. As the value-adding operational activities become mature, human energy is continuously spent through the skills of human atoms to collectively transform raw materials into finished products.

7.2.3 Kaizen–Lean Six Sigma Model

A study in North America identified a number of best practices, techniques, and major groups involved in improving manufacturing flexibility, while keeping broader organisational strategies such as lean (Boyle and Scherrer-Rathje, 2009). Together with other complementary tools such as six sigma and total quality management

(TQM), a complete set of executable tools are available to support continuous improvement in successful enterprises. The core essence of the Kaizen–Lean Six Sigma Model (KLSSM) is about change through transformation to their best practices (Vella et al., 2009). The model has a layered flexible enterprise architecture with built-in life-cycle phases driven by the three pillars, namely, skills, resources, and information.

7.2.4 GLOBALLY DISPERSED MODEL

This global dispersed model (GDM) represents a holistic approach for virtual manufacturing in a global setting (Chattopadhyay et al., 2010). The human-centric and eco-friendly approach in line with the global trend in manufacturing in the last decade suggests more IT requirements enabling collaborative decision environments and incubating multi-enterprise business network delivery. This model proposes a typical regional production system connected through globally dispersed networks. The regional production system enumerates a manufacturing capability on the shop floor that operates through an integrated and optimised combination of cellular, manufacturing execution system (MES) and hybrid of cellular and MES mode.

7.2.5 DISAGGREGATED VALUE CHAIN MODEL

The disaggregated value chain model (DVCM) is perceived as a value-adding collaborative partnership amongst people who work closely together on trust, to manage the flow of goods and services along the entire value-adding chain (Chattopadhyay et al., 2012). In a DVCM, structured value-generating activities take place across various units within or between many independent organisations acting as suppliers, distributors, and producers, moving asynchronously towards its final outcomes. This model requires fast transitions of people to new roles and relationships. The DVCM therefore, represents a synchronised material and material flow cycle, servicing customer needs and forming a closed loop "demand-design-develop-deliver" cycle.

7.2.6 INTEGRATED ARCHITECTURE – THE PENTATOMIC ORGANISATION

While the five portraits of an organisation focus on people's roles, responsibilities, and reactions, the link to consultation processes is the key characteristic of modern organisations, the evolutionary ability is missing in these models. From the point of view of enterprise architecture, the five portraits are different views of an enterprise. The pentatomic organisation model (POM) is a federation model that encapsulates different enterprise modelling requirements by enabling or disabling its constituent sub-models (Chattopadhyay and Mo, 2011) (Figure 7.1).

Each of the five foregoing models apparently looks different but has the same vibrant characteristics. As individual human actors are involved, emotion plays a very significant part in decisions which is usually catalysed by personal and external stimuli. Despite the uniqueness of each human sphere of influence, emotion can largely influence the dynamics of interaction and hence the outcome of a transaction. This unpredictability makes it very hard to model or simulate human interactions, which is

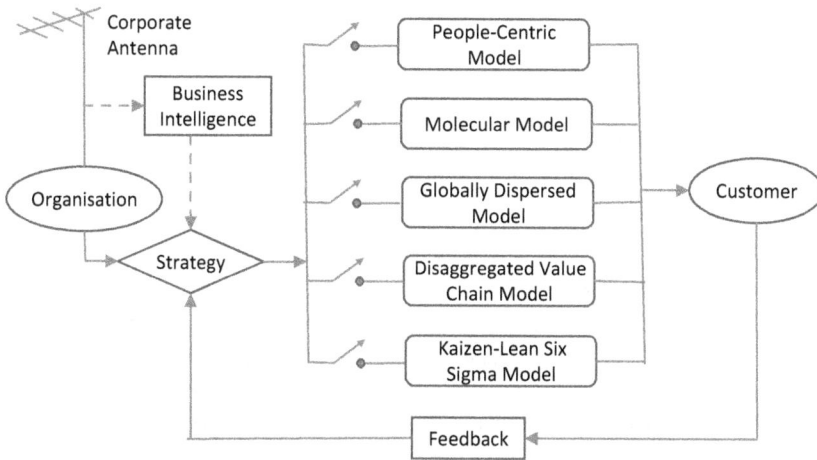

FIGURE 7.1 The pentatomic organisation model.

key to the organisation's or alliance's success. The POM allows the adaption of the organisation's structure with a combination of the five models on a need-based deployment. It should be remembered that modelling is a continuous effort and not a one-off exercise. Success is governed largely by the accurate assessment of the external business environment, business needs, customer requirements, strategy, cost, and so on.

7.3 MODELLING ALLIANCES

Forming an alliance between several potentially competing organisations brings new challenges into the alliance system and creates new risks. The coalition relationship is not binding and the "enterprise" is not a cohesive group. Every partner in the alliance has its own goals and agenda, not even mentioning the differences in organisational culture and practices. In such circumstance, trying to develop or implement strategies within the "alliance" is extremely difficult without a shared understanding and flow of information between parties.

The pentatomic organisation model could be the closest enterprise model but there is a fundamental discrepancy in the application. While the POM can switch to different enterprise models flexibly, each of the enterprise models represents a monolithic structure of the alliance. On the contrary, due to the nature of alliances and multiple characterisations of alliance mechanisms, there is no absolute authoritative relationship between partner organisations. The operation of an alliance under the POM model is obviously different from reality, and thus a new theoretical construct is required.

7.3.1 ENTERPRISE NETWORK AND VIRTUAL ENTERPRISE

The international research programme "Global Intelligent Manufacturing" code-named "Globemen" investigated the formation of enterprises at different stages of the global manufacturing supply chain (Vesterager et al., 2000). The research group uses

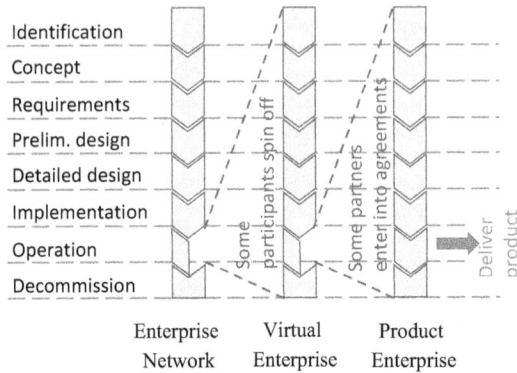

FIGURE 7.2   The Globemen three phases product enterprise evolution model.

the term "virtual enterprise" to describe the working relationships of organisations at different stages of developing and executing project works separately. In Figure 7.2, the Globemen model has three phases of evolution from "enterprise network", "virtual enterprise" to "product enterprise". The Globemen research found that there are three enterprises co-existing in the global business environment trying to form project enterprises aiming at different products.

The three different forms of enterprise serve the needs of the group with different evolutionary stages of relationship. Each form has its own enterprise life cycle. The "enterprise network" is practically not an enterprise. It is a voluntary collaborative arrangement among the partners in the network. Anyone can join or leave without penalty. Therefore, although new participants can create additional risks, the system can adjust itself quickly. Information shared is not sanitised due to the lack of security protection. The sole aim of an "enterprise network" is to explore business opportunities.

Eventually, some of the organisations in the "enterprise network" might be able to define a business opportunity more precisely among themselves. These organisations can work more closely with better-defined protocols; for example, confidentiality agreement and memorandum of understanding can be signed among the participating organisations. The aim of a "virtual enterprise" is to restrict participation to maximise potential benefits to those involved.

When the "virtual enterprise" secures a real contract, it enters into an obliged environment. The "product enterprise" is established with a formal structure of the relationship(s) among the partner organisations. Legally binding contracts are required to clearly define the division of work, responsibilities, and targets.

The Globemen model provides a conceptual framework of how alliances are evolved from interactions between enterprises. Alliances are expected to exhibit sufficient integrity towards the completion of the project. Hence, they can be thought of as the "product enterprise" but its evolution has gone through the virtual enterprise stage. Therefore, to represent the alliance structure, a different approach is necessary.

7.3.2 Enterprise DNA

DNA (deoxyribonucleic acid) is the foundation building block for all living cells. Baskin (1995) was among the first few researchers using DNA as an analogy for exploring enterprise characteristics. In this early model, the enterprise DNA was represented by two interwoven "strands": Identity and Procedure. Since then, researchers including Sireesh (2006), Spear and Bowen (1999), and Towill (2007) explored the use of DNA as a modelling construct to explain the inheritance of characteristics of organisations. Mo and Nemes (2010) defined the enterprise DNA as three foundation elements: data, people, and assets (machines). In such DNA thinking, these building blocks are itemised to enable free combinations in any order for creating new enterprise architectures. With these building blocks, the anatomy of the enterprise can be constructed using three interconnected strands: control, knowledge, and processes. Subsystems of the enterprise are therefore characterised by the genes in the form of departments, products, services, and other tangible entities (Figure 7.3).

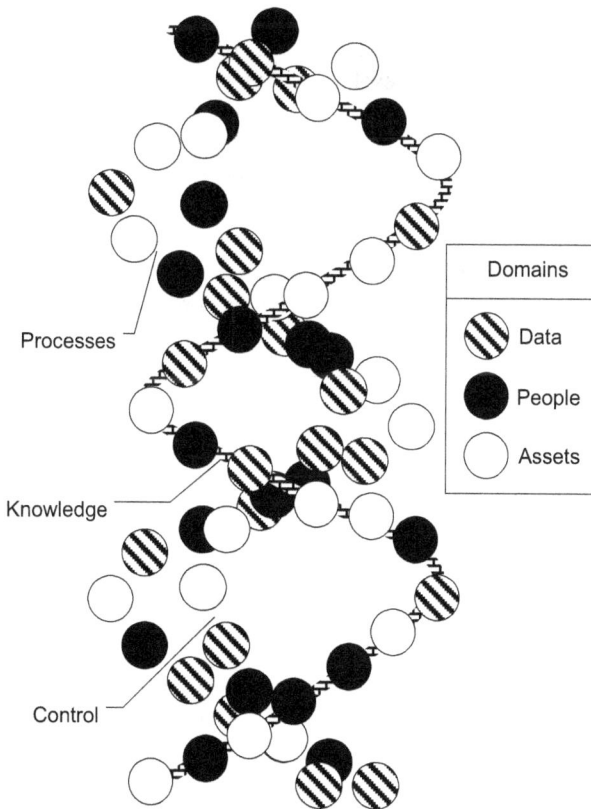

FIGURE 7.3 DNA constructs for enterprise architecture modelling. (Mo, J.P.T., & Nemes, L., Issues using EA for merger and acquisition. In G. Doucet, J. Gøtze, P. Saha, & S. Bernard (Eds.), *Coherency Management: Architecting the Enterprise for Alignment, Agility, and Assurance*, pp. 235–262, AuthorHouse, Chapter 9, 2010. ISBN 978-143899-60783. Used with permission.)

The DNA model provides a traceable granular method to represent subsections of an enterprise that have differentiating characteristics. As the enterprise continues to participate in enterprise networks, virtual enterprises, and possibly a consortium of some sort, the enterprise will undergo transformations that enable it to adapt to a new environment. According to the DNA model, changes are brought about by different permutations of the bases.

The DNA theory offers a flexible way of representing the nature of organic transformations such as merger and acquisition. When changes are required, some of the enterprises can be nurtured with different DNA, for example, the people and assets combination can be replaced causing a change of teams and infrastructure support. It is also possible to create plug and play (i.e. by replacing genes) enterprise sub-systems that fit the newly emerged enterprise requirements. It is also worth noting that the inheritance of genes may not be entirely controllable by the enterprise designer, for example, one may not have the freedom to select products or services due to historical reasons.

7.3.3 THREE INTERACTING ELEMENTS IN AN ENTERPRISE

While the DNA model offers the touch points of implementing change, it does not predict outcomes at the alliance level from changes externally. Research has shown that a more generic representation of the enterprise can be defined with three foundation elements: product, process, and people (Mo, 2012). This earlier model asserts that the interaction of these elements could be cooperative, but equally speaking, they could be in conflict, depending on how well the "system" is managed. These elements are made (if managed properly) to align towards a common goal within an environment that is imposed onto the "system" (Figure 7.4). Any socio-technical entity with a unique line of authority can be modelled including commercial companies.

The single enterprise 3PE model has three elements:

P = People
C = Process
D = Product

The interaction links are indicated by P – C, P – D, C – D.

An important concept in the 3PE model is that activities and outcomes in the system are the result of interactions between the three elements. The notion of people, process, product, and their situation within an environment is the background for any system to operate. Operation of the system depends on the extent of the interactions. Active, high-performing systems obviously have a lot of inter-P's interactions. Performance of the system then depends on what effect these interactions have within the constraints of the environment.

Furthermore, the 3PE model recognises that a successful working enterprise will not need to change. Change is only required if the enterprise detects changes in the environment in which the enterprise lives in. In other words, an enterprise changes because it has to adapt to a new environment. Figure 7.5 illustrates this concept.

In Figure 7.5, the expanded environment demands more from the enterprise. The enterprise responds by changing its people and process, but keeping the product

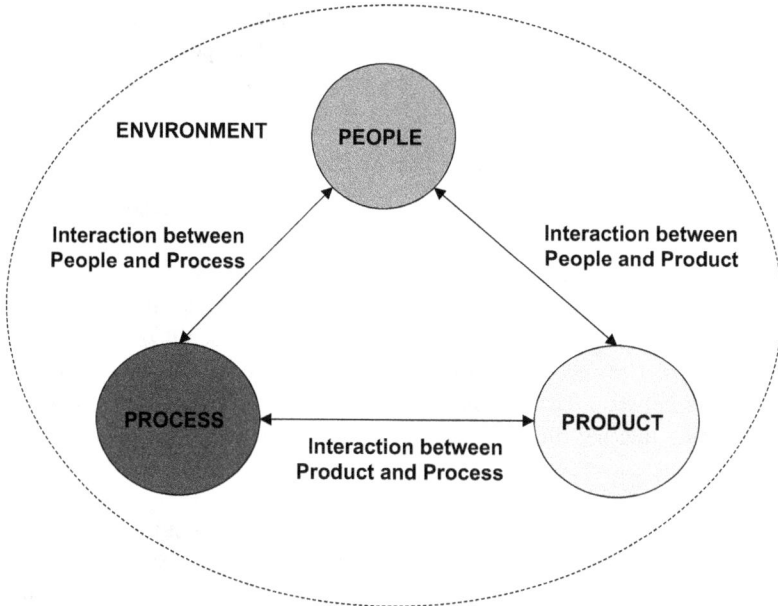

FIGURE 7.4 Product Process People Environment (3PE) model forming an organisation.

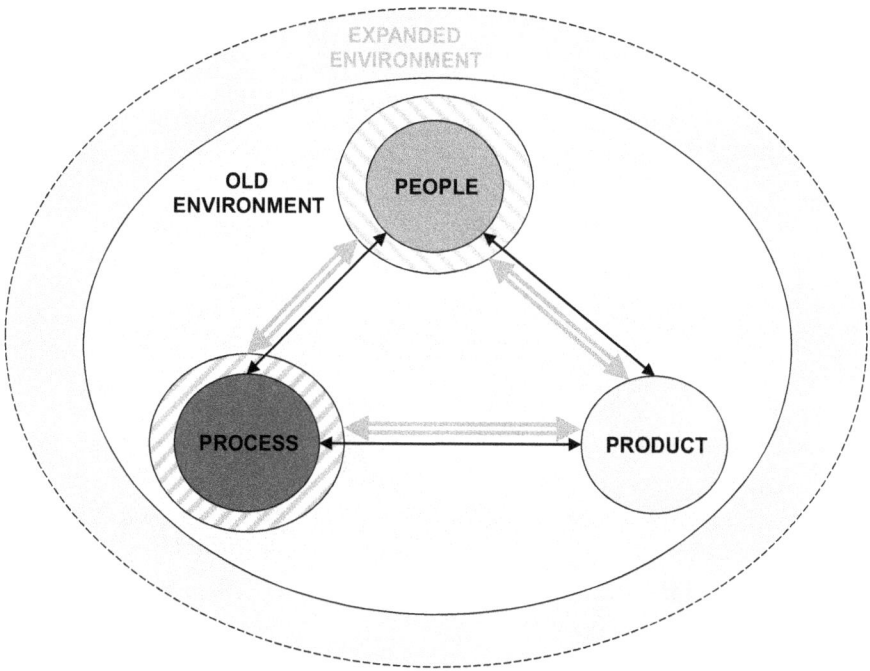

FIGURE 7.5 3PE organisation reacting to changes of environment.

unchanged, i.e., offering the same product. Due to changes in people and process, all interactions in the 3PE model have changed. It is the responsibility of the system leader to ensure a smooth transition of the enterprise to its new form.

7.3.4 3PE–SOS Alliance Interaction Model

When two or more organisations enter into a partnership based on the aforementioned reasoning, the 3PE model offers a readily explicit model to work with. Modelling of interactions in the alliance environment, including internal organisation environments (systems), can be logically represented by an extended formulation of the 3PE model to a system of systems (SOS) structure (Cook and Mo, 2019). In a two systems situation (System 1 and System 2), there are nine interaction links as shown in Figure 7.6. The links between the elements have now increased from 3 (per organisation) to 16 (including internal links). Forming an alliance potentially represents an enormous increase in risk factors that will need to be controlled, managed, and mitigated throughout the project life cycle.

However, if we examine the links carefully, there are in fact only six types of interaction links between systems as shown in Table 7.1. The notion of System 1 and System 2 are interchangeable. The links that are labelled "Replicated with (n)" are technically indistinguishable from the link (n) between the two systems because the nature of interaction does not change, except that the source of the element might come from the other side of the interaction. The analysis approach and potential solution space are similar.

This number of risk interfaces can be generalised to any number of systems in the alliance. Suppose there are n companies forming an alliance. Each company's 3PE

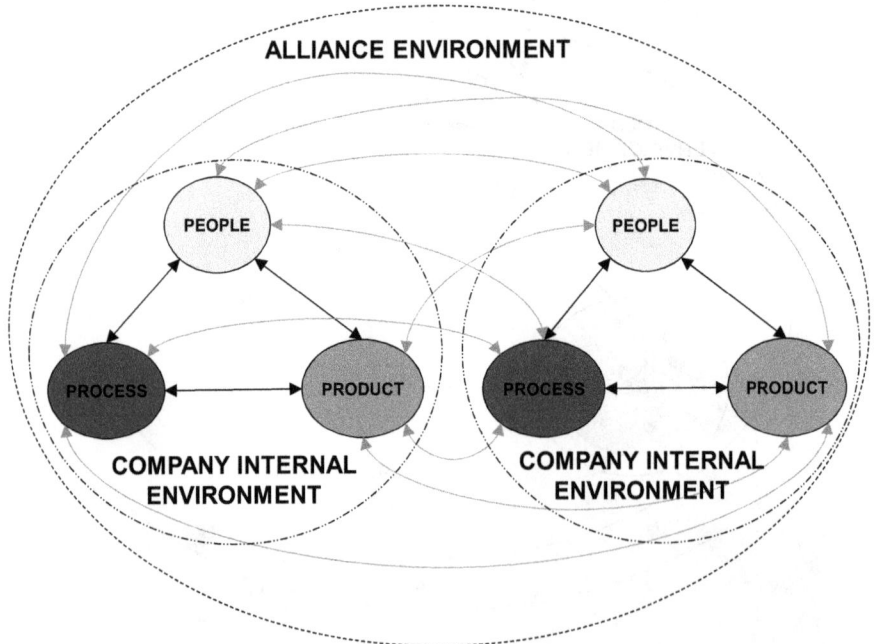

FIGURE 7.6 3PE–SoS model with two partner organisations.

TABLE 7.1

Inter-Company's Interaction Matrix

		System 1		
		People	**Process**	**Product**
System 2	People	(1) P – P	(2) P – C	(3) P – D
	Process	Replicated with (2)	(4) C – C	(5) C – D
	Product	Replicated with (3)	Replicated with (5)	(6) D – D

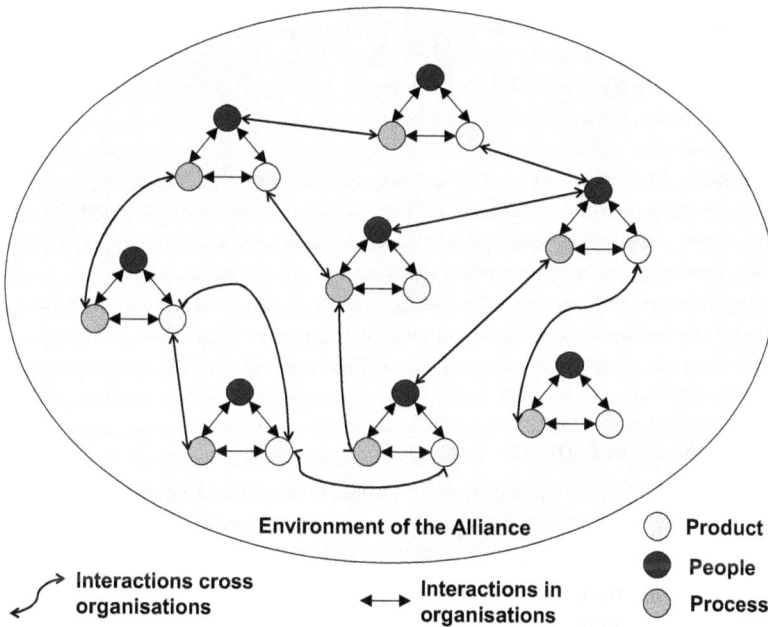

Environment of the Alliance

◯ Product
● People
◔ Process

Interactions cross organisations

Interactions in organisations

FIGURE 7.7   3PE expanded into system of systems modelling (3PE–SOS).

system is represented as a column or row in the same way as in Table 7.1. The number of interaction matrix is given by:

$$\frac{n^2 - n}{2} \tag{7.1}$$

In each interactive matrix, there are 6 inter-company interactions. Hence, for a n company alliance, the number of links is given by:

$$l_h = 6\left(\frac{n^2 - n}{2}\right) \tag{7.2}$$

This situation can be represented in Figure 7.7. It should be noted that not every pair of P's have interactions. Some P-pairs may not have any business or organisational relationship.

Equation (7.2) calculates the maximum number of links in the alliance. Some links could be problematic, for example, two companies in the alliance may offer competing products. It is therefore crucial that a strategy mitigating potential issues when companies entering into an alliance needs to be developed.

7.4 STRATEGIES FOR PARTICIPATION IN ALLIANCES

It is time to define more precisely the term alliance in the context of this chapter. An alliance is the formal and agreed inter-enterprise structure that binds several enterprises, with clearly specified relationships and responsibilities towards defined strategic goal(s) of the project. Section 7.3 highlights the importance of people-centric organisation models together with skills and assets. The Globemen model outlines the evolution of product enterprise. Practically, the product enterprise is an alliance with the goal of making the "product". It should be noted that there are also alliances for "services".

In addition, the DNA model that makes use of the analogy of an enterprise with biological systems is reviewed. The 3PE model consolidates these research findings into a generic engineering management architecture applicable to any organisation. It is postulated that any organisations must have these three foundation elements that intertwine like DNA in biological systems, but can be replaced or restructured under constraints and influences of the business environment. With the 3PE model as the basis, it is now possible to explore how an alliance could operate successfully.

7.4.1 ALLIANCES IN INDUSTRY 4.0

With the advancement in the Internet of Things (IoT), global business networks rely heavily on information technology infrastructure to do business. The change in business processes triggers typical issues in Industry 4.0 operations that include shorter product life cycle, more supply variability, difficult collaboration, risk to confidentiality, conflicts in intellectual property, opportunity loss, capacity constraints, and others. This phenomenon is generalised as X4.0 system development where X represents a specific industry sector (Mo and Beckett, 2019). X4.0 systems have inherent complexity, societal and technological challenges as IoT technologies and associated data assets become the main platform to do business. Through IoT connection, people with different background knowledge and potentially different cultural norms, together with other stakeholders such as financial institutions, governments, and certification authorities, will have a strong influence on the development.

The complexity challenge of X4.0 can be explored with a system of systems modelling approach. The dispersed environment nature of X4.0 essentially evolves into different types of business environments. Due to high volume of data transmission in IoT, the data-driven X4.0 paradigm is centred around the decision system context and supported by a knowledge network with collaborating knowledge agents (which are systems themselves). The knowledge network supports the performance of tasks and is supported by the collection and distribution of information.

System of systems models depend on a clear definition of the single system model and the interconnection among single system models. Mo and Beckett (2018)

TABLE 7.2
3PE Element Relationships with Industry 4.0 Functional Blocks

Elements Of 3PE (Both Product And Service)	Industry 4.0 Functional Blocks			
	Cyber-Physical Systems	Data Analytics	Data Integrity	Work 4.0
Product/ service	Smart sensors and integrated autonomous robotic agents	Machine learning software, "Big Data" management and analysis services	System security and data quality monitoring software, Massive on-line storage, clean data, secure storage services	Decision support. VR, learning management systems, educational products and services
Process/ procedure	System development and operational processes and procedures	Data collection from multiple sources, data organisation, analysis, and interpretation procedures	Data quality screening, data access, and distribution procedures	Agile Human Resources and Intellectual Property Management
People/ stakeholders	Engineering orientation. Cross-disciplinary collaboration	Information Systems orientation. Cross-functional collaboration	Computer Science orientation. Cross-disciplinary collaboration	Knowledge Management orientation. Knowledge capture/ sharing collaboration
Environment/ context	Industry sector automation	Industry problem-solving and continuous improvement	Industry sector and community data security requirements	Industry and regional socio-technical norms

Mo, J.P.T., & Beckett, R.C., *Engineering and Operations of System of Systems*, Taylor and Francis, 269 p., 2018. ISBN 978-113-863473-2. With permission.

elaborated the X4.0 system as superimposed on top of 3PE–SOS model, making use of network-centric operations for connections and information sharing, i.e., interactions. Whilst frequent and recurring events may be described as workflows, it is the actors (people) who are driving the system's performance.

The relationships between the elements of 3PE and the Industry 4.0 system can be mapped with details as shown in Table 7.2.

It is clear from the mapping that an alliance in Industry 4.0 environment will be able to maximise its performance by "big data", multi-directional interactions, knowledge sharing, and cross-disciplinary collaboration.

7.4.2 INNOVATION STRATEGY IN ALLIANCES

Shenhar et al. (2016) developed a model that suggests high-tech projects must include at least three cycles of design, build, and test. It also suggests that such projects need to allocate about 30% of the time and budget, as contingent resources, beyond a typical traditional plan. This was supported by a study into financial service firms by Das et al. (2018), the research highlighted if an innovation strategy, active management support, and a separate governance structure for innovation are in place, projects get stimulated at the exploration phase and do not experience a lack of appropriate resources or competition with traditional projects.

Striving to survive in an ever-changing world, as well as maintaining the ability to innovate has become increasingly crucial (Zhuang et al., 2018). Understanding the link between project complexity and innovation is highly pertinent. However, many challenging projects are further complicated by the introduction of partners, with the formation of an alliance traditionally viewed as decisive for success (Young et al., 2016). However, how can alliances benefit from interactions creating innovative solutions while trying to avoid the pitfalls of complexity as modelled in 3PE–SoS?

From their research, Zhu et al. (2019) found that in an alliance it is essential that the lead firm (prime) and usually the initiator of the project, should understand the exploratory nature of the project as well as foster innovation-related capabilities and network-related capabilities as a pre-condition. Furthermore, Trappey et al. (2017) reviewed research in collaborative systems and concluded that the concept of collaborative systems is not just a collection of enabling technologies but also a fundamental business philosophy requiring strategic thinking for a variety of applications at different stages of collaboration, including management, dissemination, use of data, information, and knowledge throughout the entire life cycle of product development. What strategies and system architecture should be adopted in joining an alliance?

Clearly, these questions are applicable across all industries and fields with innovative advancements and the introduction of digital technology being ubiquitous. Strategic options to control and minimise risk through systematic modelling offer some benefits, if implemented and managed comprehensively from the outset (Cook and Mo, 2022).

As an example, the automotive industry is seeing a watershed moment with the move from traditional petrol and diesel vehicles to hybrid and to fully electric powertrains. Electric cars have been a consideration for many decades, and previous attempts at development have seen little success. The domination of hydrocarbon fuels and automotive manufacturers' reluctance to innovate in new areas of powertrain design and technology is a large factor. From the 3PE–SOS model's perspective, these can happen by interactions in P – D, D – D, and P – P.

Many alliances are formed based on the instant perception of financial or technical feasibility, without considering much broader implications, such as individual enterprises having to adapt to the merger and enlarged business environment. Crucially, the 3PE–SOS model is used to highlight potential interaction links that may be exploited to foster innovations. The strategic alliance should, therefore, be formed with careful consideration and systematic analysis of each interaction link in search for maximising the benefits of forming the alliance.

7.4.3 MANAGING RISKS IN AN ALLIANCE

As previously mentioned, the formation of an alliance could be thought of as a risk reduction strategy that could share technical challenges, tap into appropriate resources, ensure competitive edge, spread the financial burden, and schedule pressure of large complex projects. However, Cook and Mo (2018) provided evidence that all is not well with the alliance strategy as a method for mitigating risk. Their research detailed how introducing partners to a project actually increases risk pathways and the chances of success are limited without a systematic holistic approach to risk management. Many factors contributed to failure in alliances including the complexity of controlling partnership risks, the process of how individual partners will work, emerging behaviours of alliance partners, and lack of ability to learn and adapt as they go (Cui et al., 2018).

In recent times, co-innovation has emerged as a popular concept for how organisations may create partnerships to develop products. Bugshan (2015) defined the term "co-innovation" as innovation deriving from the collaboration of two or more parties. Of course, the reasons driving companies to co-innovate are manifold, spanning from accessing and co-producing new knowledge, to designing new products and services and decreasing time to market. Through co-innovation, partners increase their competitiveness by creating and sharing knowledge, resources, improve production, create new commercial opportunities. Ombrosi et al. (2019) specifically highlighted two major sets of drivers that can be recognised for co-innovation: relationship-based reasons on the one side and technology-based reasons on the other side. Hence, there are great opportunities for forming alliance in terms of co-creation, but equally speaking, these opportunities do not come without risks.

In their study on co-innovation risk, Abhari et al. (2018) found co-innovation actors (external co-innovators) perceived four different individual risks: time, social, intellectual property right, and financial. The empirical results demonstrate a high degree of confidence in both translation validity and criterion-related validity. Negative effects of perceived co-innovation risk on actors' continuous intention to ideate, collaborate, and communicate in co-innovation were evident, but prior experience moderated these relationships.

Cook and Mo (2020) further researched the severity of risks in any engineering developments including projects that do not normally bear significant risks such as system upgrade and engineering modifications. As long as there is a need for organisations to form an alliance when undertaking the project, risks within the alliance emerge. The research highlights how risk pathways increase with the introduction of partners and thus need to be carefully managed. As innovation is desirable in alliance interactions, these risks should be anticipated and mitigated well in advance.

7.4.4 SUMMARY

It has been well established that innovation is laden with risk and presents extreme challenges for any organisation. Forming an alliance is seen as a strategic method for spreading the risk of innovation and development. However, this strategy is a double-edge sword due to the substantial increase of challenges in managing interactions among partners. A new system architecture approach that can expose the origin of innovation risks in complex (alliance) projects and provide an investigative direction for identifying these risks is required.

7.5 ALLIANCE INNOVATION LIFE CYCLE ASSESSMENT METHODOLOGY

This section outlines management strategies for alliances working in conjunction with an agreed mission. Within the alliance, every unique system has its own goals and agenda. Therefore, there are no fixed rules or authoritative drivers to govern the alliance. However, if the interactions are understood and managed intelligently (and in many respects, dynamically and diligently), the alliance and the individual systems in the system of systems can have a good chance to migrate to a win-win situation.

The Alliance Innovation Lifecycle Assessment Methodology (AILAM) is the outcome of research conducted into characterising operating principles of alliances with the aim to outline a strategic approach to develop desirable, although may not be effective, alliances.

7.5.1 Systems Engineering Life Cycle

Complex projects in areas such as defence and aerospace are usually coordinated by a technical process built on a backbone of Systems Engineering (SE) methodology (INCOSE, 2007). This methodology has been well established for many decades and is structured around the SE V life-cycle model (Figure 7.8). Broadly speaking, the

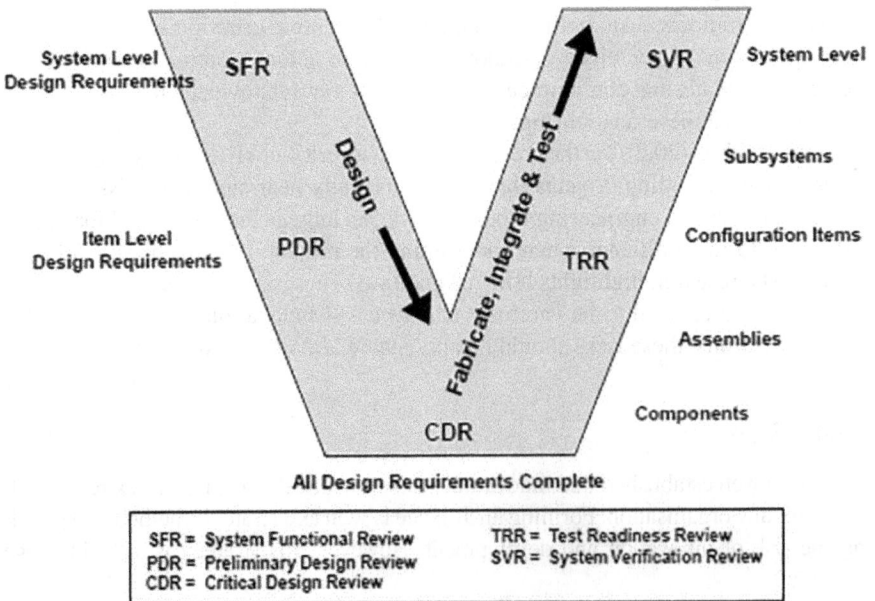

FIGURE 7.8 Typical systems engineering V life cycle.

SE methodology encourages innovative ideas to be proposed according to a set of requirements that are determined at the infancy of the project, followed by specific phases of the design and realisation process (verification and validation).

Although the successful application of innovative ideas can be highly rewarding, it is inherently risky. How can the project team decide that a certain innovative idea (at the SFR phase), has a good chance of success and will produce great benefit(s) later in the project? The SE V life cycle has a good theoretical foundation that has been applied with varying degrees of success. This problem of making "the right choice" for an innovation becomes increasingly apparent when significant complexity is introduced, such as the formation of partnerships and alliances. By applying 3PE–SOS, as more partners join the alliance, there is a significant jump in possible risk pathways, and as a result, without very robust and comprehensive risk analysis and mitigation strategy, the project can become overwhelmed.

7.5.2 Prioritising Strategic Actions

To address these challenges, the analytic hierarchy process (AHP) is used to determine both the significance and priority that risks need to be addressed as a snapshot assessment at certain points in the SE V life cycle. The 3PE–SOS model provides a structure where all risks (including anticipated innovation) can be located within the topology.

Using the 3PE methodology, the enterprise responsible for innovation can be modelled as part of the product element in that enterprise. This modelling construct clarifies that each organisation can have its own innovation reflected in its product offering. If there is co-innovation in an alliance, the relevant part of the co-innovation in each organisation will need to be separately represented in each of the 3PE models.

It is worth noting that if innovation is in the process of using an existing product, or new procedure to manage the project team, innovation can also be identified in the process element of the leading organisation.

7.5.3 Continuous Improvement Approach

Having defined the architectural model of an organisation, this chapter proposes an iterative approach to assessing the effect of innovation risks on long-term projects under alliance arrangement, known as Alliance Innovation Lifecycle Assessment Methodology (AILAM). The iterative approach is a significant enhancement to the combined methodology, to enable a thorough assessment of risks at the infancy of the project. The AILAM can be illustrated in Figure 7.9.

After the initial set of risks has been established, the output requires evaluation by the project team. Post this evaluation, a new 3PE system model can then be launched, and the initial set of risks can then be refreshed and expanded as updated scenarios are materialised with the new 3PE-SOS model and re-considered innovative ideas.

With the new 3PE-SOS risks, an AHP matrix can be created and a prioritisation process can be done to refine the mitigation plan. Theoretically, this cyclic process can continue until there is an acceptable set of risks for all innovative ideas being proposed. Research into past innovative projects has shown that each

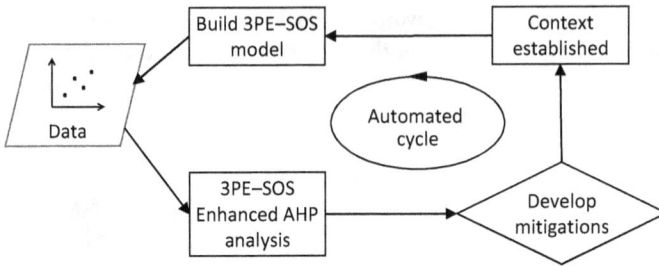

FIGURE 7.9   AILAM.

round of assessment can be more effective by setting improvement goals. This systematic approach ensures completing risk assessments of innovation projects in circa three rounds. The initial round of assessment (or first iteration) will focus on a single organisation and the risk of innovation within that organisation's own environment.

A second-round iteration will be focused on interactions among partners within the alliance environment. It is important to highlight here that the elements of the 3PE model (people, product, and process) are established within each of the organisation's own environments, during the first iteration. The second iteration models interactions between the elements of all the partners, within the alliance environment.

A third and final iteration of the 3PE model is now run where risks that are located within the individual organisational environments and the overall alliance environment and re-evaluated for the final time, and it is at this point that a definitive list of all project risks is captured.

7.6   CASE STUDY OF AILAM FOR A LONG-TERM DEFENCE PROJECT

To illustrate how AILAM could be used to assess innovative work in complex engineering projects, a case study is provided that highlights the challenges and disturbances forming an alliance and managing innovation can introduce to a project. A historical summary of the case study is provided, and an analysis of how the 3PE–SOS model could have improved the outcome is presented.

7.6.1   CONTEXT ESTABLISHED

In the late 1990s, the UK began the development of a new naval surface destroyer known as Type-45 or Daring class, with the first ship, HMS Daring, planned to enter service in 2007. Among a whole array of advanced and cutting-edge systems that were integrated into Type-45, the platform benefited from the introduction of a new state-of-the-art innovative engine package known as WR-21, designed to meet fuel efficiency and endurance requirements.

The engine package was developed by partners in an alliance including Rolls-Royce, Westinghouse (Northrop Grumman), BAE Systems, and the UK Ministry of Defence (MoD). However, midway through the design phase, Westinghouse was purchased by Northrup Gruman, and upon assessment of the WR-21 project, Northrop Grumman made the decision to leave the project. Consequently, Rolls-Royce inherited the unfinished design and development work but significantly was offered little relief on the programme schedule and cost. To achieve critical delivery milestones, the engine package underwent minimal analysis and testing, before being hastily finalised and built, in order to be ready for the first-of-class integration deadline. The results of WR-21/Type-45 project are well documented in the media (Weiler and Chiprich, 1997), with the consequences still being felt to this day.

Many classes of naval ship use a gas turbine(s) as their prime mover, as these engines offer incredible power-to-weight attributes. However, improvements in fuel efficiency have been desired for some time in both the aviation and maritime sectors. The WR-21 engine package mainly offered two innovative technologies:

1. An intercooler at the compression stage
2. A recuperator at the exhaust stage

These innovations ensured the WR-21 engine package would be unique in the world, with the major advantage being significantly increased fuel efficiency and thus increased endurance for the ship. Overall, the technical theory behind the WR-21 engine package remains sound, albeit an extremely difficult and challenging technology to develop. However, the subsequent issues and problems surrounding the project can arguably be traced back to poor management when introducing such an innovation via an alliance business model.

Some news feed documents have been consulted to construct the major events of the Type 45 ships systems engineering life cycle and are listed chronologically in Figure 7.10 (Writer, 2016; Trevithick, 2018; Allison, 2021).

7.6.2   AILAM FOR WR-21 PROJECT

Using the 3PE–SOS methodology, each of the companies within the WR-21 project alliance can be modelled according to the 3PE elements. Next, interactions between the P's in the alliance are characterised. For reference, this is the "Build 3PE–SOS Model" step of AILAM in Figure 7.9. Risks within the individual companies can be identified and suitable mitigations can be planned.

In practice, there can be hundreds or even thousands of risks in the 3PE–SOS model. This is the "Data" step in Figure 7.9. The data are then analysed, with AHP applied to determine the priority that risks need to be addressed. This is the "3PE–SOS Enhanced AHP analysis" step in Figure 7.9. Finally, for each risk, an appropriate mitigation is developed and modelled, and as the final step of AILAM, the model cycle should be run three times to generate three iterations as follows:

1. Iteration 1 – Solve for an initial risk profile
2. Iteration 2 – Include additional scope like innovation, alliance, etc.
3. Iteration 3 – Final pass to including everything in the final risk profile

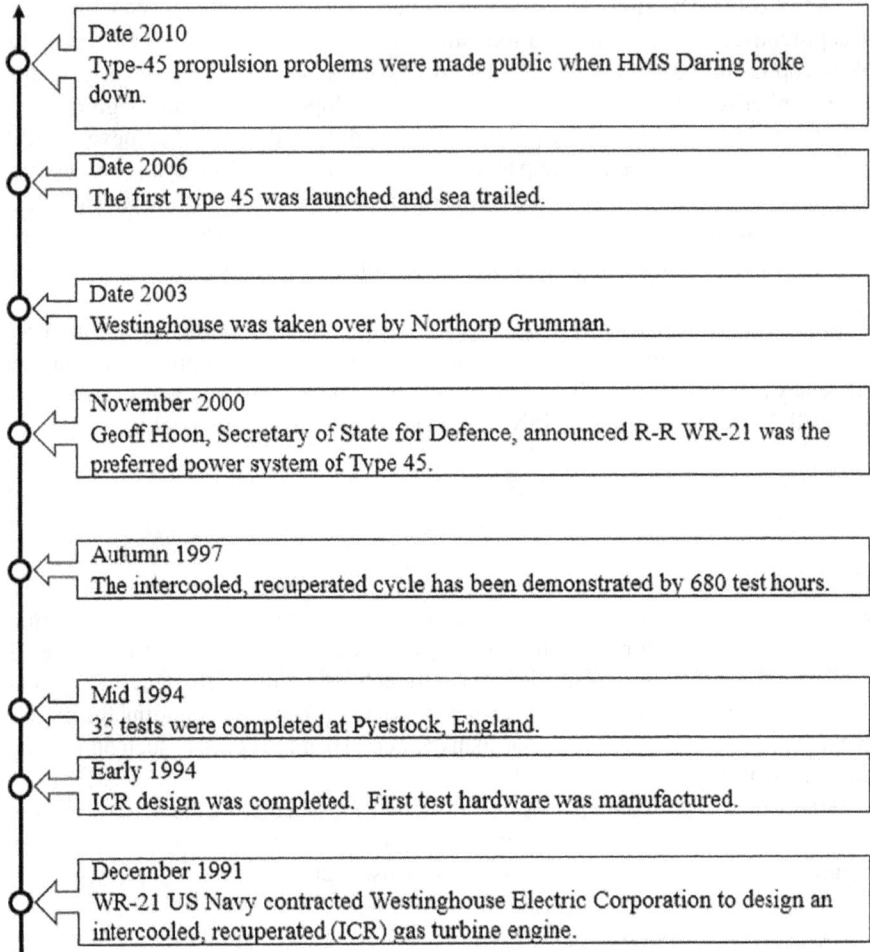

FIGURE 7.10 Major events in the systems engineering life cycle of WR-21.

The AILAM model will now be applied using the historical information as explained in the context-established stage.

7.6.3 First Iteration

Within the WR-21 alliance, both Westinghouse and Rolls-Royce needed to develop new technology as their part of the WR-21 engine package project. Any organisation attempting to bring a new innovative product to market will see a significant increase in the risks relating to the product within their organisation's own environment regardless of whether the project is part of an alliance or not.

By applying the first iteration of the 3PE-SOS methodology solely to Westinghouse as a single organisation, an emphasis on the product element has been highlighted by the inclusion of such a challenging innovation, i.e., designing and building an intercooler and recuperator. It should be noted that organisations rarely bring a completely new product to market, and this is due mainly to the severity of the risk it carries. In the majority of cases, organisations will generally only make small incremental changes/updates to an extant product to minimise their risk exposure. However, in the case of the WR-21 project, both Westinghouse and Rolls-Rolls needed to develop significant innovative technical engineering product solutions themselves, these were to be brought together to form the final engine package.

The first iteration of the 3PE model was used to identify risks that Westinghouse would face developing the intercooler and recuperator for the WR-21 engine package. By defining a topology framework, the 3PE-SOS model drives and robust and accurate risk capture process (see Figure 7.11).

Post the risk identification process, it was found that significant numbers of risks were clustering around the Product/Process elements and their interactions, this can be seen in Figure 7.11. This is clearly not unexpected considering the type of project and its technical nature. In order to give an example of the identified risks, but also maintain this section to a manageable length, three risks for each of the 3PE elements and interactions have been provided in Figure 7.11.

With a comprehensive set of risks now identified and associated with either an element or interaction within the 3PE-SOS model topology, the next challenge for Westinghouse is to establish the priority for managing risks through the life cycle. It is important to note that this first iteration is taken at the start of the project, so it is this priority modelling that will establish the initial baseline risk profile.

This phase of the modelling primarily incorporates the analytic hierarchy process (AHP). Each risk is essentially assessed with consideration of the likelihood and consequence of the risk in three values of optimistic, normal, and pessimistic situations. A graphical representation of the results for WR-21 project and how the risks are spread across the 3PE elements and interactions can be seen in Figure 7.12. This level of capture and fidelity of risks has only been possible with the use of the 3PE model, as the structure of the 3PE framework provides a systemic methodology that ensures consideration is given to all areas within the project where risks could be present.

The integration of AHP with the 3PE-SOS model is a novel approach and has established a method to assess risks for the priority of mitigation and management across different strands in the project. This is only possible because the framework of the 3PE-SOS model has provided a structure that would have allowed Westinghouse to really understand where the extreme, high medium, and low risks are located or clustering within their development of the WR-21 technology. As a result, Westinghouse should have been able to direct its efforts in mitigating, or at minimum, control the right risks initially and on through the project life cycle.

For the WR-21 project, the most significant risks reside around requirements, design process, performance, and testing. These are all fundamental to an engine development project and as this risk analysis has shown, essential to be managed and controlled from the outset. Further complications were to befall the project (which are detailed in the following section), which would bring increased challenges and further emphasise the need to robustly control and manage project risks.

Objective

Group

Sub-Group

Project Risks

Product

RID-1
Developing new technology presents technical challenges beyond the organisation's ability.

RID-2
Component design has not fully considered manufacture, build and integration.

RID-3
The size and weight of the package is greater than the initial estimate.

Risk 115

Process

RID-4
The project requirements are not clear and fully defined.

RID-5
The difference disciplines across the business are not aligned correctly to deliver the project.

RID-6
The schedule is unrealistic and cannot be achieved.

Risk 72

People

RID-7
The main SME (Subject Matter Expert) leaves the project during a key stage.

RID-8
Team dynamics are generally poor due to lack of positive leadership.

RID-9
The remuneration packages offered to staff are too low to attract or keep good talent.

Risk 27

Process-People

RID-10
There is no succession planning should key staff leave the project.

RID-11
The recruitment process takes too long to meet the project schedule.

RID-12
The company design process is too complicated and convoluted for staff to follow

Risk 22

People-Product

RID-13
The introduction of new technology is not mature enough to meet the programme schedule.

RID-14
The process to achieving product certification is not understood.

RID-15
Any purchased equipment of components does not meet project requirements.

Risk 28

Product-Process

RID-16
Key staff do not fully understand their organisations product

RID-17
There is no legacy knowledge remaining with the organization

RID-18
Staff do not have a holistic view of the product and only focus on their areas.

Risk 48

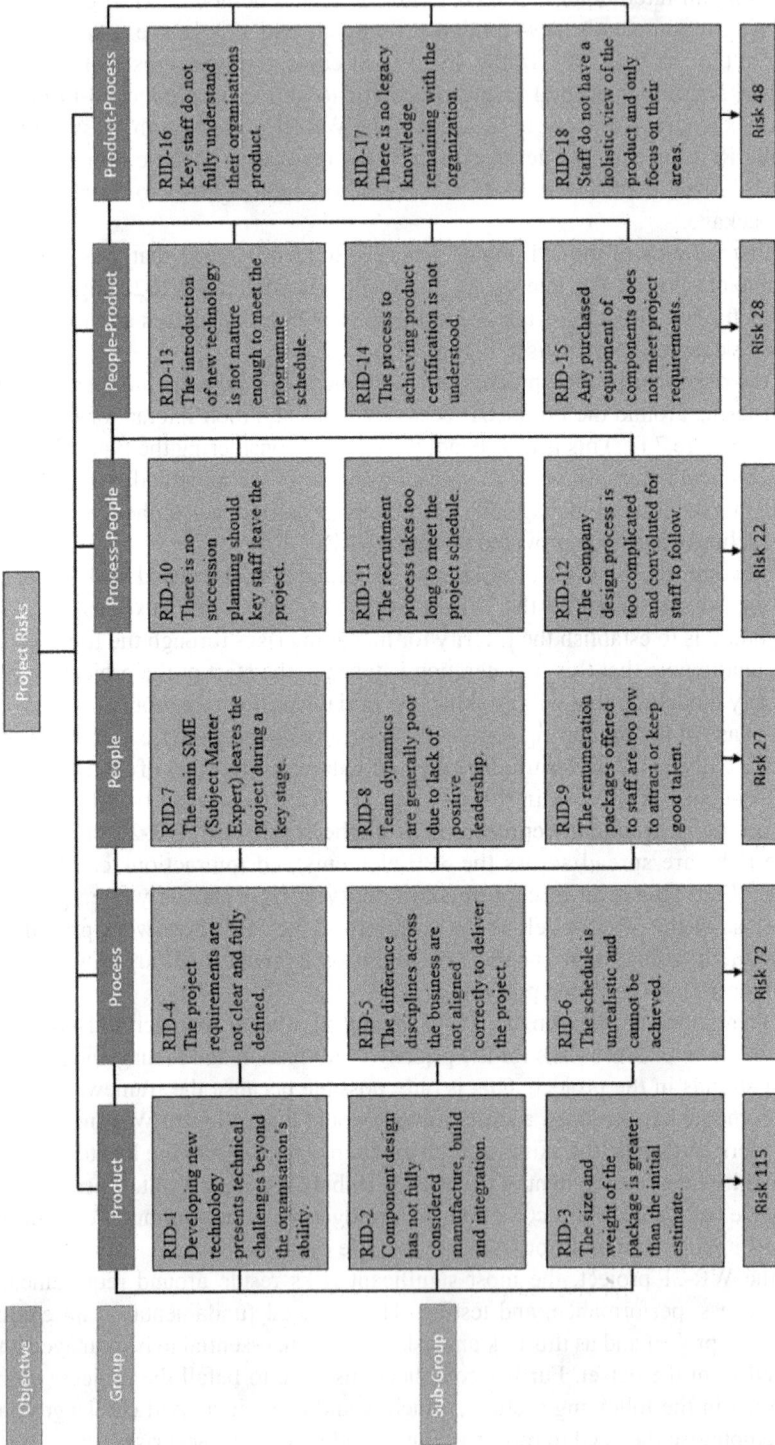

FIGURE 7.11 3PE framework of risk for Westinghouse.

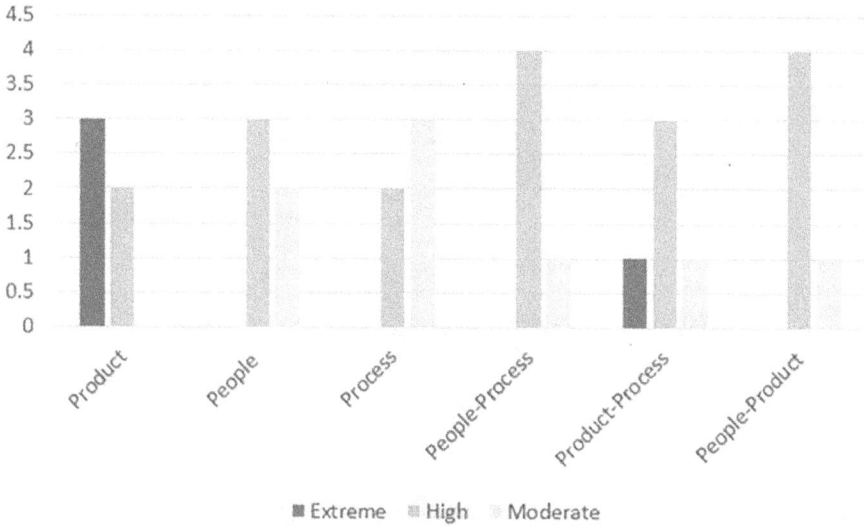

FIGURE 7.12 Distribution of Westinghouse risk levels for the WR-21 project across the 3PE framework.

7.6.4 SECOND ITERATION

Due to the complexity of the WR-21 engine package, Westinghouse, Rolls-Royce, and BAE Systems formed an alliance to develop this technology. This three-way partnership was formed to ensure that organisations with specific skills, such as gas turbine technology, were engaged and responsible (it should be noted that BAE Systems was the prime contractor to the Ministry of Defence for the overall Type-45 destroyer). However, when this alliance is examined more closely by applying the 3PE-SOS model, it becomes apparent that the potential risk pathways have increased dramatically, and this can be seen visually in Figure 7.13.

According to the 3PE-SOS model, by introducing two or more partners into an alliance environment, it is the interactions between the 3PE elements that will expand significantly within the alliance environment, whereas the risks associated with the elements themselves remain static within each of the organisation's environments. This is defined in Table 7.3.

As before, the 3PE-SOS model framework was applied to identify a set of project risks. However, this time the risk analysis included potential risks that forming an alliance has introduced to the WR-21 engine project. The unique 3PE-SOS structure defines interactions between elements or risk pathways across the alliance partners and ensures consideration is given to all possible risks. Figure 7.14 presents a comparison between the distribution of risk for a single organisation, in this case Westinghouse, and the formation of an alliance by Westinghouse, Rolls-Royce, and BAE systems.

As described earlier, the main motivation to form an alliance is to reduce and spread risk across several organisations. From Figure 7.14, it can be seen that both the elements of product and process have seen some reduction in the

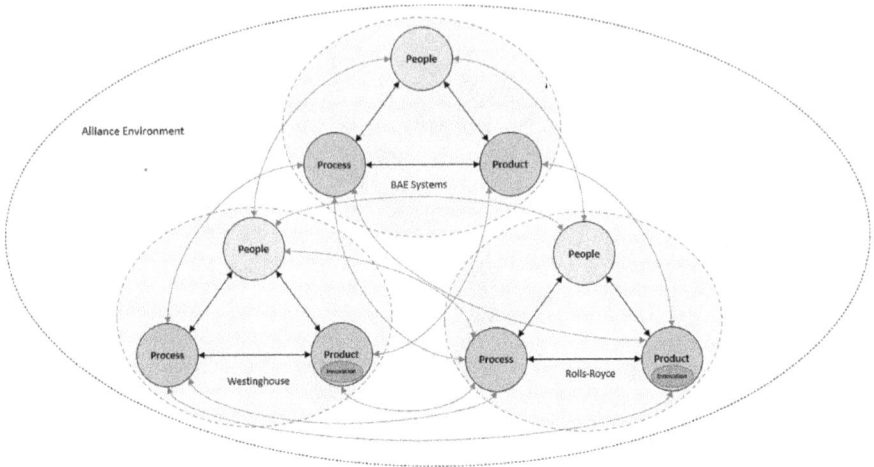

FIGURE 7.13 Alliance 3PE model that includes significant product innovation.

TABLE 7.3
Identification of Alliance Risks by 3PE-SOS Models of Westinghouse and Rolls-Royce

		Westinghouse			Rolls-Royce			BAE Systems		
		People	Product	Process	People	Product	Process	People	Product	Process
Westinghouse	People	Internal interactions			1	2	3	7	8	9
	Product				4	5			10	11
	Process					6				12
Rolls-Royce	People	Replicated			Internal interactions			13	14	15
	Product							16	17
	Process									18
BAE Systems	People	Replicated			Replicated			Internal interactions		
	Product									
	Process									

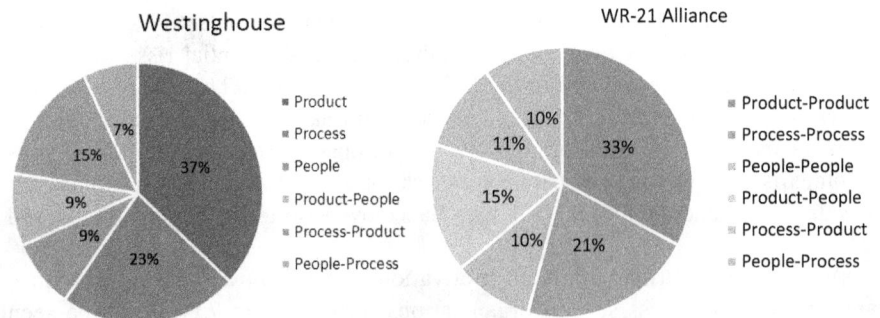

FIGURE 7.14 Distribution of risk, single organisation verses an alliance.

identified risks. However, by using the novel 3PE-SOS model, the analysis highlights an actual increase in risks located in the interactions between the elements due to the number of risk pathways introduced by the formation of an alliance. There are now circa 459 risks identified against the WR-21 alliance identified at the beginning of the project, while the sole organisation of Westinghouse had a total of 312 risks.

The results of the 3PE modelling for the WR-21 engine project as an alliance throw up a set of initial risks that are somewhat different to a sole organisation attempting to complete the project. While clearly the challenges of the product and the introduction of a new technology remain dominant, the difficulties of working with partners have become significant with the elements of process, people, and their interactions being more pronounced.

The next stage of the AILAM model was then applied to the risk set to again determine the priority for tackling the project risks. As before, the novel topology of the 3PE-SOS model highlights where risks are tending to cluster and thus allows more precise mitigations to be developed. Once the model has been run and Program Evaluation and Review Technique (PERT) has been initiated to define the severity of the risks, AHP is then deployed to determine the priority. The AHP output is shown in Figure 7.15.

With the AHP modelling complete, the priority for managing/mitigating risks within the WR-21 alliance can be determined. Again, this is a set of risks that have been defined at the infancy of the project, clearly the priority will evolve as the project moves through the life cycle. Some risks will be mitigated, some will be realised, and there will also be emergent risks to account for, and hence, the model should be applied frequently. A list of the initial risks that have been determined as high importance for the WR-21 project is presented in Table 7.4.

For the WR-21 engine package alliance, there was a watershed moment just over halfway through the project that would have significant consequences. In the early 2000s, Westinghouse was bought by the US defence giant Northrup Grumman. Upon reviewing all of Westinghouse's live projects, Northrup Grumman concluded that the WR-21 project was unsatisfactory and therefore decided to cut their losses and pull out. This decision clearly had devastating consequences for the project.

From the results of AILAM presented in Table 7.4, it can be seen that "Risk 13" identified the possibility of a partner leaving the alliance. There are also a number of other risks that will be realised, should a partner exit the alliance. For a project like WR-21, where new and innovative technology is being developed, this risk is increased as organisations face significant ongoing technical and development issues, slippage in schedule, and challenges controlling costs. If the WR-21 alliance had used the AILAM model, these risks would have been identified at the infancy of the project and mitigations/conditions imposed on cost, schedule, and technical shortfalls.

		RID1	RID2	RID3	RID4	RID5	RID6	RID7	RID8	RID9	RID10	RID11	RID12	RID13	RID14	RID15	RID16	RID17	RID18
		Product-Product			Process-Process			People-People			Process-People			People-Product			Product-Process		
Product-Product	RID1	1.000	5.000	3.000	5.000	5.000	1.000	1.000	0.333	0.333	0.333	0.333	0.333	3.000	3.000	0.200	0.200	5.000	1.000
	RID2	0.200	1.000	0.200	0.200	1.000	0.143	0.111	0.143	0.143	0.143	0.143	0.143	0.200	0.200	0.333	0.333	0.333	0.333
	RID3	0.333	5.000	1.000	0.333	5.000	0.333	0.200	0.333	0.333	0.200	0.200	0.200	0.333	0.333	0.333	0.333	0.333	1.000
Process-Process	RID4	0.200	5.000	3.000	1.000	3.000	0.200	0.143	0.143	0.333	0.200	0.200	0.200	0.333	0.333	0.333	0.333	1.000	0.333
	RID5	0.200	1.000	0.200	0.333	1.000	1.000	0.200	0.143	0.200	0.143	0.143	0.143	0.333	1.000	0.200	0.333	0.333	0.200
	RID6	1.000	3.000	3.000	5.000	5.000	1.000	1.000	3.000	5.000	3.000	3.000	1.000	1.000	3.000	3.000	1.000	0.333	1.000
People-People	RID7	1.000	9.000	5.000	7.000	7.000	5.000	1.000	1.000	1.000	1.000	1.000	1.000	5.000	1.000	5.000	3.000	1.000	1.000
	RID8	3.000	7.000	3.000	7.000	7.000	0.333	1.000	1.000	5.000	1.000	1.000	1.000	-3.000	1.000	3.000	0.333	3.000	0.143
	RID9	3.000	7.000	3.000	3.000	5.000	0.200	1.000	0.200	1.000	0.333	0.333	0.333	1.000	3.000	3.000	0.333	3.000	1.000
Process-People	RID10	3.000	7.000	5.000	5.000	7.000	0.333	1.000	1.000	3.000	1.000	0.333	0.333	3.000	3.000	3.000	0.333	5.000	5.000
	RID11	3.000	7.000	5.000	5.000	7.000	1.000	0.333	1.000	3.000	3.000	1.000	1.000	3.000	3.000	3.000	1.000	5.000	1.000
	RID12	3.000	7.000	5.000	5.000	7.000	1.000	0.333	1.000	3.000	3.000	1.000	1.000	3.000	3.000	3.000	1.000	5.000	1.000
People-Product	RID13	0.333	5.000	3.000	3.000	3.000	1.000	0.200	-0.333	1.000	0.333	0.333	0.333	1.000	1.000	1.000	0.333	3.000	1.000
	RID14	3.000	3.000	3.000	3.000	1.000	0.333	1.000	1.000	1.000	0.333	0.333	0.333	1.000	1.000	1.000	1.000	3.000	1.000
	RID15	5.000	3.000	3.000	3.000	5.000	0.333	0.333	0.333	3.000	0.333	0.333	0.333	1.000	1.000	1.000	0.333	5.000	3.000
Product-Process	RID16	5.000	5.000	3.000	3.000	1.000	1.000	0.333	3.000	3.000	3.000	1.000	1.000	3.000	3.000	3.000	1.000	5.000	5.000
	RID17	0.200	3.000	3.000	1.000	3.000	0.333	1.000	3.000	0.333	0.200	0.200	0.200	0.200	0.333	0.200	0.200	1.000	0.200
	RID18	1.000	3.000	1.000	3.000	5.000	1.000	7.000	0.200	1.000	1.000	0.200	1.000	1.000	1.000	0.333	0.200	5.000	1.000
Sum		33.467	90.000	52.400	59.867	80.000	14.743	16.197	13.829	31.676	17.752	11.086	11.886	24.400	24.667	28.267	10.800	56.000	32.210

FIGURE 7.15   AHP for the WR-21 Alliance

TABLE 7.4
Alliance risks for WR-21

Element or Interaction (3PE-SOS)	Risk Description There is a risk that...
Product-Product	When all components (products) are brought together and built into one system, the propulsion package does not perform as expected.
	One of the partners produces a substandard product.
Process-Process	Partner internal processes do not align.
	The timescales the partners are working to, do not align.
People-People	There is a personality clash between partner senior managers (e.g. CEOs).
	Staff from alliance partners take no ownership or responsibility.
	Staff try to undermine the reputation of alliance partners.
	Certain staff do not want to communicate with the alliance partners.
People – Process	Organisations lack control or direction over partner staff.
	Suitably qualified and experienced people (SQEP) cannot be resourced to work on the project.
	Partner staff are unable or unwilling to work across the alliance partners.
Process – Product	The lack of an integrated process leads to lack of holistic product understanding by partners in the alliance.
	One or more of the partners leaves the alliance and a replacement product is needed.
Product – People	Staff are unwilling to share information about their organisation's product.
	Staff from the alliance do not really understand their partner's product.

7.6.5   THIRD ITERATION

The exit of a partner from the WR-21 alliance was identified by the 3PE-SOS model at the initial stage of the WR-21 project. Mitigations for this risk should have been established before the project commenced. Westinghouse (Northrup Grumman) was developing an innovative and complicated set of parts for the engine package that are essentially bespoke (i.e., the alliance could not simply reach out to the industry for a similar commercial-off-the-shelf product). The alliance was now facing the technical challenge of being able to progress the intercooler and recuperator development, concurrently with huge pressure to complete the WR-21 engine package on time to meet the ship schedule and control costs.

This risk(s) had clearly not been addressed by the alliance to a satisfactory level, and consequently, Rolls-Royce inherited the partially complete design from Northrup Grumman and became solely responsible for the WR-21 package. This can be seen in Table 7.5, where the alliance has now reduced to two partners.

TABLE 7.5
Identification of Alliance Risks by AILAM for BAE Systems and Rolls-Royce

		Rolls-Royce			BAE-Systems		
		People	Product	Process	People	Product	Process
Rolls-Royce	People	Internal interactions			1	2	3
	Product				5		
	Process				6		
BAE systems	People	Replicated			Internal interactions		
	Product						
	Process						

While clearly the number of potential risk pathways within the alliance has reduced with the exit of Northrup Grumman, risks around technical challenges of the engine package, schedule, and cost have all dramatically increased across all the interactions. The 3PE-SOS model has identified that without dramatic intervention to mitigate these risks, the WR-21 alliance is in significant danger of failure. This means if AILAM had been available for risk assessment prior to commencing the WR-21 project, the risk of dramatic change in the alliance could have been mitigated, possibly by one of the following plans:

- A contractual agreement could have been set up such that any innovation developed as part of WR-21 should remain part of the WR-21 project. This could include all IP including the part Westinghouse was responsible for innovating.
- BAE Systems and Rolls Royce could have defined a backup plan in case the innovation was considered a failure. This backup plan might have deterred innovation development, but it could have ensured the successful completion of the project.

If the above were missed, there was still an opportunity to set up legal conditions such that the sale of Westinghouse to Northrop Grumman was allowed only if Northrop Grumman was forbidden from pulling out of the WR-21 project, otherwise heavy penalties would apply.

AILAM does not discourage innovation; instead it takes into account potential risks that might hinder innovation success and ensures the alliance is prepared and ready to take up the challenges.

7.7 CONCLUSION

This chapter has highlighted the challenges and complexity of forming or joining an alliance. As opposed to the conventional thinking of sharing risks in an alliance, it transpires that significant increases in risks are the result of the interactions between people, process, and product among the partners in the alliance. To develop the strategy for joining an alliance, one should develop the model of 3PE–SOS in two stages. First model, the individual enterprises by the 3PE-SOS model, so that internal risks are clearly identified and understood. When the individual companies join the alliance, the P's in each pair of companies are then characterised and innovation sources are identified and supported.

The favoured mitigation by organisations is to attempt to spread the risk by entering into co-innovation partnerships. While on the surface this seems reasonable, this research has detailed how this strategy actually introduces significant numbers of risk pathways the more partners that are introduced to the alliance.

AILAM for assessing innovation risks in complex product(s) (system) is an innovative methodology using 3PE–SOS development. The 3PE–SOS model has the ability to identify risks within a sole organisation's environment as well as clustering across the three main elements of the 3P model (People, Product, and Process) and the associated interactions, to ensure effort and resources are applied to the right areas for targeted mitigation.

AILAM then goes further to assess projects that are developing challenging innovations under an alliance structure. As each partner enters the alliance, AILAM demonstrates an expansion in potential risk pathways. The integrated 3PE-SOS model in AILAM is unique in identifying these pathways and this crucially offers enhanced risk identification fidelity. Furthermore, by applying AHP capability to the model, each risk can be assessed and ranked for mitigation priority throughout the project life cycle.

REFERENCES

Abhari, K., Davidson, E. J., Xiao, B. (2018). A risk worth taking? The effects of risk and prior experience on co-innovation participation. *Internet Research*, 28, 804–828.

Allison, G. (2021). All Type 45s to have received engine repairs by mid-2020s. *UK Defence Journal*.

Barmeyer, C., Mayrhofer, U. (2008). The contribution of intercultural management to the success of international mergers and acquisitions: An analysis of the EADS group, *International Business Review*, 17(1), 28–38

Baskin, K. (1995). DNA for corporations: Organizations learn to adapt or die. *The Futurist*, 29(1), Jan–Feb, 68.

Bernus, P., Nemes, L. (1996). A framework to define a generic enterprise reference architecture and methodology. *Computer Integrated Manufacturing Systems*, 9(3), 179–191.

Boyle, T.A., Scherrer-Rathje, M. (2009). An empirical examination of the best practices to ensure manufacturing flexibility: Lean alignment. *Journal of Manufacturing Technology Management*, 20(3), 346–366.

Bugshan, H. (2015). Co-innovation: The role of online communities. *Journal of Strategic Marketing*, 23, 175–186.

Chattopadhyay, S., Chan, D.S.K., Mo, J.P.T. (2010). Business model for virtual manufacturing – A human-centered and eco-friendly approach. *The International Journal of Enterprise Network Management*, 4(1), 39–58.

Chattopadhyay, S., Chan, D.S.K., Mo, J.P.T. (2011). Analysis of disaggregated corporations of the 21st century using molecular structure. *International Journal of Management*, 28(3, Pt.2), 849–866.

Chattopadhyay, S., Chan, D.S.K., Mo, J.P.T. (2012). Modelling the disaggregated value chain – The new trend in China. *International Journal of Value Chain Management*, 6(1), 47–60.

Chattopadhyay, S., Mo, J.P.T. (2010). Modelling a global EPCM (Engineering, Procurement and Construction Management) enterprise. *International Journal of Engineering Business Management*, 2(1), 1–8.

Chattopadhyay, S., Mo, J.P.T. (2011) The Pentatomic face of organizations, Ch.5, in *The New Faces of Organizations in the 21st Century*, Eds.: Mohammad Ali Sarlak, Payam Noor University, NAISIT Publishers. ISBN: 978-0-9865335-0-1, pp. 188–233.

Cook, M., Mo, J.P.T. (2019). Architectural modelling for managing risks in forming an alliance. *Journal of Industrial Integration and Management*, 4, 1–17.

Cook, M., Mo, J.P.T. (2020). Determination of the severity of risks in engineering projects using a system architecture approach. *TMCE 2020: Thirteenth International Tools and Methods of Competitive Engineering Symposium*. Dublin, Ireland.

Cook, M.C., Mo, J.P.T. (2022) Architectural approach for analysing and managing innovation in complex system design projects. *Product*, 20(1), e20210019.

Cui, V., Yang, H., Vertinsky, I. (2018). Attacking your partners: Strategic alliances and competition between partners in product markets. *Strategic Management Journal*, 39, 3116–3139.

Das, P., Verburg, R., Verbraeck, A., Bonebakker, L. (2018). Barriers to innovation within large financial services firms. *European Journal of Innovation Management*, 21, 96–112.

Goa, S., Zhang, S. (2008). Opportunism and alliance risk factors in asymmetric alliances. *IEEE Xplore Conference Series*, 1109, 680–685.

Gregor, S., Hart, D., Martin, N. (2007). Enterprise architectures: Enablers of business strategy and IS/IT alignment in government. *Information Technology and People*, 20(2), 96–120.

INCOSE (2007). Systems engineering handbook. *International Council on Systems Engineering*, 3, 32–34.

Janssen, M., Klievink, B. (2012). Can enterprise architectures reduce failure in development projects? *Transforming Government: People, Process and Policy*, 6(1), 27–40.

Keller, P. (2016). *Alliance at Risk: Strenthening European Defence in an Age of Turbulence and Competition*. Atlantic Council. https://www.atlanticcouncil.org/event/alliance-at-risk-strengthening-european-defense/

Mo J.P.T., Zhou M., Anticev J., Nemes L., Jones M., Hall W. (2006). A study on the logistics and performance of a real 'virtual enterprise'. *International Journal of Business Performance Management*, 8(2–3), 152–169.

Mo, J.P.T. (2012). Performance assessment of product service system from system architecture perspectives. *Advances in Decision Sciences*, 2012, 19, Article ID 640601.

Mo, J.P.T., Beckett, R.C. (2019) Architectural modelling of transdisciplinary system with inherent social perspectives. *Journal of Industrial Integration and Management: Innovation and Entrepreneurship*, 4(4), 1950012 (19 p.).

Mo, J.P.T., Beckett, R.C. (2018). *Engineering and Operations of System of Systems*, Taylor and Francis, ISBN 978-113-863473-2, 269 p. https://www.taylorfrancis.com/books/mono/10.1201/9781315206684/engineering-operations-system-systems-john-mo-ronald-beckett

Mo, J.P.T., Nemes, L. (2010). Issues using EA for merger and acquisition, in *Coherency Management: Architecting the Enterprise for Alignment, Agility, and Assurance*, Eds.: Gary Doucet, John Gøtze, Pallab Saha and Scott Bernard, AuthorHouse, Chapter 9, ISBN 978-143899-60783, pp. 235–262.

Oberg, C., Henneberg, S.C., Mouzas, S. (2007). Changing network pictures: Evidence from mergers and acquisitions. *Industrial Marketing Management*, 36(7), 926–940.

Ombrosi, N., Casprini, E., Piccaluga, A. (2019). Designing and managing co-innovation: The case of Loccioni and Pfizer. *European Journal of Innovation Management*, 22, 600–616.

Shenhar, A.J., Holzmann, V., Melamed, B., Zhao, Y. (2016). The challenge of innovation in highly complex projects: What can we learn from Boeing's Dreamliner experience? *Project Management Journal*, 47, 62–78.

Sireesh, S. (2006). Embedding enterprise compliance and risk management culture into organizational DNA. *InFinsia*, 120(1), Feb–Mar, 48–51.

Spear, S., Bowen, H.K. (1999). Decoding the DNA of the Toyota Production System. *Harvard Business Review*, 77(5), Sep–Oct, 96–106.

Towill, D.R. (2007). Exploiting the DNA of the Toyota Production System. *International Journal of Production Research*, 45(16), Aug, 3619–3637.

Trappey, A.J.C., Elgh, F., Hartmann, T., James, A., Stjepandic, J., Trappey, C.V., Wognum, N. (2017). Advanced design, analysis, and implementation of pervasive and smart collaborative systems enabled with knowledge modelling and big data analytics. *Advanced Engineering Informatics*, 33, 206–207.

Trevithick, J. (2018). Royal Navy will retrofit Type 45 destroyers to keep them from breaking down. *The Drive*. https://www.thedrive.com/the-war-zone/19509/royal-navy-will-retrofit-type-45-destroyers-to-keep-them-from-breaking-down

Vella, R., Chattopadhyay, S., Mo, J.P.T., (2009). Six sigma driven enterprise model transformation. *International Journal of Engineering Business Management*, 1(1), 1–8.

Vesterager, L.B. Larsen, J.D. Pedersen, M. Tølle, P. Bernus (2000) Use of GERAM as basis for a virtual enterprise framework model. *Fourth International Working Conference on the Design of Information Infrastructure Systems for Manufacturing (DIISM 2000)*, IFIP TC5 WG5.3/5.7/5.12, 15-17 November, 2000, Melbourne, Victoria, Australia, pp. 75–82.

Watson, J.D., Crick, F.H.C. (1953). Molecular structure of nucleic acids. *Nature*, 171(4356), April 25, 737–738.

Weiler, C., Chiprich, J. (1997). WR-21 intercooled recuperated gas trubine system overview & update. *The ASME ASLA '97 Congress and Exhibition*, Singapore.

Writer, T. (2016). Putting the Type 45 propulsion problems in perspective. *Navy Lookout – Independent Royal Navy News and Analysis*. https://www.navylookout.com/putting-the-type-45-propulsion-problems-in-perspective/

Young, B., Hosseini, A., Lædre, O. (2016). The characteristics of Australian infrastructure alliance projects. *Energy Procedia*, 96, 833–844.

Zhu, F., Jiang, M., Yu, M. (2019). The role of the lead firm in exploratory projects. *International Journal of Managing Projects in Business*, 13, 312–339.

Zhuang, L., Williamson, D., Carter, M. (2018). Innovate or liquidate – Are all organisations convinced? A two-phased study into the innovation process. *European Journal of Innovation*, 21, 96–112.

8 Digital Manufacturing

Annelize Botes
Nelson Mandela University

This chapter serves as an introduction to digital manufacturing for the management of technology practitioners. The main concepts associated with digital manufacturing are explained with some examples of systems, industry standards, and definitions given. The aim of this chapter is to educate the reader about the importance of digital manufacturing within the engineering design and manufacturing environment as well as its associated technologies. With the international drive for the digitalisation of industries, it is of utmost importance to consider digital manufacturing and its human, technological, and economic challenges from a strategic perspective, both from within manufacturing organisations and at an industry-wide level. The topics covered in this chapter are graphically displayed in Figure 8.1.

8.1 INTRODUCTION

Conventional manufacturing methods are in-line processes in which the product is designed, and the drawings are forwarded to the production line for manufacturing of a prototype, whereas digital manufacturing is a cyclic process in which the product is designed and refined using computer-aided design and prototyping simulation software. During the simulation process, it is relatively easy to identify required componentry and material resources, production processes, and associated technology, to determine the feasibility of manufacturing the product. It also enables the design of an efficient supply chain for effective inventory control (Paritala et al., 2017).

According to TWI (The Welding Institute, 2022b), digital manufacturing can be explained as the application of computer systems to:

- Manufacturing services,
- Supply chains,
- Products, and
- Processes.

Digital manufacturing is an area within product lifecycle management (PLM) that establishes collaboration between various phases of the product lifecycle (Paritala et al., 2017). It is more pronounced during the development stages of a product, i.e., new product development, and incorporates concepts such as design for manufacturability, computer-integrated manufacturing and flexible manufacturing, potentially

 DOI: 10.1201/9781003374879-8

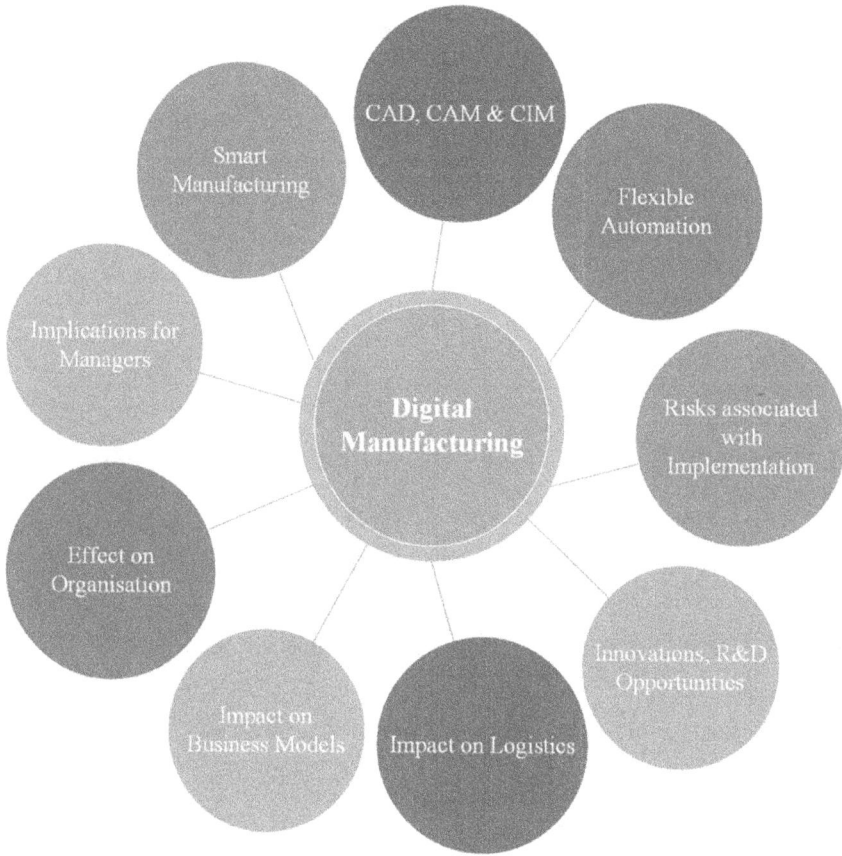

FIGURE 8.1 Sections covered in Chapter 8.

using artificial intelligence, 3D printing, human-machine interaction, automation, and robotics. In essence, digital manufacturing technologies link systems and processes across all areas of manufacturing to create an integrated approach (from design to production, servicing, and maintenance of the final product/s with the potential for efficient recycling of materials in the context of a future circular economy). These technologies also allow companies to model and simulate manufacturing processes and ensure required product quality before production begins. They expedite decision-making that results in cost savings and reduced time to market for the final product.

The three types of digital manufacturing foci according to their application field and purpose are:

1. Design-centred (product engineering),
2. Production-centred (process engineering), and
3. Control-centred (production/manufacturing engineering) (Choi et al., 2015).

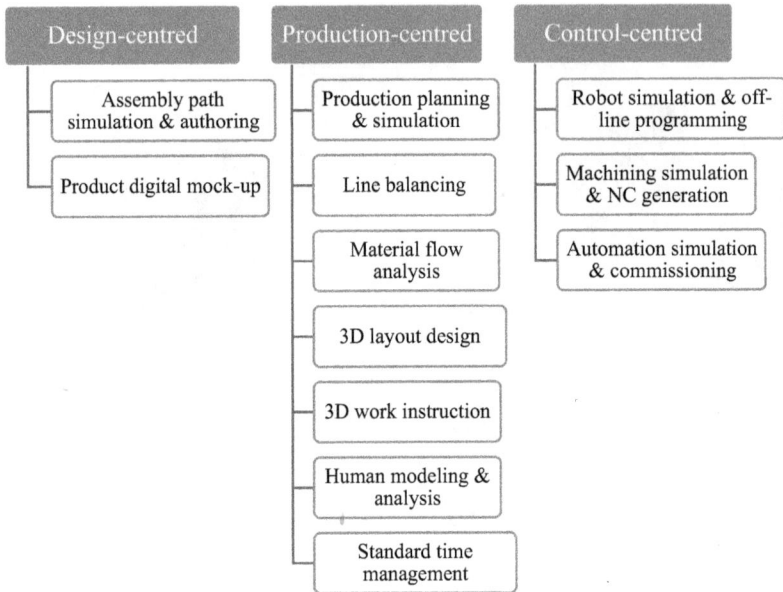

FIGURE 8.2   Digital manufacturing classification according to function.

Figure 8.2 displays the main tasks of the various functional categories of digital manufacturing.

The main driving forces for companies to adopt digital manufacturing technologies are (Paritala et al., 2017):

- Time to market (short product development time),
- Improved productivity (improved quality and reduced scrap rate),
- Managing cost, and
- Product customisation.

According to the European Factories of the Future Research Association (EFFRA), digital manufacturing platforms are described as:

> manufacturing platforms that are enabling the provision of services that support manufacturing in a broad sense. The services that are enabled by digital manufacturing platforms are associated to collecting, storing, processing, and delivering data. These data are either describing the manufactured products or are related to the manufacturing processes and assets that make manufacturing happen (material, machine, enterprises, value networks and – not to forget – factory workers.

(European Factories of the Future Research Association, 2016)

A simplified generic digital manufacturing framework is shown in Figure 8.3, which shows the links between the historical factory, current factory, and envisaged future virtual factory.

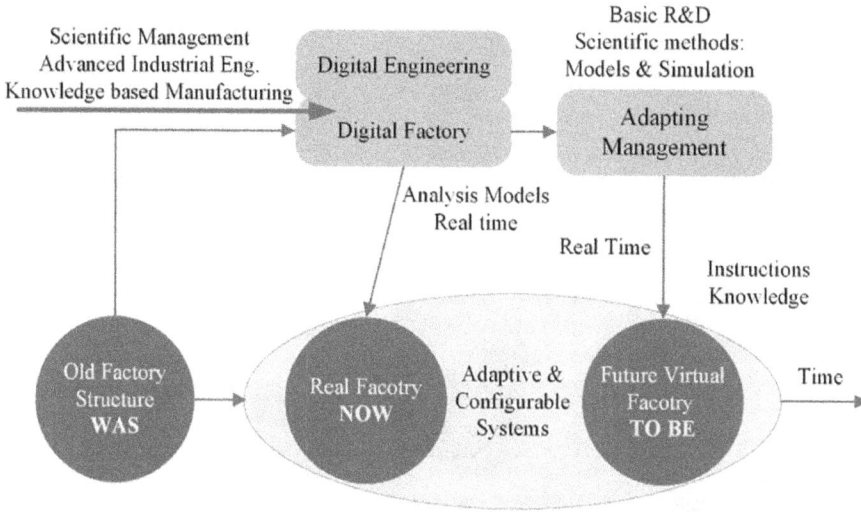

FIGURE 8.3 Example of a generic digital manufacturing framework. (Westkämper, E., Digital manufacturing in the global era. In P. Cunha & P. Maropoulos (Eds.), *Digital Enterprise Technology*. Springer US, 2007. https://doi.org/10.1007/978-0-387-49864-5_1. With permission.)

8.2 CAD, CAM, AND CIM

Traditionally, computer-aided design (CAD) and computer-aided manufacturing (CAM) systems were used for the design of virtual models and physical products with the use of numerical control (NC) machines. The automotive and aviation industries were the leaders in the development of computer-integrated manufacturing (CIM) to control production processes. This included the use of data communication, robotics, and automation. The integration of total quality management (TQM), just-in-time (JIT) manufacturing, concurrent engineering (CE), and lean manufacturing (LM) into CIM leads to a revolution in the manufacturing sector (Zhou et al., 2011). Figure 8.4 shows the interaction between CAD, CAM and CIM in a typical factory operation (Mourtzis et al., 2022).

8.3 SMART MANUFACTURING (INTELLIGENT MANUFACTURING)

Smart manufacturing generally refers to manufacturing systems that are completely integrated and collaborative and respond in real time to comply with the changing environments of the industry landscape, their supply networks, and customer requirements (Yang et al., 2023). The most significant enabling technologies associated with smart manufacturing are:

1. Immersive technologies,
2. Additive manufacturing (AM),
3. Big data analytics,

CAM
Cost Estimation
Computer Aided Process
Planning
Numerical Control Part
Programming
Computerised Work
Material Resource Planning
Capacity Planning

CAD
Geometric Modelling
Engineering Analyis
Design Review & Evaluation
Automation Drafting

Design

Manufacturing
Planning

Factory
Operation

Computer
Integrated
Manufacturing
System (CIM)

Business
Function

Manufacturing
Control

Computerised Business
System
Order Engty
Accounting
Payroll

CAM
Process Control
Process Monitoring
Shop Floor Control
Computer-aided Inspection

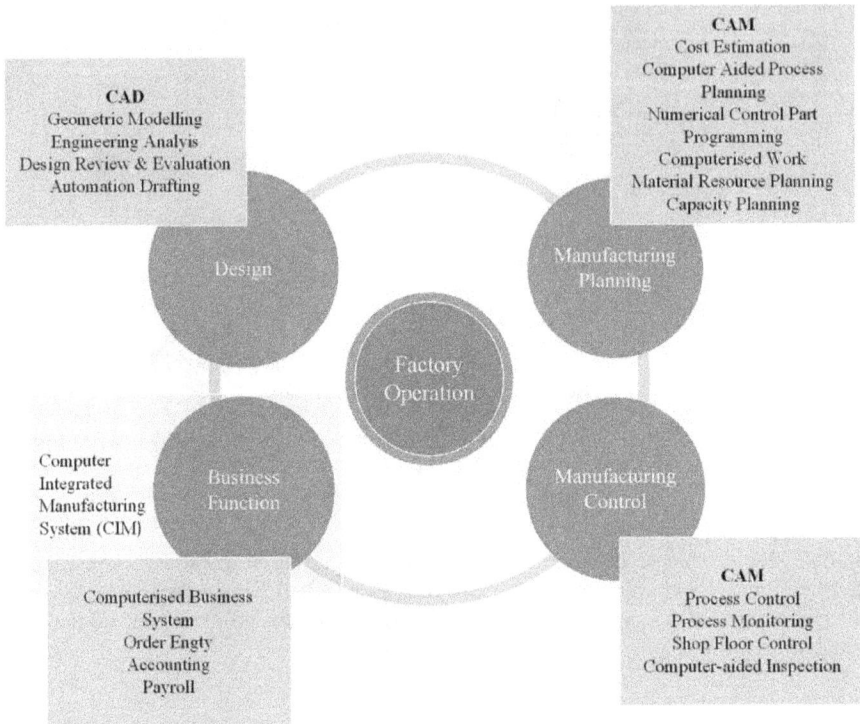

FIGURE 8.4   Interaction between CAD, CAM, and CIM in a factory operation. (Mourtzis, D., Angelopoulos, J., & Panopoulos, N., Digital manufacturing. In *The Digital Supply Chain.* Elsevier, 2022. https://doi.org/10.1016/B978-0-323-91614-1.00002-2. With permission.)

4. Industrial Internet of Things (IIOT),
5. Artificial intelligence (AI),
6. Digital Twin (DT) (Sahoo & Lo, 2022), and
7. Machine Learning (ML) (Küfeoğlu, 2022a).

Industry 4.0 is based on cyber-physical systems (CPS), and the integration of physical processes, storage systems, and production facilities coordinates and exchanges information and monitors each other (Colombo et al., 2017).

8.3.1   IMMERSIVE TECHNOLOGIES

The use of immersive technologies refers to techniques that combine the senses of sight, sound, and touch into the digital experience of product design. There are currently three types of immersive technologies available in the market, i.e., virtual reality, augmented reality, and mixed reality. Virtual reality (VR) is defined as the computer-generated simulation of a three-dimensional image or environment that can be interacted with in a seemingly real or physical way by a person using special electronic equipment, such as a helmet with a screen inside or gloves fitted with sensors, whereas augmented reality (AR) involves technologies that superimpose a computer-generated image on a user's view of the real world, thus providing a composite view (Oxford University Press, 2022). Mixed reality (MR) is a computer-generated environment in which elements of a physical and a virtual environment are combined.

VR helps to visualise the product design from the initial stage without a physical product, which reduces the prototyping time and costs, as the design engineer can make product customisations before manufacturing starts. AR, on the contrary, acts as an interface that improves human-machine interactions, whereby operators can evaluate and react appropriately to the manufacturing process in real time. MR devices are often used to train operators on the machinery they will be working on, resulting in a highly skilled workforce prior to applying their skills to the real-world machine (Sahoo & Lo, 2022).

8.3.2 ADDITIVE MANUFACTURING (AM)

Also called additive layer manufacturing (ALM) or 3D printing, AM is a computer-controlled method that creates three-dimensional objects by depositing materials in layers (The Welding Institute, 2022). Due to the technologies involved (e.g., robot arm and external axis), the process lends itself to be remotely controlled online. Since AM only uses the material required to build the object with very limited post-processing machining required, the process leads to time and cost saving during the R&D and prototyping stages of designs (Song et al., 2022).

According to ISO/ASTM 52900 *Additive Manufacturing – General Principles – Terminology*, AM processes can be classified into seven main categories as listed in Table 8.1 (American Society for Testing and Materials (ASTM), 2015). Of all the AM processes, powder-bed fusion has shown the greatest potential for AM of metallic components due to the ability to be accurately controlled to manufacture components in intricate detail (Korpela et al., 2020).

In AM processes, component complexity does not add to the cost of manufacture compared to subtractive manufacturing processes. Therefore, from a material usage point of view, AM processes are more economical, although some design and programming software-related costs might add cost in the initial stages of the manufacturing process. Another advantage of using AM in component manufacture is the minimisation of waste, thereby contributing to the sustainability of raw materials. AM also allows for the ease of design customisation without going through a time-consuming setup and production line change as with conventional subtractive manufacturing processes.

According to the Report on Global Markets for 3D Printing (McWilliams, 2021), consumer goods manufacturing was the largest end-user of AM technologies in 2020, accounting for revenues of US$2.7 billion (17.4% of the total), followed by the automotive sector at US$2.6 billion (16.6%), with medical and dental products ranked third, with sales of almost US$2.6 billion (16.4%). They expect the market to grow at a compound annual growth rate (CAGR) of 23.5% and to exceed US$56.1 billion by the end of 2026 (McWilliams, 2021).

TABLE 8.1

AM Process Categories According to ISO/ASTM 52900:2015

AM Process Categories	Explanation
Binder jetting	Process in which a liquid bonding agent is selectively deposited to join powder materials.
Directed energy deposition	Process in which focussed thermal energy is used to fuse materials by melting as they are being deposited.
Material extrusion	Process in which material is selectively dispensed through a nozzle or orifice.
Material jetting	Process in which droplets of build material are selectively deposited.
Powder bed fusion	Process in which thermal energy selectively fuses regions of a powder bed.
Sheet lamination	Process in which sheets of material are bonded to form a part.
Vat photopolymerisation	Process in which liquid photopolymer in a vat is selectively cured by light-activated polymerisation.

American Society for Testing and Materials (ASTM), *Standard Terminology for Additive Manufacturing – General Principles – Terminology1*. ASTM International, 2015. With permission.

Although AM is a promising manufacturing technology, there are some limitations and challenges that need to be considered when looking at incorporating it into a production system. The main limitations and challenges include the following:

- Low production volumes. Traditional subtractive manufacturing is still better suited for high-volume production of components. AM is better suited for low- to medium-volume production.
- Restricted material options compared to traditional manufacturing processes.
- Limited size of objects that can be manufactured. AM is generally better suited to produce smaller and more complex parts. The size of the components is constrained by the AM equipment currently available in the market.
- High investment cost of AM equipment, specialised design software, and raw materials.
- Limited standards for AM parts to ensure quality and acceptable level of component performance, especially in safety-critical applications.

8.3.3 BIG DATA ANALYTICS

Big data analytics involves the collection and analysis of large amounts of data from a wide range of sources including the various production units, customer feedback, market analyses and product request systems, material specifications, and performance characteristics, which assists in real-time decision-making for smart manufacturing. This benefits the manufacturer in the identification of resources required for quality production and the ongoing assessment of product performance and potential causes of product failures in real time. The application of big data analytics can also provide enhanced data-driven marketing for predictive manufacturing (Phuyal et al., 2020). There is no single definition for the term "Big Data", but it is generally accepted to comprise structured data found in organisational databases and unstructured data created by communication technologies. Big data encompass huge collections of complicated data sets that are too immense for traditional data processing tools to evaluate, manage, and record in the needed time scale (Küfeoğlu, 2022a). Traditionally, there were three V's associated with big data, i.e., volume, velocity, and variety, but recently some more V's have been added to adequately characterise big data (Bigelow & Botelho, 2022). Table 8.2 gives a summary of the six V's associated with big data.

With the development and integration of information technologies, the manufacturing industry has seen many improvements in connectivity in every stage of the production process making it easier for managers to implement process control and quality assurance (Küfeoğlu, 2022a).

8.3.4 INDUSTRIAL INTERNET OF THINGS (IIoT)

IIoT is a sub-division of IoT (Internet of Things) with a focus on the manufacturing and production environment. IoT architecture is still being developed but has already several characteristic layers/levels, that is, business layer, application layer, middleware layer, network layer, and perception layer. A brief description and function of each layer are summarised in Table 8.3.

IIoT enables the inter-connection of multiple physical and electronic components such as sensors and actuators, potentially located across multiple sites, using internet cloud computing and communications technology. This enables

TABLE 8.2
The Six V's Associated with Big Data

	Traditional			New	
Volume	**Variety**	**Velocity**	**Veracity**	**Value**	**Variability**
The amount of date from myriad sources.	The types of data: Structured, semi-structured, and unstructured.	The speed at which big data is generated.	The degree to which big data can be trusted.	The business value of the data collected.	The ways in which the big date can be used and formatted.

Bigelow, S., & Botelho, B., *Data Management*. The Ultimate Guide to Big Data for Businesses, 2022. https://www.techtarget.com/searchdatamanagement/definition/big-data. With permission.

TABLE 8.3
Summary of Functions of IoT Layers

Business layer	Connects with the IoT network and generates business models using information from the application layer.
Application layer	Global management of the whole program and receives information from the middleware layer.
Middleware layer	Transfers information from the network layer to the database. Regular data analysis and computations. Managing service between IoT and apps. Communicates with the local data hub. Makes automatic judgements based on the results obtained.
Network layer	Wired or wireless communication. Transfer information from the sensors to the processing unit reliably.
Perception layer	Categorise things and collect information about them using sensor tools. Artefacts of the physical environment and sensor equipment, e.g., RFID tags, barcodes, and infrared sensors.

Küfeoğlu, S., Emerging technologies. In *Emerging Technologies. Sustainable Development Goals Series.* Springer, Cham, 2022a. https://doi.org/10.1007/978-3-031-07127-0_2. With permission.

interaction, cooperation, and control of each connected component in a production system to reach a common goal such as production planning, predictive maintenance, fault finding, improved human-machine interaction, and intelligent process control (including material optimisation). This system can also be used in digital presentations of factories, products, and processes for marketing purposes (Phuyal et al., 2020).

8.3.5 ARTIFICIAL INTELLIGENCE (AI)

Artificial intelligence is defined as the ability of a digital computer or computer-controlled robot to perform tasks normally undertaken by intelligent beings (Copeland, 2022). The key components of AI are shown in Figure 8.5 (Kanade et al., 2022).

Artificial intelligence (AI) is facilitating higher value-added production volumes by accelerating the integration of manufacturing and information communication technologies. AI was in essence developed to support human intelligence by AI techniques, such as perception, machine learning, and reinforcement learning, and AI-enabled applications, such as computer vision, natural language processing, and intelligent robotics (Wan et al., 2021).

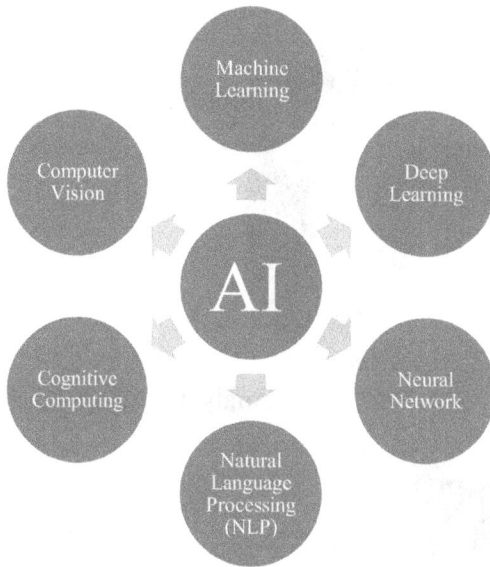

FIGURE 8.5 Key components of AI. (From Kanade et al., 2022. Used with permission.)

8.3.6 DIGITAL TWINS (DT)

A virtual model designed to precisely reflect a physical object is known as a digital twin. An object is equipped with various sensors in areas that are critical to its functionality. These sensors then produce data sets of the physical object's performance, which is then transmitted to a processing system and applied to the digital copy of the object (IBM, 2023). According to the International Organization for Standardization (ISO), a digital twin is defined as "a fit-for-purpose digital representation of an observable manufacturing element (OME) with synchronization between the OME and its digital representation. OMEs include personnel, equipment, materials, manufacturing processes, facilities, environment, products in a manufacturing environment" (International Organization for Standardization, 2021).

Recent advances in digital technologies have led to the realisation of using digital twins in manufacturing; however, there is a lack of protocols and implementation frameworks, which has become a hurdle to the wide adoption of digital twin technology (Shao et al., 2023). Figure 8.6 shows the basic working principle of a digital twin (Küfeoğlu, 2022a).

The motivation to use digital twins (to reach business goals) is focussed around five main areas (Arnautova, 2020):

- Risk evaluation and manufacturing times are both accelerated.
- Accurate predictive maintenance regimes.
- Synchronised monitoring remotely.
- Enhanced association.
- Making profitable financial choices.

Table 8.4 gives a concise summary of the most common questions regarding digital twins (Küfeoğlu, 2022a).

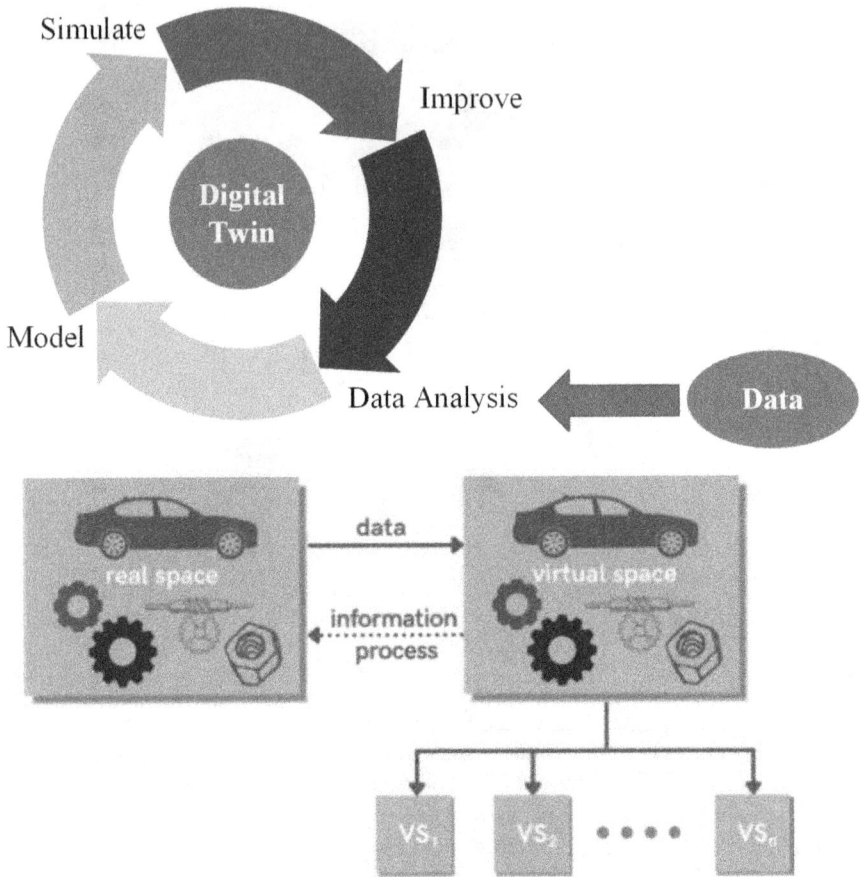

FIGURE 8.6   Basic working principle of a digital twin.

TABLE 8.4
Questions and Answers Pertaining to Digital Twins

What is a digital twin?	A collection of processes that simulate the behaviour of a physical system in a virtual system that receives real-time input to update itself throughout its lifespan.
	A digital twin duplicates the physical system to detect failures and modify opportunities, suggest real-time measures for optimising unpredictable situations, and monitor and evaluate the operational profile system.
Where is it used?	Healthcare, city management, maritime and shipping, manufacturing, aerospace, and AR/VR.
Why should it be used?	Digital twins can help businesses enhance their date-driven decision-making processes substantially. Business utilises digital twins to evaluate the capabilities of physical assets, adapt to changes, enhance operations, and add value to systems by connecting them to their real-world versions at the edge.
Who is doing digital twins?	Microsoft Azure, Ansys twin builder, Siemens PLM, Akselos, GE Predix, Aveva.

8.3.7 Machine Learning (ML)

Machine learning is a subset of artificial intelligence and can be explained as the use and advancement of computer systems that can learn and adjust without following specific and clear instructions, by using algorithms and statistical models to evaluate and make judgements from patterns in data (Oxford University Press, 2022).

Research in the field of machine learning has undergone rapid growth in the last decade and has changed the way data-driven decision-making is done in the fields of physical and social sciences. A large part of data-driven decision-making is the need to forecast the future behaviour of a system based on historical time-series data. Machine learning methods aim to learn a non-linear function mapping of stochastic historic input data to a forecasted output value. Methods such as support vector regression (SVR), random forest (RF), and artificial neural networks (ANN) are the simplest existing methods that have shown abilities in learning these non-linear functions (Wicaksono et al., 2021). The relationship between artificial intelligence (AI), machine learning (ML), neural networks (NN), and deep learning (DL) is shown in Figure 8.7 (Küfeoğlu, 2022a).

Machine learning techniques are currently being employed in various digital manufacturing techniques such as directed energy deposition (DED). DED is a suitable process for the manufacture of complex components, usually associated with high value, because the process has high build rates, manufactures near-net-shape parts, can create strong dense parts, and can handle a wide range of materials. As examples, ML techniques were used to predict the tensile strength for samples produced by DED (Cooper et al., 2023), while other researchers have focused on developing a strategy involving machine control for error compensation, as well as domain expert knowledge to identify input-output relations and supervise machine learning in order to optimise the DED process (Gröning et al., 2023).

ARTIFICIAL INTELLIGENCE (AI)

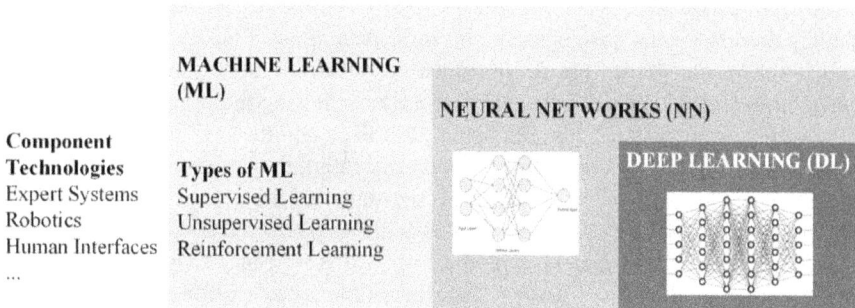

FIGURE 8.7 Relationship between AI, ML, NN and DL. (Küfeoğlu, S., Emerging technologies. In *Emerging Technologies. Sustainable Development Goals Series*. Springer, Cham, 2022a. https://doi.org/10.1007/978-3-031-07127-0_2. With permission.)

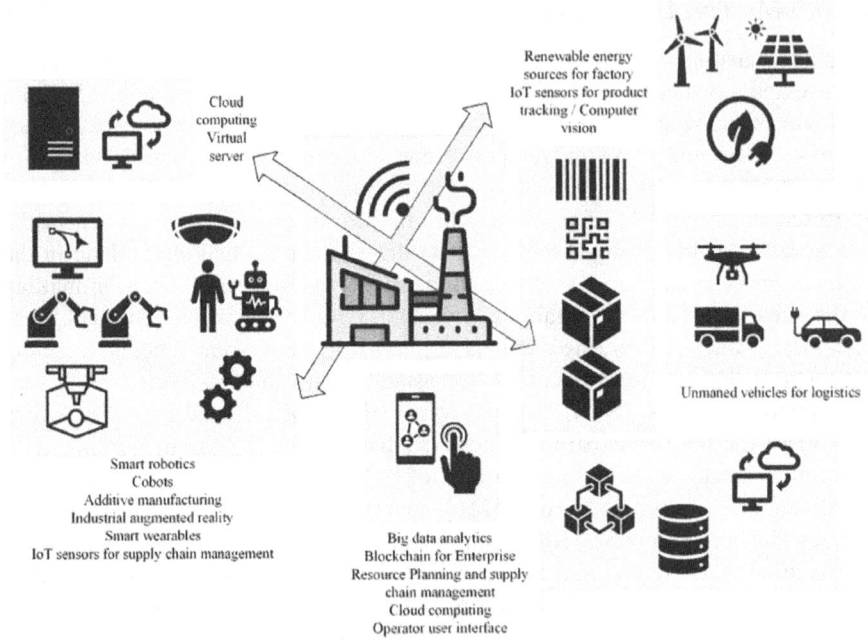

FIGURE 8.8   Elements of a smart factory.

8.3.8 THE SMART FACTORY/FACTORY OF THE FUTURE

Simply stated, the aim of smart manufacturing is to take advantage of new technologies to make processes more economical and productive. Figure 8.8 is an illustration of the various intelligent systems associated with a smart factory with descriptions of elements of the factory of the future also given.

8.4   FLEXIBLE AUTOMATION

The key drivers for the development and implementation of flexible automation for manufacturing companies are to minimise the downtime associated with product changeovers and to keep a diverse range of products flowing through their production lines (Miller & Hannifin, 2017). As opposed to fixed automation (which is designed to produce a single product repetitively and efficiently), flexible automation involves the seamless changeover in a process by the touch of a button. Flexible automation uses electromechanical automation that achieves positional control for quick and repeatable process changeovers, which allows a diverse range of products to flow through the production line with very little downtime. The cost-effectiveness of fixed automation vs. flexible automation is shown in Figure 8.9. The IIoT and the access to new robotic technologies, such as collaborative robots, allow for flexible innovative automation designs.

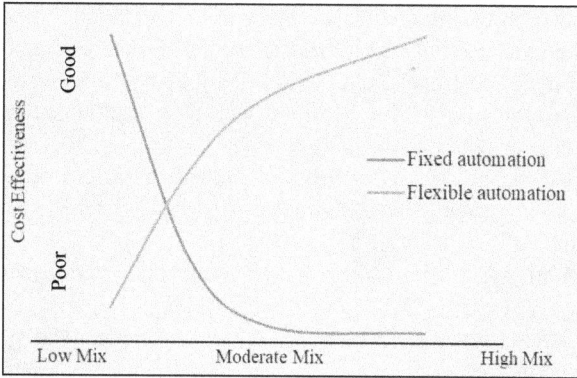

FIGURE 8.9 Comparison of cost-effectiveness of fixed vs. flexible automation.

8.4.1 New Robotic Technologies

According to ISO 8373:2021 Robotics – Vocabulary, an industrial robot is defined as an "automatically controlled, reprogrammable, multipurpose manipulator, programmable in three or more axes, which can be either fixed in place or mobile for use in industrial automation applications" (International Organization for Standardization, 2021). Robotics is a fundamental element in digital manufacturing and has played a key role in the development and use of automation. The motivation to use robots as part of a production line is due to their excellent capabilities in terms of speed, accuracy, repeatability, and flexibility, which leads to the manufacture of high-quality products. A further benefit when using robotic systems during manufacturing is their ease of online monitoring and control (Altus Market Research, 2022).

Recent advances include incorporating artificial intelligence towards semi- and fully autonomous robotic systems, e.g., transfer learning and imitation learning. Robotic systems have been in use in industry for many years, and it has recently become critical to have digital twins for robots, especially in practical scenarios where there are multi-robot setups or those that require safe human-robot interaction (HRI) or complex human-robot collaboration (HRC) (Huang et al., 2021).

The implementation of digital technologies within factories allows the use of autonomous mobile robots (AMRs) on assembly lines. AMRs are equipped with sensors and cameras to navigate their environments. The information gathered by AMRs can be used in predictive data systems and allows manufacturers to make more informed decisions when developing their plants. Cobots are collaborative robots that can safely interact with humans. Robotic systems that do not interact with humans are usually contained within an enclosure due to the high speed at which they operate. Due to the presence of humans in a cobot environment, there is a need for high signalling, high bandwidths, low latency, and rapid decision-making capabilities, which are needed to ensure a safety-critical environment. Further advances in robotics include methods to develop new operations more reliable and secure. Some of the new generations of robotics include autonomous robots, cobots, interactive autonomous smart robots, humanoids, mobile robots, and cloud robots (Javaid et al., 2021). Although logistics robots are outside the scope of industrial robotics, they play an integral role in international trade. Currently, logistics robots are used in large warehouses due to the ease with which their workflows can be set up and changed with the assistance of robotic data cloud systems (Li et al., 2020).

Figure 8.10 shows the interrelationship between the various advanced robotics with Table 8.5 summarising the advantages of robotic systems in manufacturing (Küfeolu, 2022a).

Advanced robotic systems are comprised of elements on various levels, such as:

- Physical system that is defined by materials and mechanical systems such as gears and motors.
- Control and measuring systems.
- Electronic systems that connect various sensors, actuators, and controllers.
- Computational systems, e.g., real-time operating systems.

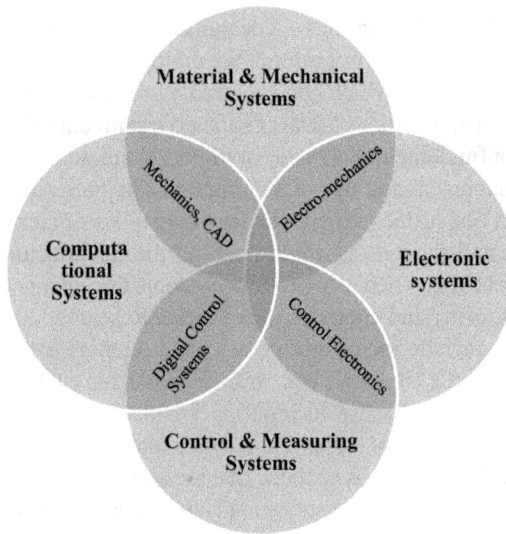

FIGURE 8.10   Interrelationship between robotic subsystems.

TABLE 8.5
Advantages of Robotic Systems

Advantages of Robotics		
Productivity	Safety	Savings
• Higher quality	• Doing hazardous jobs	• Time saving
• Fewer errors	• Carrying heavy loads	• Consuming less material
• More precise		

TABLE 8.6
Core Properties of Industrial & Advanced Robotics

		Industrial Robotics	Advanced Robotics	New Hazards Examples
Autonomy	Robot control	Automatic	Decisional autonomy	Hazardous decisions
	Workspace	Structured	Non-structured (uncertainties)	Adverse situations/ uncertainties in perception
Collaboration	Motion	No robot motion in human presence	Simultaneous motion (human & robot)	Bad synchronisation between human & robot Non-human-legible movements
	Human-robot closeness	Human is far	Human is close/ physical interaction	Collisions, contact forces too high
	Human-robot communication	Remote device	Advanced interaction (cognitive)	Mode confusion/ communication errors
Task	Mechanical architecture	Heavy/stiff/powerful	Light/compliant/ limited power	Precision hazards/energy storage due to compliance
	Task complexity	Mono-function	Multi-function	Safety rules not adapted (diverse and evolving rules)

Guiochet, J., Machin, M., & Waeselynck, H., *Rob. Auton. Syst.*, 94, 2017. https://doi.org/10.1016/j.robot.2017.04.004. With permission.

8.4.2 Productivity, Quality, and Safety

Robots have been used in the manufacturing industry for many decades, and subsequent development has led to streamlined processes by using intelligent robots that operate with great precision and speed. In the current manufacturing climate, robotic demands require highly adjustable systems that allow for product changes at low cost (Javaid et al., 2021). Safety in robotic applications is not a new concept and has been studied for many years in manufacturing applications. However, the development and implementation of advanced robotics with new abilities, such as decision-making autonomy and physical interaction with humans, necessitate the re-evaluation of traditional safety regulations concerning industrial robots. Many studies focus on safety-related robot functions such as collision avoidance control and human-aware motion. Safety involves not only humans in the vicinity of robots but also the environment within which the robot operates (Guiochet et al., 2017).

Table 8.6 gives a summary of the core properties of industrial and advanced robotics as well as examples of new hazards (Guiochet et al., 2017).

8.5   EFFECT ON ORGANISATION

8.5.1   OPERATIONAL PERFORMANCE

There is a positive relationship between the digital maturity of an organisation and its operational performance. This positive relationship can be partly explained by organisations having access to new digital technologies that offer new ways to promote, communicate, and analyse the market. Positive operational performance does not imply that the organisation will experience profit growth. This can be attributed to digital technologies that require large initial investments that have not yet made a return on investment (ROI) (Grooss et al., 2022). It is also evident that larger organisations are more prone to show a positive relationship between digital maturity and corporate performance when measured in monetary terms. There is evidence that companies that can use big data through business analytical tools and decision-making experience a significant improvement in their operational performance. Thus, the deduction can be made that if organisations can increase their digital maturity, they will also experience greater operational performance advantages.

It is imperative for companies, especially small and medium-sized enterprises (SME's), to increase their rate of implementation and integration of digital technology initiatives to remain competitive. The large long-term investment into non-instantaneous value-adding activities can be unattractive to companies that run lean operations (Grooss et al., 2022).

The fast development and implementation of digital manufacturing technologies had dramatically changed business practices leading to a disruptive digital transformation of the whole manufacturing industry value chain. Many large organisations had to completely redesign their business processes and models to achieve the benefits associated with these new paradigms. Not all organisations have made the necessary advances by developing high digital capabilities to obtain a competitive advantage (Savastano et al., 2018).

Recent studies have shown that the concurrent use of digital manufacturing technologies and lean manufacturing leads to the largest increase in operational performance (Buer et al., 2021). It was found that digital manufacturing technologies do not contribute significantly to improved operational performance but that these technologies are enablers of lean manufacturing (Hahn, 2020; Kamble et al., 2020). The relationship between digital manufacturing technologies and lean manufacturing is synergistic, i.e., combined they have a bigger impact on organisational performance than alone. The relationship between factory digitalisation and the effect of lean manufacturing on operational performance is shown graphically in Figure 8.11.

8.5.2   EMPLOYEE COLLABORATION

In the modern manufacturing environment, organisations are increasingly making use of enterprise social media platforms to support a digital work environment. The expected benefits of introducing a digital work environment into an organisation include improved employee performance (Dittes & Smolnik, 2019). When implementing digitalisation strategies, organisations should ensure that their staff find

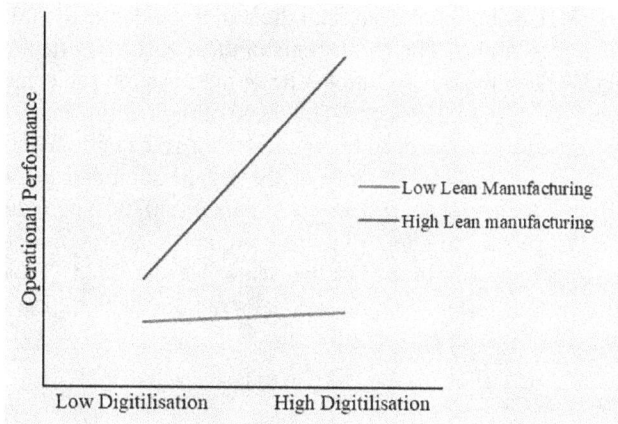

FIGURE 8.11 Relationship between factory digitalisation and lean manufacturing.

the technologies valuable and appropriate. It is also important to ensure adequate and suitable skills and competencies of staff to maintain a high productivity rate. The organisation must, therefore, facilitate digitalisation, so that staff at all levels (operators, team leads, managers, etc.) understand the need for it and can take greater responsibility in their organisations (Thun et al., 2022).

Digital work environments will encourage work from home and the more frequent use of technology products as more employees will interact with each other by using hybrid communication channels that can be accessed from anywhere and not just in the physical environment of the organisation. Care should be taken during this digital transformation, as the individual productivity of workers may be different from when they work in an office environment as usual (Abidin, 2021).

It is therefore critical that organisations use the right tools to assist their employees to collaborate and stay connected across geographies and functions/roles. Digital collaboration (teamwork) and communication tools are designed to enable employees to tap into the shared knowledge of the enterprise, solve challenges with experts remotely, and turn IoT data from the factory floor into long-term benefit (De Boer et al., 2020).

8.5.3 Organisational Structure

Digital transformation (including all aspects of digital manufacturing) is expected to have a huge impact on organisational design (structure). Although computerised systems and software have been in use in the manufacturing industry since the 1980s, the development and adoption of new technologies have seen exponential growth with the strong drive of Industry 4.0 initiatives by the German government since 2011. It should be noted that the process of organisational structure design to incorporate digitalisation is complex, with many potential interdependencies due to the complex nature of digital transformation. It is suggested that the organisational structure and organisation strategy be developed simultaneously, as the use of digital technologies increases the speed to which the organisation has to adjust to market demands/changes and hence will affect the organisation strategy directly (Kretschmer & Khashabi, 2020).

New leadership roles have emerged with the aim to guide the digital transformation process of organisations. The incumbents in these roles are often referred to as "digital leaders". The new roles (positions) in an organisation's management teams include the role of chief digital officer (CDO)/head of digital transformation/director of digital transformation, and/or head of digital strategy. The main functions of these roles are to inform the business strategy of the organisation and guide the implementation of digital technologies (Engesmo & Panteli, 2020). The industrial sector was one of the last sectors to adopt digitalisation, and many organisations' CDOs have only been appointed within the last five years (Visnjic, 2023). Table 8.7 gives a summary of the various forms of organisational structures with digital technologies emphasised in bold (Xiang et al., 2022).

TABLE 8.7
Comparison of Various Forms of Organisational Structures

	Structure	Benefits	Weaknesses	Scenarios
Traditional forms	Classical	Simple and effective	Lack of systematic scheduling	Small workshops
	Linear	Clear chain of command with clear authority and responsibility	Poor horizontal communication and high cost for management	Small-sized enterprises
	Functional	Professional division of labour and reasonable decentralisation	Unclear authority and responsibility	Specialised enterprises
	Linear-functional	Clear authority and responsibility, professional division of labour and stable structure	Poor horizontal communication & inflated management costs	Medium- & large-sized enterprises
	Business unit	Balanced cost and control with consideration of both individual motivation & high-level workload	Internal competitive frictions & managerial redundancies	Large global enterprises
	Matrix	High flexibility & efficiency for projects	Two lines of leadership with unclear authority & responsibility	Large- & medium-sized project-based enterprises
New forms	Network	High connectivity & complementarity	Lack of business process pull	Strategic network organisations
	Multi-team system	Multi-team synergy & responsiveness	High demands for coordinators to cross boundaries	Temporary organisations responding to emergencies
	Platform	User-oriented & high degree of digitisation	High demand for digital technology & resource allocation capabilities, higher risk, & uncertainty	Digital platform organisations act as intermediates of multiple sided markets

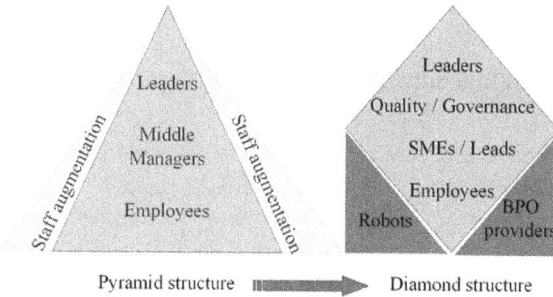

FIGURE 8.12   Changing business hierarchy.

Robotic process automation (RPA) is a software approach that uses business process automation technology to automate tasks performed by human workers, such as extracting data, filling in forms, and moving files. With the implementation of these technologies, the business hierarchy has changed from the conventional triangle organisational model to a diamond model. The diamond-shaped organisational model needs more discipline experts, quality assurance, and management to coordinate services with internal business units, RPA systems, and business process outsourcing (BPO) providers (Küfeoğlu, 2022a). Figure 8.12 shows the change in the business hierarchy from a pyramid to a diamond shape.

8.6   RISKS ASSOCIATED WITH DIGITAL MANUFACTURING TECHNOLOGIES

The current published literature has a tendency to overemphasise the positive effects of digital technologies in manufacturing and underestimate the potential risks associated with their implementation (Flyverbom et al., 2019). Some of the positive implications for the corporate risk situation are the improved traceability of intelligent products, while a strong negative implication is the product's technology dependency and the predisposition to technical failures (Arlinghaus & Rosca, 2021). Table 8.8 gives examples of an altered risk situation caused by the application of digital technologies on different supply chain levels. Key risk factors for Industry 4.0 technologies include environmental risks, industry-specific, and company-specific risk factors.

TABLE 8.8

Examples of Altered Risk Situation Due to the Application of Digital Technologies

	Products & Objects	**Processes & Factories**	**Business Models & Supply Chains**
Positive impact on risk	• Innovative business models maintain competitiveness. • Smart products are traceable and can store quality-related information.	• Increasing productivity & flexibility due to automation, failure reduction, process integration & transparency. • Predictive analytics for better MRO.	• Visibility & integration increases efficiency & service level of SCs. • Flexibility & velocity in case disruptions. • Individualisation of SCM.
Negative impact on risk	• Strong dependence on technology as crucial part of business models. • Smart products have predispositions for technical failures.	• Low inventory levels can cut both ways. • Conventional planning approaches may fail. • Negative effects of autonomous objects.	• Loss of data sovereignty. • New gateways for cyber-attacks. • Increasing complexity of SCM & planning processes.

Arlinghaus, J.C., & Rosca, E., *IFAC-Papers Online*, 54(1), 337–342, 2021. https://doi.org/10.1016/j.ifacol.2021.08.15. With permission.

8.6.1 ENVIRONMENTAL RISKS

These risks are frequently linked to legal risks that occur during the implementation stage and comprise aspects of data privacy and data protection. Most countries have data protection regulations in place which can be perceived as a risk (e.g., violations of data privacy and disadvantages for the employees). Industrial spying or the theft of corporate data using wireless or mobile connections is also a major threat to organisations. Another concern is unauthorised access to physical facilities through digital interfaces, undetected manipulation of data, or malicious encryption of data through ransomware.

8.6.2 INDUSTRY-SPECIFIC RISKS

The risk associated with the integration of Industry 4.0 solutions has led to direct and indirect dependencies on service providers (technologies), as well as a lack of flexibility in certain conditions. Many of the digital technologies incorporated into the manufacturing process require provider-specific hardware, software, and expertise, thus making the users directly dependent on the service provider for issues such as fault-finding, maintenance of systems, and hardware replacements. These types of risks are more complicated when international companies are used as service providers that do not have a local technical support team.

8.6.3 COMPANY-SPECIFIC RISKS

From a company perspective, it was found that the most important risks relate to workforce (human) adaptation, mistakes, and attacks. The digital workflow might not give

TABLE 8.9

Risk Factors and Mitigation Strategies

Risk Factor	Risk Management Approach	Possible Risk Mitigation Strategies
Missing technology expertise	Rejecting projects, maintaining the status quo	Screening of provider markets Use of external consulting services
Legal risk	Consultation of the works council (workers' union) as part of the introduction	Inclusion of works council/unions Adjustment of employment contracts Technical adjustment/certification
Employee adoption	Trian-and-error	Involvement of affected employees skills training (upskilling) Intuitive solution design
Cyber risk	Blind trust in existing protective measures such as firewalls	Sensitisation/training of employees Technical measures such as IDS/IPS Segmentation of networks Backup strategy

workers the freedom to change their operation freely and spontaneously as they did before and therefore cannot react to operation incidents due to new digital technology restrictions (Arlinghaus & Rosca, 2021). Table 8.9 provides a summary of risks and mitigation strategies found in a study conducted by Arlinghaus and Rosca (2021).

8.7 CONSIDERATIONS FOR DIGITAL MANUFACTURING IMPLEMENTATION

For organisations to implement digital manufacturing into their operations, some basic requirements must be met. The transformation to digital manufacturing necessitates a holistic change regarding the digital capabilities of the entire organisation. This implies that several different digital technologies must be implemented across the organisation to benefit from digital manufacturing. This requires a detailed digital transformation strategy for the organisation, as well as a digital transformation implementation plan.

Other considerations include:

- Organisation digital risk assessment,
- Digital transformation strategy development,
- Sustainability of digital manufacturing,
- Human resource management (skills development, etc.).

Additional complications that might arise during the implementation of digital manufacturing technologies are when the organisation is operating on a global scale. Cultural differences, educational level, available infrastructure, etc., are all important considerations for the management of global operations. Often, developing countries do not have a stable and steady electricity supply for the operation of digital manufacturing (sensitive electronics) equipment, and other challenges such as the safety and security of equipment and staff all influence the digital manufacturing implementation strategy in these countries.

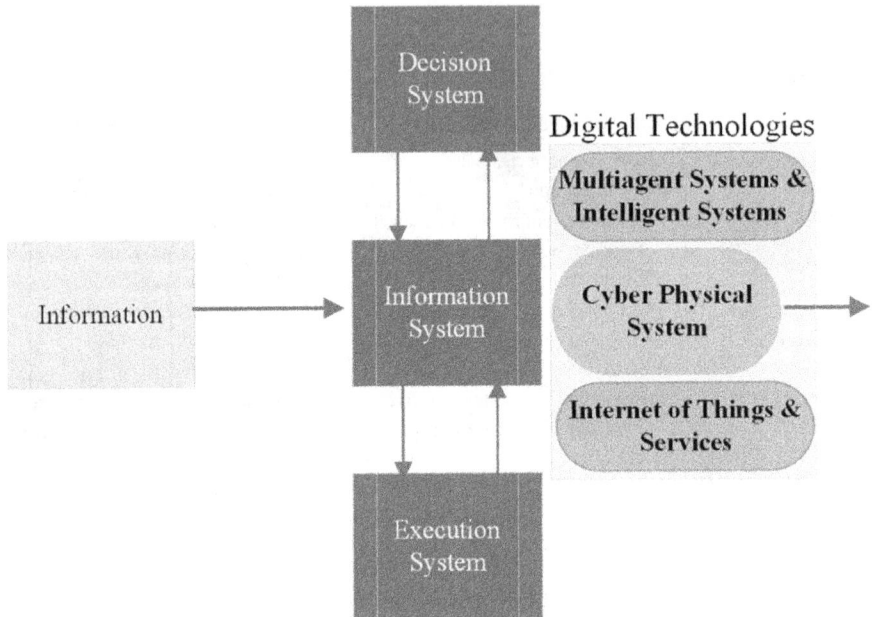

FIGURE 8.13   Overview of how digital technologies influence logistics systems (holistic view).

8.8   IMPACT ON LOGISTICS AND COMPLEX PROJECTS

The most used digital technologies in logistics are big data, artificial intelligence, blockchain, the Internet of Things, and robotics. And they are mainly used to reduce the cost of supply chains by optimising logistics (Tajudeen et al., 2022). Since logistics is an interdisciplinary field between engineering (manufacturing, mechanical engineering), natural sciences (mathematics, computer science), and economics, it is evident that digital manufacturing technologies will have a direct influence on the complexity of logistics. The implementation of digital technologies is leading to new and innovative products and processes, such as small lot sizes, variable customer demand, flexible response, and customised mass production. Figure 8.13 gives an illustration on where digital technologies are affecting the holistic logistics system (Herzog & Timm, 2021).

Multiagent systems model the information flow of autonomous logistic processes where material flow is represented using logistics sensor data (such as location, speed, stops, and temperature) which is then combined with the information flow. Research has shown that the distributed artificial intelligence technologies of intelligent agents and multi-agent systems are suitable to build comprehensive models (simulations) for complex autonomous cooperating logistics processes (Herzog & Timm, 2021).

The drive for the Industry 4.0 initiative has led to the development of the Logistics 4.0 paradigm to accommodate the elements of digital manufacturing. The building blocks of Logistics 4.0 is shown in Table 8.10 (Knapp et al., 2021).

TABLE 8.10
Building Blocks of Logistics 4.0

Big Data	Data collection & processing
	Logistics control tower
	Augmented reality
New methods of physical transportation	Driverless transport systems
	Robots
	Drones
Digital platforms & marketplaces	Big cross-border platform
	Shared transport capacity
	Shared warehouse capacity
	Crowdsourcing
New production methods	Additive manufacturing (3D printing)
	Digital manufacturing techniques (direct energy deposition, etc.)

One of the largest logistics networks is associated with the automotive industry as it involves the movement of materials, components, and complete vehicles around the globe. The logistics chain involves several stakeholders, as well as dealers, which requires seamless coordination between them. It is in this complex network that digitalisation can add tremendous value in securing a high level of transparency among the stakeholders. It also provides high-tech solutions for developing effective and tailor-made support and control systems that help processes along the supply chain (Hoff-Hoffmeyer-Zlotnik et al., 2021).

A good example of the benefits of digitalisation in automotive supply chain logistics is during peak periods of unusually high demand. The coordination between the release and execution of an order can be improved by an intelligent network of different logistics units. A concept of a cyber-physical logistic system was developed by researchers from BIBA (Bremer Institut für Produktion und Logistik), which uses tailored autonomous control methods to attain an efficient demand-orientated material supply in manufacturing (Schukraft et al., 2021).

The availability of IT systems and increased computational power has led to a renewed interest in using a network modelling approach to create models of complex networks from various domains in the manufacturing system (Becker & Wagner-Kampik, 2021). Modelling complex networks in manufacturing and logistics requires a huge amount of input data, which is normally collected from shop floor control systems. These records contain information such as order ID, machine/s that processed a certain operation, and time data. The advent of digitalisation in the manufacturing industry will lead to the availability of massive amounts of system data that can be used in network modelling. The use of artificial intelligence methods in the processing of the system data will increase the quality of logistic network models. Figure 8.14 shows two complex network graphs for two production scene scenarios, using a network modelling approach for logistics in manufacturing (Becker & Wagner-Kampik, 2021).

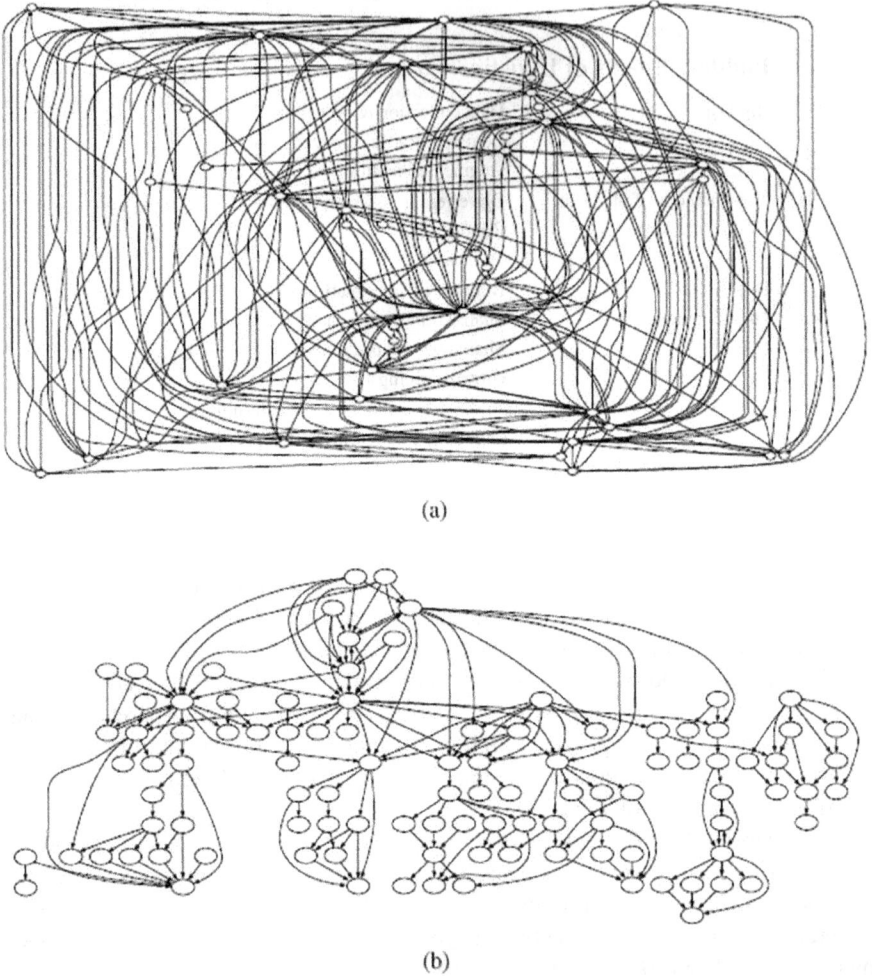

FIGURE 8.14 (a) Network model of a shop floor production of machine parts and (b) Network model from a process industry.

8.9   IMPACT ON BUSINESS MODELS

To understand the impact digital manufacturing has on the business model of an organisation, it is important to look at the core components of a business model. Figure 8.15 shows the core components of a business model (Küfeoğlu, 2022b). For a company to conduct its business (profitably), all four components need to be resolved.

A business model can be seen as a qualitative method of planning how business should be conducted. Due to the changes in the way in which business is being conducted with the implementation of digital manufacturing technologies, more diverse and some controversial concepts and approaches to business models have emerged (Seidel et al., 2017).

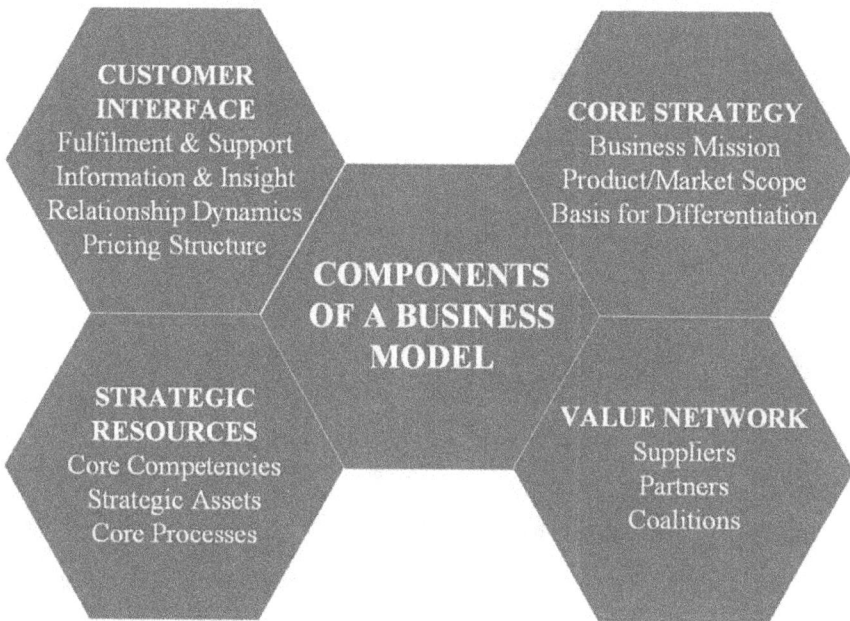

FIGURE 8.15 Core components of a business model.

Business models that combine products and services (or manufacturing as a service) are commonly referred to as product service systems (PSSs). Business models that are examples of wide-ranging transformative models include PSS-based business model and circular economy-based business model (CBM). These models include the product's entire lifespan and are particularly well suited for the manufacturing industry. These models are viewed as the most effective and sustainable business models that require a shift from profit-orientated to enhanced benefits or reduced negative effects on the environment and society. Both models include new and innovative ideas for manufacturing processes (Seidel et al., 2017). Some of the effects of including environmental and societal needs on the traditional business model components are shown in Figure 8.16.

The main difference between additive manufacturing-based business models compared to conventional models is that production can be made on-demand and the necessity to store an inventory is further reduced. AM also lends itself to the fact that production can be done on location, nearer to location, and production of ready-to-use products (Savolainen & Collan, 2020).

8.10 IMPLICATIONS FOR ENGINEERING MANAGERS

Since digitalisation entails a considerable organisational change that is driven by digital technologies as well as changes in business strategy, managers need to understand the impact of digitalisation, especially on the current workforce. This implies that managers may need to move faster to avoid losing ground to competitors, which is especially true for smaller organisations that might not have the necessary resources (Kretschmer & Khashabi, 2020).

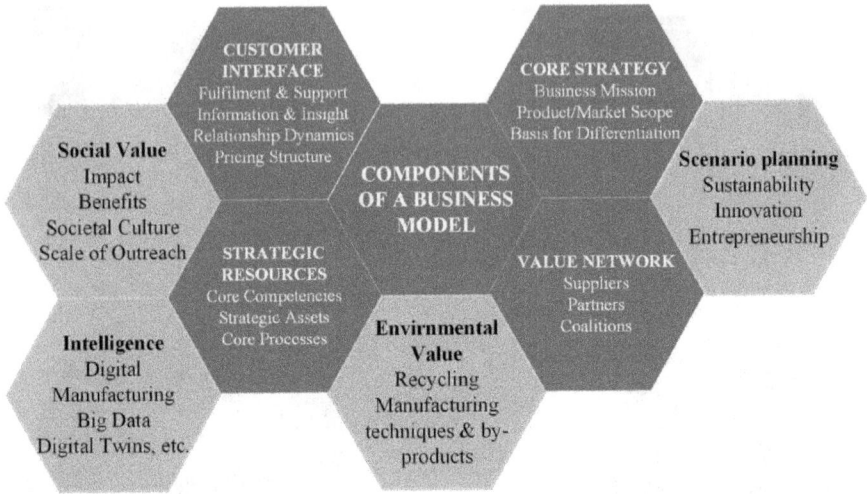

FIGURE 8.16 Effect of environmental and societal needs on business model components.

Also of importance is the management's ability to foster new digital intrapreneurial opportunities for staff. It has been shown that intrapreneurs are essential to corporate innovation. An organisation's ability to nurture intrapreneurial talent affects its ability to positively respond to opportunities and disruptions caused by digital transformation. Digital intrapreneurs can be described as employees who use their entrepreneurial spirit for the benefit of their employer and concurrently to give meaning to their work by implementing their ideas to produce impactful digital innovations. If managers can accommodate the needs of digital intrapreneurs, organisations can function more effectively in a digitally transforming environment. The successful implementation of digital technologies within an organisation requires a change in managerial attitudes and building employee trust (Pinchot & Soltanifar, 2021). Figure 8.17 shows a digital intrapreneurship model to assist rapid digitalisation within an organisation.

Engineering managers in large manufacturing industries need to understand the value of domain-specific knowledge and devise a growth strategy for domain experts that aligns with the new required capabilities. This will ensure that skills and expertise that have been accumulated and perfected over time do not get lost in the digital transformation process (Szalavetz, 2022).

Managers, especially manufacturing companies, need to ensure that the needs of the staff (all levels) and social systems are respected and balanced with the advantages that digital technologies offer. Large transformative changes are often met with resistance from employees, especially when they feel threatened with the idea of being replaced by digital technologies. Therefore, a deeper understanding of the digital technologies' implementation process and the effect on the interaction between humans, digital technology, and organisations are required (Thun et al., 2022).

Other challenges facing engineering managers working with modern production technologies are technology-related factors such as technical problems (software freezes and errors, system crashes, etc.), poor usability, low situation awareness,

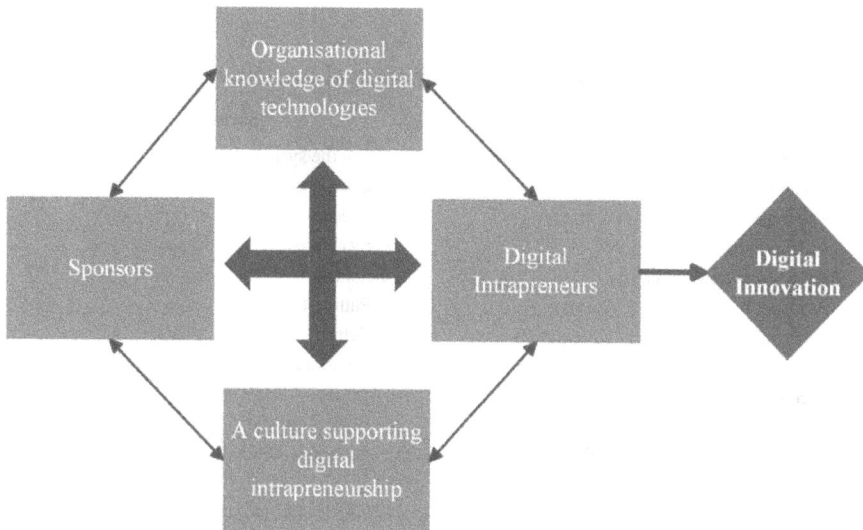

FIGURE 8.17   Digital intrapreneurship model. (Pinchot, G., & Soltanifar, M., Digital intrapreneurship: The corporate solution to a rapid digitalisation. In M. Soltanifar, M. Hughes, & L. Göcke (Eds.), *Digital Entrepreneurship, Future of Business and Finance*, 2021. https://doi.org/10.1007/978-3-030-53914-6_12. With permission.)

and increased qualification requirements of workers (team leaders, operators, etc.). These challenges have a negative effect on the production (interrupted workflow, added time pressure, etc.) and add to the perceived work stress of managers and workers. One of the benefits of digital manufacturing is its ability to adapt to rapid changes in the production process (agility), and the negative impact of this is the added complexity of the production system to be managed. Another challenge faced by engineering managers when the digital transformation process has a top-down approach is that operators are less likely to be satisfied with new digital tools, i.e., digitalisation fails to reach the operational level. It is crucial that the aim of a digital intervention is well communicated and that there is a focus on continuous evaluation during the implementation (Thun et al., 2022). The key enablers and barriers to the development and implementation process of digital technologies are summarised in Table 8.11.

8.11   INNOVATIONS AND OPPORTUNITIES FOR RESEARCH AND DEVELOPMENT

The implementation of digital manufacturing technologies has led to many changes in the design and production of components and systems. Digital manufacturing technologies have opened the development of a new range of novel materials, as well as new ways of designing structures and complex components. Advanced materials, advanced robotics, and the exploitation of digital manufacturing information are all areas for further research and innovation. These areas can be seen as complex multi-level changes that will affect how industrial systems operate. Furthermore, these

TABLE 8.11

Enablers and Barriers to the Development and Implementation Process

Enablers	Barriers
Shared trust	Trusting the system
• Build trust	• Compatibility with existing systems
• Extended collaboration with unions	• Speed and stability of networks
• Transparency of purpose	• System access and data security
Shared visual understanding	Understanding the benefits
• Visual mapping and structuring	• Putting the old tools away
• Visualisation facilities participation	• Return on investment (ROI)
	• Effect measures, ROI and business cases
Shared user perspectives	Perspective of economics
• Flexible and versatile involvement	• Budget change
• Rapid release of functionality	• Economic conditions
• Communicating user needs	
Shared learning	Learning to manage scope
• Train the trainer	• Large-scale implementation and training
• Digitalisation as continuous improvement	• Managing user feedback

technological changes propel changes in the business models that organisations adopt which will cause a shift in the architecture of businesses to become more networked in niche areas. The advanced technologies associated with digital manufacturing also challenge the current social and economic structures with the aim to create a more efficient, more productive, and more sustainable workforce and products.

Due to globalisation, international expansion, and the demand for collaboration of manufacturing organisations, future research should test the inter-organisational coordination mechanisms among organisations (suppliers, partners, etc.) (Xiang et al., 2022).

Expanded multidisciplinary studies are needed on the development and implementation processes of organisations in digital transformation to understand the social and technical aspects of work systems. Some of the biggest barriers to the implementation of digital applications that need further evaluation are as follows:

* Trusting the system.
* Understanding the benefits.
* Perspectives of economics (Thun et al., 2022).

Furthermore, areas identified for future research are performance measures and the correlations between each manufacturing level in the factory. There is a need for standard reference models and data schemes for digital manufacturing system development. More attention is needed for the adoption of virtual reality technology into manufacturing systems (Choi et al., 2015).

Another important factor that needs further study is the relationship between digital leaders, IT capabilities, and digital technology programmes. Some organisations choose to integrate digital technologies into their IT department, while others prefer to differentiate between them, and research is needed to ascertain why organisations make these different choices (Engesmo & Panteli, 2020).

There seems to be no consensus on the future effective capacity of robots to fully substitute human labour, because some skills are only associated with human beings (for now), such as judgement and common sense of the ability to identify the purposiveness of objects. More studies are needed on the effect of digital technologies on the labour market, that is, new skills that will be in demand in the future in manufacturing (Freddi, 2018).

A number of traditional theories where the underlying assumption is that industry boundaries are stable need to be re-evaluated due to the change that digital technologies bring to manufacturing companies (e.g., changing structure of industries and value chains) (Szalavetz, 2022).

REFERENCES

Abidin, A. Z. (2021). The influence of digital leadership and digital collaboration on the digital skill of manufacturing managers in Tangerang. *International Journal of Artificial Intelligence Research*, 6(1), 1027–1041. https://doi.org/10.29099/ijair.v6i1.330

Altus Market Research. (2022). *Role of Robotics in Digital Manufacturing!* Robotics Tomorrow. https://www.roboticstomorrow.com/news/2022/07/25/role-of-robotics-in-digital-manufacturing/19148

American Society for Testing and Materials (ASTM). (2015). *Standard Terminology for Additive Manufacturing – General Principles – Terminology1*. ASTM International. https://www.iso.org/standard/69669.html

Arlinghaus, J. C., & Rosca, E. (2021). Assessing and mitigating the risk of digital manufacturing: Development and implementation of a digital risk management method. *IFAC-Papers Online*, 54(1), 337–342. https://doi.org/10.1016/j.ifacol.2021.08.159

Arnautova, Y. (2020, June 11). *Digital Twins Technology, Its Benefits & Challenges to Information Security*. GlobalLogic. https://www.globallogic.com/insights/blogs/if-you-build-products-you-should-be-using-digital-twins/

Becker, T., & Wagner-Kampik, D. (2021). Complex networks in manufacturing and logistics: A retrospect. In *Dynamics in Logistics*. Springer International Publishing. https://doi.org/10.1007/978-3-030-88662-2_3

Bigelow, S., & Botelho, B. (2022). *Data Management*. The Ultimate Guide to Big Data for Businesses. https://www.techtarget.com/searchdatamanagement/definition/big-data

Buer, S.-V., Semini, M., Strandhagen, J. O., & Sgarbossa, F. (2021). The complementary effect of lean manufacturing and digitalisation on operational performance. *International Journal of Production Research*, 59(7), 1976–1992. https://doi.org/10.1080/00207543.2020.1790684

Choi, S., Jun, C., Zhao, W. Bin, & Do Noh, S. (2015). Digital manufacturing in smart manufacturing systems: Contribution, barriers, and future directions. In S. Umeda, M. Nakano, H. Mizuyama, H. Hibino, D. Kiritsis, & G. von Cieminski (Eds.), *Advances in Production Management Systems: Innovative Production Management towards Sustainable Growth. APMS 2015. IFIP Advances in Information and Communication Technology* (Vol. 460, pp. 21–29). Springer, Cham. https://doi.org/10.1007/978-3-319-22759-7_3

Colombo, A. W., Karnouskos, S., Kaynak, O., Shi, Y., & Yin, S. (2017). Industrial cyberphysical systems: A backbone of the fourth industrial revolution. *IEEE Industrial Electronics Magazine*, 11(1), 43–47. https://doi.org/10.1109/MIE.2017.2648857

Cooper, C., Zhang, J., Huang, J., Bennett, J., Cao, J., & Gao, R. X. (2023). Tensile strength prediction in directed energy deposition through physics-informed machine learning and Shapley additive explanations. *Journal of Materials Processing Technology*, *315*, 61–67. https://doi.org/10.1016/j.jmatprotec.2023.117908

Copeland, B. (2022). *Artificial Intelligence*. Encyclopedia Britannica. https://www.britannica.com/technology/artificial-intelligence

De Boer, E., Luse, A., Mangla, R., & Trehan, K. (2020). *Digital Collaboration for a Connected Manufacturing Workforce*. McKinsey & Company. https://www.mckinsey.com/capabilities/operations/our-insights/digital-collaboration-for-a-connected-manufacturing-workforce

Dittes, S., & Smolnik, S. (2019). Towards a digital work environment: The influence of collaboration and networking on employee performance within an enterprise social media platform. *Journal of Business Economics*, *89*(8–9), 43–48. https://doi.org/10.1007/s11573-019-00951-4

Engesmo, J., & Panteli, N. (2020). Digital transformation and its impact on IT structure and leadership. In K. Ahlin, P. Mozelius, & L. Sundberg (Eds.), *11th Scandinavian Conference on Information Systems (SCIS2020)*. https://core.ac.uk/download/pdf/351022286.pdf

European Factories of the Future Research Association. (2016). *Factories 4.0 and Beyond*. https://www.effra.eu/sites/default/files/factories40_beyond_v31_public.pdf

Flyverbom, M., Deibert, R., & Matten, D. (2019). The governance of digital technology, Big Data, and the Internet: New roles and responsibilities for business. *Business & Society*, *58*(1), 21–29. https://doi.org/10.1177/0007650317727540

Freddi, D. (2018). Digitalisation and employment in manufacturing. *AI & Society*, *33*(3). https://doi.org/10.1007/s00146-017-0740-5

Gröning, H., Zenisek, J., Wild, N., Huskic, A., & Affenzeller, M. (2023). Method of process optimization for LMD-processes using machine learning algorithms. *Procedia Computer Science*, *217*. https://doi.org/10.1016/j.procs.2022.12.350

Grooss, O. F., Presser, M., & Tambo, T. (2022). Balancing digital maturity and operational performance – Progressing in a low-digital SME manufacturing setting. *Procedia Computer Science*, *200*, 32–43. https://doi.org/10.1016/j.procs.2022.01.247

Guiochet, J., Machin, M., & Waeselynck, H. (2017). Safety-critical advanced robots: A survey. *Robotics and Autonomous Systems*, *94*, 12–22. https://doi.org/10.1016/j.robot.2017.04.004

Hahn, G. J. (2020). Industry 4.0: a supply chain innovation perspective. *International Journal of Production Research*, *58*(5), 43–47. https://doi.org/10.1080/00207543.2019.1641642

Herzog, O., & Timm, I. J. (2021). Intelligent agents for social and learning logistics systems. In M. Freitag, H. Kotzab, & N. Megow (Eds.), *Dynamics in Logistics* (pp. 21–29). Springer International Publishing. https://doi.org/10.1007/978-3-030-88662-2_5

Hoff-Hoffmeyer-Zlotnik, M., Teucke, M., Oelker, S., & Freitag, M. (2021). Automobile Logistics 4.0: Advances through digitalization. In *Dynamics in Logistics* (pp. 197–226). Springer International Publishing. https://doi.org/10.1007/978-3-030-88662-2_10

Huang, Z., Shen, Y., Li, J., Fey, M., & Brecher, C. (2021). A survey on AI-driven digital twins in Industry 4.0: Smart manufacturing and advanced robotics. *Sensors*, *21*(19), 32–39. https://doi.org/10.3390/s21196340

IBM. (2023). *How Does a Digital Twin Work?* https://www.ibm.com/topics/what-is-a-digital-twin

International Organization for Standardization. (2021). *ISO 8373:2021 Robotics – Vocabulary*. ISO. https://www.iso.org/standard/75539.html

Javaid, M., Haleem, A., Singh, R. P., & Suman, R. (2021). Substantial capabilities of robotics in enhancing Industry 4.0 implementation. *Cognitive Robotics*, *1*, 58–75. https://doi.org/10.1016/j.cogr.2021.06.001

Kanade, A., Bhoite, S., Kanade, S., & Jain, N. (2022). Artificial intelligence and morality: a social responsibility. *Journal of Intelligence Studies in Business*, *13*(1). doi: 10.37380/jisib.v13i1.992

Kamble, S., Gunasekaran, A., & Dhone, N. C. (2020). Industry 4.0 and lean manufacturing practices for sustainable organisational performance in Indian manufacturing companies. *International Journal of Production Research*, *58*(5), 32–25. https://doi.org/10.10 80/00207543.2019.1630772

Knapp, F., Kessler, M., & Arlinghaus, J. C. (2021). The influence of cognitive biases in production logistics. In *Dynamics in Logistics* (pp. 183–193). Springer International Publishing. https://doi.org/10.1007/978-3-030-88662-2_9

Korpela, M., Riikonen, N., Piili, H., Salminen, A., & Nyrhilä, O. (2020). Additive manufacturing-past, present, and the future. In *Technical, Economic and Societal Effects of Manufacturing 4.0* (pp. 17–41). Palgrave Macmillan, Cham. https://doi.org/10.1007/978-3-030-46103-4_2

Kretschmer, T., & Khashabi, P. (2020). Digital transformation and organization design: An integrated approach. *California Management Review*, *62*, 000812562094029. https://doi.org/10.1177/0008125620940296

Küfeoğlu, S. (2022a). Emerging technologies. In *Emerging Technologies. Sustainable Development Goals Series* (pp. 349–369). Springer, Cham. https://doi.org/10.1007/978-3-031-07127-0_2

Küfeoğlu, S. (2022b). Innovation, value creation and impact assessment. In *Emerging Technologies. Sustainable Development Goals Series* (pp. 349–369). Springer, Cham. https://doi.org/10.1007/978-3-031-07127-0_1

Li, M., Milojević, A., & Handroos, H. (2020). Robotics in manufacturing: The past and the present. In M. Collan & K. Michelsen (Eds.), *Technical, Economic and Societal Effects of Manufacturing 4.0* (pp. 85–95). Springer International Publishing. https://doi.org/10.1007/978-3-030-46103-4_4

McWilliams, A. (2021). *Global Markets for 3D Printing* (Report No. MFG074B).

Miller, J., & Hannifin, P. (2017). *The Drivers of flexible Automation*. Valin. https://www.valin.com/resources/articles/drivers-of-flexible-automation

Mourtzis, D., Angelopoulos, J., & Panopoulos, N. (2022). Digital manufacturing. In *The Digital Supply Chain* (Vol. 112, pp. 45–50). Elsevier. https://doi.org/10.1016/B978-0-323-91614-1.00002-2

Oxford University Press. (2022). *The Oxford English Dictionary*. Oxford University Press, Oxford.

Paritala, P. K., Manchikatla, S., & Yarlagadda, P. K. D. V. (2017). Digital manufacturing: Applications past, current, and future trends. *Procedia Engineering*, *174*, 982–991. https://doi.org/10.1016/j.proeng.2017.01.250

Phuyal, S., Bista, D., & Bista, R. (2020). Challenges, opportunities and future directions of smart manufacturing: A state of art review. *Sustainable Futures*, *2*, 23–32. https://doi.org/10.1016/j.sftr.2020.100023

Pinchot, G., & Soltanifar, M. (2021). Digital intrapreneurship: The corporate solution to a rapid digitalisation. In M. Soltanifar, M. Hughes, & L. Göcke (Eds.), *Digital Entrepreneurship, Future of Business and Finance* (pp. 233–262). https://doi.org/10.1007/978-3-030-53914-6_12

Sahoo, S., & Lo, C.-Y. (2022). Smart manufacturing powered by recent technological advancements: A review. *Journal of Manufacturing Systems*, *64*, 32–39. https://doi.org/10.1016/j.jmsy.2022.06.008

Savastano, M., Amendola, C., & D'Ascenzo, F. (2018). *How Digital Transformation is Reshaping the Manufacturing Industry Value Chain: The New Digital Manufacturing Ecosystem Applied to a Case Study from the Food Industry*. https://doi.org/10.1007/978-3-319-62636-9_9

Savolainen, J., & Collan, M. (2020). Industrial additive manufacturing business models: What do we know from the literature? In *Technical, Economic and Societal Effects of Manufacturing 4.0*. Springer International Publishing. https://doi.org/10.1007/978-3-030-46103-4_6

Schukraft, S., Teucke, M., Freitag, M., & Scholz-Reiter, B. (2021). Autonomous control of logistic processes: A retrospective. In *Dynamics in Logistics* (pp. 115–130). Springer International Publishing. https://doi.org/10.1007/978-3-030-88662-2_1

Seidel, J., Barquet, A.-P., Seliger, G., & Kohl, H. (2017). Future of business models in manufacturing. In R. Stark, G. Seliger, & J. Bonvoisin (Eds.), *Sustainable Manufacturing, Sustainable Production, Life Cycle Engineering and Management* (pp. 34–43). Springer, Cham. https://doi.org/10.1007/978-3-319-48514-0_10

Shao, G., Hightower, J., & Schindel, W. (2023). Credibility consideration for digital twins in manufacturing. *Manufacturing Letters*, *35*, 24–28. https://doi.org/10.1016/j.mfglet.2022.11.009

Song, X. T., Kuo, J.-Y., & Chen, C.-H. (2022). Design methodologies for conventional and additive manufacturing. In *Digital Manufacturing*. INC. https://doi.org/10.1016/b978-0-323-95062-6.00007-3

Szalavetz, A. (2022). The digitalisation of manufacturing and blurring industry boundaries. *CIRP Journal of Manufacturing Science and Technology*, *37*, 32–39. https://doi.org/10.1016/j.cirpj.2022.02.015

Tajudeen, F. P., Nadarajah, D., Jaafar, N. I., & Sulaiman, A. (2022). The impact of digitalisation vision and information technology on organisations' innovation. *European Journal of Innovation Management*, *25*(2), 81–98. https://doi.org/10.1108/EJIM-10-2020-0423

Thun, S., Bakås, O., & Storholmen, T. C. B. (2022). Development and implementation processes of digitalization in engineer-to-order manufacturing: Enablers and barriers. *AI & Society*, *37*(2), 76–79. https://doi.org/10.1007/s00146-021-01174-4

Visnjic, I. (2023). *How Industrial Organizations Are Leading the Digital Transformation*. Innovation & Techology. https://dobetter.esade.edu/en/industrial-sector-digitalization?_gl=1*s9p4e8*_up*MQ..*_ga*MTIzNDQ1ODgxNi4xNjc5NTc2ODk0*_ga_S41Q3C9XT0*MTY3OTU3Njg5NC4xLjEuMTY3OTU3NzQwNy4wLjAuMA

Wan, J., Li, X., Dai, H.-N., Kusiak, A., Martinez-Garcia, M., & Li, D. (2021). Artificial-intelligence-driven customized manufacturing factory: Key technologies, applications, and challenges. *Proceedings of the IEEE*, *109*(4), 23–31. https://doi.org/10.1109/JPROC.2020.3034808

The Welding Institute. (2022a). *What Is Additive Manufacturing?* https://www.twi-global.com/technical-knowledge/faqs/what-is-additive-manufacturing

The Welding Institute. (2022b). *What Is Digital Manufacturing? (A Definitive Guide)*. https://www.twi-global.com/technical-knowledge/faqs/what-is-digital-manufacturing

Westkämper, E. (2007). Digital manufacturing in the global era. In P. Cunha & P. Maropoulos (Eds.), *Digital Enterprise Technology*. Springer. https://doi.org/10.1007/978-0-387-49864-5_1

Wicaksono, H., Boroukhian, T., & Bashyal, A. (2021). A demand-response system for sustainable manufacturing using linked data and machine learning. In M. Freitag, H. Kotzab, & N. Megow (Eds.), *Dynamics in Logistics* (pp. 155–181). Springer International Publishing. https://doi.org/10.1007/978-3-030-88662-2_8

Xiang, Q., Zhang, Y., Zhong, J., Wang, G., & Long, L. (2022). Construction of digital 3D magic-cube organization structure for innovation-driven manufacturing. *Frontiers of Engineering Management 2*, 12–19. https://doi.org/10.1007/s42524-022-0237-x

Yang, L., Zou, H., Shang, C., Ye, X., & Rani, P. (2023). Adoption of information and digital technologies for sustainable smart manufacturing systems for Industry 4.0 in small, medium, and micro enterprises (SMMEs). *Technological Forecasting and Social Change*, *188*(March 2022). https://doi.org/10.1016/j.techfore.2022.122308

Zhou, Z., Xie, S., & Chen, D. (2011). *Management of Technology in Digital Manufacturing Science*. Springer, London, Dordrecht, Heidelberg, New York. https://doi.org/10.1007/978-0-85729-564-4_7

9 Future Fuels

Colin A. Scholes
The University of Melbourne

9.1 INTRODUCTION

The transition of the global economy toward renewable and green energy requires a suitable fuel to replace petroleum and natural gas. This transition is being driven by concerns about climate change and the environmental impact of fossil fuels, as well as the strategic need for many countries to gain energy independence. This transition is, therefore, being driven both at a consumer level and at an institutional/government level and will impact all aspects of a regional economy. The unprecedented growth in low-emissions technology for electricity generation has ensured that photovoltaics and wind turbines are now recognised as the cheapest form of electricity (Graham, Hayward, Foster, & Havas, 2022; Stefani, Hallam, & Wright, 2022). For the transportation sector, which includes vehicles, rail and shipping, the future fuel is clearly hydrogen and hydrogen base carriers (COAG Energy Council Hydrogen Working Group, 2019). Hydrogen-based fuels are also suitable for small-scale thermal requirements, such as cooking and domestic heating, to replace or be mixed with natural and petroleum gases. Liquified natural gas (LNG) has also been touted as a future fuel and is already used in vehicles and large-scale power generation. As such, an industry already exists to support this fuel. However, LNG's high fugitive emissions, those associated with the production and processing of the fuel, as well as methane having 21 times the global warming potential of carbon dioxide means LNG should only be viewed as a transition fuel. That is LNG, will assist in the transition of the global economy, but not the final fuel that our futures will be based upon.

Hydrogen's strongest advantage as a future fuel is the emissions are almost exclusively water, with negligible carbon. For example, Hyundai hydrogen fueled cars can travel up to 100 km on 1 kg of hydrogen while producing only 0.2 kg of CO2-e per km. The standard automobile to undertake that distance requires the equivalent of 7.5 L of diesel or 9.3 L of petrol and emits 20 kg of CO2-e per km (COAG Energy Council Hydrogen Working Group, 2019). Hydrogen can also compete with battery-based vehicles in the transport sector, as hydrogen carrier fuels contain much more energy than the equivalent weight of batteries, and the recharge time is a fraction of that required for modern electric vehicles. As such, hydrogen-based vehicles are likely to dominate the large transport sector of buses, trucks, and shipping, which are required to transport heavy loads over long distances. Already the world produces ~70 million tonnes of hydrogen per year

DOI: 10.1201/9781003374879-9

(*The Future of Hydrogen*, 2019). Primarily, this hydrogen is used in petrochemical refineries for desulfurisation as well as in the Haber-Bosch process for reaction with nitrogen to produce ammonia, the bases for the global industry in fertilisers and explosives. Hydrogen also has niche applications associated with metal processing, fine chemical production as well as glass fabrication and the electronic industry. Hence, the safe handling of hydrogen is known, and hazard minimisation has come a long way since the Hindenburg disaster (Najjar, 2013; Yang et al., 2021). There are however some disadvantages in using hydrogen as a fuel for the transport and energy sector. The most significant is the low energy density of hydrogen compared to petroleum-based fuels, which means that a larger volume of hydrogen must be stored within a vehicle to travel the same distance. This comparison of energy density can be seen in Table 9.1, which indicates that even liquid hydrogen and highly compressed hydrogen gas have relatively low energy density. The need to store liquid hydrogen or highly compressed hydrogen gas leads to the other significant disadvantage, the storage requirements of this fuel and the difficulty in handling compared to liquid petroleum. Hence, hydrogen-based carriers that include ammonia, methanol and dimethyl ether are alterative future fuels (Dalena et al., 2017; MacFarlane et al., 2020), as shown in Figure 9.1. Importantly, these hydrogen carriers have higher energy densities than hydrogen and can be stored as liquids at or near ambient conditions, which correlates with safer handling. However, the production of these carrier fuels from hydrogen requires additional processing stages which add to the cost of production. In addition, for methanol and dimethyl ether, carbon dioxide is a by-product of their combustion, though at a lower level than petroleum.

This chapter is focused on hydrogen and hydrogen carriers as future fuels, the energy economics that will be constructed around them in the coming decades, the processes available to generate these fuels, focusing on low emission approaches. In addition, a discussion on the transition approaches that will need to be undertaken as economics shift to hydrogen-based fuels will be presented.

TABLE 9.1
Energy Density of Existing and Future Fuels Based on Hydrogen

	Energy Density (MJ/L)
Liquid H_2	8.53
Gaseous H_2	1.91 (200 bar)
Methanol	15.6
Dimethyl ether	19.1
Ammonia	11.5
Natural gas	22.2 (liquified)
Gasoline	34.2
Diesel	38.6

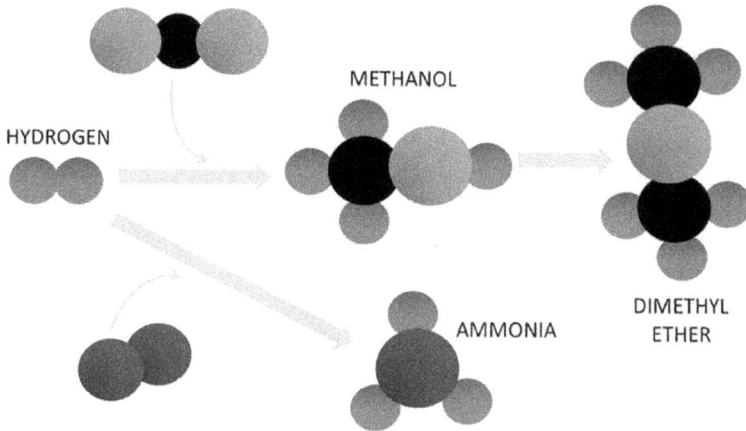

FIGURE 9.1   Future fuel hydrogen carriers based on hydrogen.

9.2   HYDROGEN ECONOMY

The hydrogen economy is based around the concept that hydrogen powers all major aspects of a regional economy. This includes the electricity and transport sectors. Hydrogen generated electricity through turbines and fuel cell technology, with the analogous for hydrogen-based vehicles being based around fuel cell technology or converted internal combustion engines.

The combustion of hydrogen for heating purposes, like natural gas, can be used to heat industry and domestic dwellings. One challenge for the hydrogen economy is replacing natural gas in domestic appliances, such as burners, given the different radiative and convective heat transfer properties of hydrogen compared to natural gas. The current strategy is to blend hydrogen with natural gas, and therefore reduce the carbon footprint of the domestic sector. The purpose of the blends is to ensure that existing appliances can still be safely used and therefore minimise the economic burden of replacing every natural gas-based appliance for hydrogen-based appliances. In addition, the existing natural gas network can safely accommodate the bends with minimal degradation in performance and loss of fuel.

Large scale demonstrates in the United Kingdom have demonstrated that a 20% blend of hydrogen in natural gas can be safely distributed through a natural gas network and appliances safely operated on the blend (Isaac, 2019). Furthermore, domestic appliances performed at the same level with the hydrogen blend as expected for natural gas. Importantly, customers reported no difference in using the hydrogen blend in cooking and heating appliances. The UK Gas Appliance Directive now requires all approved appliances to be verified for a 23% hydrogen blend. In Australia, appliances have been tested up to 10% hydrogen, and there are no regulatory constraints in distributing 20% hydrogen blends through the existing natural gas network. The distribution of hydrogen through gas networks is not a new process, with original town gas (based on gasification of coal) containing up to 60% hydrogen distributed through pipeline infrastructure, though at a lower pressure than modern networks.

Converting to 100% hydrogen appliances will take time, given the large investments required as almost every household will be affected. There are already appliances available in the United Kingdom that operate on 100% hydrogen, including space heaters, boilers and hot water heaters as well as cooktops. These appliances have additional safety features, given the additional hazards of hydrogen, and will need to re-educate the public about appliance use, given the different nature between hydrogen and natural gas. For example, on cooktops, hydrogen flame is orange and lower temperature than the natural gas blue flame, which influences the heat supplied during cooking. Regulations and mandating hydrogen ready appliances will become standard in new dwellings in the near future, to support the transition to hydrogen.

9.2.1 METHANOL ECONOMY

Methanol is a key component of the hydrogen economy, as methanol is a critical precursor chemical for much of the global chemical industry (Dalena et al., 2017). Methanol is involved in the synthesis of acetic acid and formaldehyde, as well as used in industries such as paints, adhesives, plastics, solvents and cleaning products, along with pharmaceutical products. Currently, methanol is synthesised from natural gas, through reforming to syngas, and hence methanol usage has a strong carbon footprint. Methanol can be synthesised from carbon dioxide with hydrogen, through CO_2 hydrogenation, which generates a process that can be carbon neutral (Lee et al., 2020). In this manner, CO_2 can be recycled and enables methanol dependent industries to significantly reduce their carbon footprint. The source of CO_2 can be any fossil fuel industry, such as conventional power stations, cement kilns and petrochemical refineries. Direct air capture from the atmosphere has also been suggested. The CO_2 must be captured from the industrial process, purified and compressed for the CO_2 hydrogenation reaction to methanol. Dedicated carbon dioxide capture technologies already exist, based on solvent absorption, membrane separation and adsorbents (Thambimuthu, Soltanieh, & Abanades, 2005). Generally, the more concentrated the CO_2, the lower the energy demand of the capturing process and more favourable economics.

Methanol can be used as fuel and hydrogen carrier. At room temperature methanol is a liquid with higher energy density than hydrogen, and hence represents a viable method for storing and transporting hydrogen for energy purposes. Methanol is already used in limited capacities for fuel purposes, but only in niche applications as part of a burner-boiler process. Importantly, the transportation of methanol is well known in the industry and represents lower safety risk when compared to hydrogen.

Dimethyl ether is another hydrogen carrier, which is produced by the dehydrogenation of methanol (Arcoumanis, Bae, Crookes, & Kinoshita, 2008). The advantage of dimethyl ether is that it has similar characteristics to light hydrocarbons in terms of storage and combustion; hence, dimethyl ether is a potential petroleum substitute (Catizzone, Bonura, Migliori, Frusteri, & Giordano, 2018; Chang, 1983; Semelsberger, Borup, & Greene, 2005). It is recognised that the similarity between dimethyl ether and in particular propane represents an early adoption scenario for hydrogen carriers, as dimethyl ether produced by low carbon means through methanol can replace many hydrocarbon fuels, such as liquified petroleum gas (LPG).

Safety is a major advantage of the methanol economy, as both methanol and dimethyl ether are less flammable than hydrogen and hence can be stored and transported under less stringent conditions. This is important, as any transport fuel will need to be handled by non-skilled personnel when filling up vehicles. The public already have experience with using gasoline pumps for vehicles and methanol can easily be transferred in a same manner.

9.2.2 AMMONIA ADVANTAGE

Ammonia is another hydrogen carrier, as it is synthesis from the direct reaction between hydrogen and nitrogen through the Haber-Bosch process (Erisman, Sutton, Galloway, Klimont, & Winiwarter, 2008). The advantage of ammonia is the well-established industry that already exists for this chemical, given ammonia is currently the largest single user of hydrogen globally and one of the most traded chemical commodities (Giddey, Badwal, Munnings, & Dolan, 2017; Green, 1982). For example, ammonia is the basis of fertilisers, explosives, types of plastics, textiles, dyes, as well as pesticides and herbicides. As such, the chemical industry has developed good safety practices and standards in handling this toxic chemical (Duijm, Markert, & Paulsen, 2005). Another major advantage of ammonia is that there already exists global infrastructure for the transportation and storage of this carrier. For example, there are available tankers that can ship ammonia globally. In contrast for hydrogen there is only one demonstration transportation ship, to date. Hence, the hydrogen economy can rapidly build on this existing infrastructure and utilise ammonia as a hydrogen carrier.

The disadvantage for ammonia is that it cannot be directly combusted for power generation, due to high nitric oxides (NOx) formation, the source of acid rain. As such, ammonia must be converted back to hydrogen, which then undergoes subsequent combustion. The decomposition of ammonia to hydrogen is an energy intensive process, as it requires ammonia to be heated to high temperatures (>700°C) over a catalyst, the resulting hydrogen and nitrogen gases must then be separated (Lamb, Dolan, & Kennedy, 2019). While the overall thermodynamics still favour ammonia as an energy source, this additional decomposition process reduces the overall efficiency of ammonia compared to alternative hydrogen carriers.

Ammonia does have serious human health issues, unlike the other future fuels. Exposure to ammonia can cause damage to the lungs as well as chemical burns to skin, eyes, mouth, and the throat. Critically, exposure to ammonia above 5,000 ppm will induce rapid respiratory arrest. This means that ammonia fuel sites will need to be well ventilated, similar to gasoline service stations. Ammonia does have a very strong odor, and so achieving high concentrations undetected will not be possible. However, the accumulated effect of long-term ammonia exposure is still not fully understood, but acute sensitivity and skin irritation are known medical outcomes.

9.3 FUEL GENERATION

9.3.1 HYDROGEN

The hydrogen economy is not dependent on the source of hydrogen, but for a low carbon future it is critical that hydrogen being used is 'green'. The colour of hydrogen is a way of describing its source, with various shades corresponding to the carbon footprint of the production process, which is outlined in Table 9.2. Currently, most of the world's hydrogen is produced by steam reforming of natural gas, i.e., grey hydrogen (*The Future of Hydrogen*, 2019). The carbon footprint of this hydrogen can be significantly reduced through carbon capture and storage, which converts this hydrogen to 'blue' (Thambimuthu et al., 2005). On a large scale, the target is to power the hydrogen economy through green hydrogen, generated from the electrolysis of water through renewable energy. As this approach generates very little carbon emissions and represents a sustainable process. Ideally, it is hoped that in the coming decades all aspects of an economy will transition to be based around hydrogen.

Hydrogen is produced from water through three main chemical pathways:

1. Direct electrolysis of water
2. Coal gasification
3. Natural gas reforming

Electrolysis of water is the process by which a direct electric current decomposes water into hydrogen and oxygen, through a sustained redox reaction, as shown in Figure 9.2. If the electricity is provided by renewable energy sources, then the hydrogen produced from electrolysis is green. There are a variety of electrolysers designed for water decomposition to hydrogen, with details provided in Table 9.3. The three most common electrolysers are alkaline, proton exchange membrane (PEM), and solid oxide electrolysers (SOE), with only alkaline and PEM electrolysers being commercially available.

Alkaline electrolysis is based on a liquid electrolyte that consists of basic NaOH, KOH or KCl, typically at a concentration of 30%. When the direct current is applied, hydroxide ions (OH⁻) are transported through the electrolyte from the cathode to the

---

TABLE 9.2

Definitions of Hydrogen 'Colours' and Their Link To Production Source

Colour	Production Process
Black/Brown	Gasification of coal and biomass
Grey	Steam reforming of natural gas
Blue	Steam reforming of natural gas with carbon capture and storage
Turquoise	Thermal splitting of methane via pyrolysis
Green	Electrolysis of water powered by renewable energy

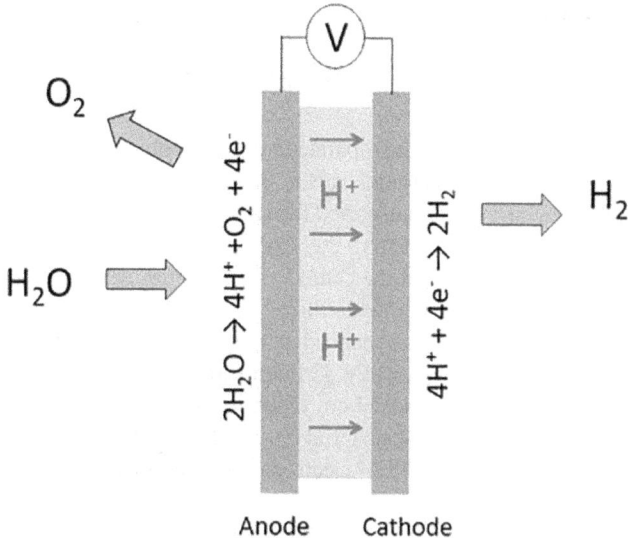

FIGURE 9.2   Electrolysis of water for hydrogen generation.

TABLE 9.3
Water Electrolyser Designs for Hydrogen Generation

Operation Principle	Low Temperature				High Temperature		
	Alkaline electrolysis		Proton exchange electrolysis		Oxygen ion electrolysis		
	Liquid	Polymer Electrolyte			Solid Oxide		
	Conventional	Solid Alkaline	H^+ – PEM	H^+ – SOE	O^{2-} – SOE	CO_2	
Charge carrier	OH^-	OH^-	H^+	H^+	O^{2-}	O^{2-}	
Temperature (°C)	20–80	20–200	20–200	500–1,000	500–1,000	750–900	
Electrolyte	Liquid	Polymer	Polymer	Ceramic	Ceramic	Ceramic	
Efficiency (%)	59–70		65–82	<100	<100		

anode, enabling hydrogen to be produced. The following chemical reactions taking place with the alkaline cell.

- Overall Reaction:

$$2H_2O \ (l) \ \rightarrow \ 2H_2 \ (g) \ + \ O_2 \ (g)$$

- Anode Reaction:

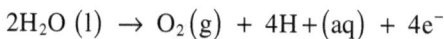

$$2H_2O \ (l) \ \rightarrow \ O_2 \ (g) \ + \ 4H + (aq) \ + \ 4e^-$$

- Cathode Reaction:

$$4H^+ (aq) + 4e^- \rightarrow 2H_2 (g)$$

In PEM electrolysis, the electrodes are separated by a solid polymer electrolyte membrane (i.e., PEM). At the anode, the water molecule is split into oxygen and protons (hydrogen ions). The resulting electrons are transported through the external circuit while the PEM allows the protons to be transported from the anode to the cathode electrode. These protons combine, with the electrons, at the cathode side to form hydrogen. The membrane is placed between two sheets of porous backing media, such as carbon paper, that acts as gas diffusion layers, providing mechanical support and an electrical pathway for the electrons. The PEM material is typically perfluorosulfonic acid (PFSA) base, which is mechanically strong and stable under both oxidative and reductive environments. To ensure proton conductivity, the PEM material must be kept hydrated, and loss of water is a limitation of the technology. The PEM electrode reactions are as follows:

- Overall Reaction:

$$2H_2O \ (l) \ \rightarrow \ 2H_2 + O_2$$

- Anode Reaction:

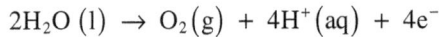

$$2H_2O \ (l) \ \rightarrow \ O_2(g) + 4H^+ (aq) + 4e^-$$

- Cathode Reaction:

$$4H^+ (aq) + 4e^- \rightarrow 2H_2 (g)$$

PEM electrolysers offer high power density, faster response times, and shorter cold-start times compared to bulker alkaline electrolysers. PEM electrolysers also require a smaller space footprint than alkaline electrolysers, providing the potential for installation in variable locations.

Solid oxide electrolysers are the least developed of the three main electrolyser types but has a potentially higher electrical efficiency as it operates at high temperatures of 700–1,000°C. Hence, the process electrolyses steam rather than liquid water. The electrolyte is based on an oxide material that enables the transportation of oxygen anions through the ceramic lattice. These require the high temperature, but the dense ceramic electrolyte prevents back diffusion of gases and good contact with the electrodes. The solid oxide electrolysis reactions are as follows:

- Overall Reaction:

$$2H_2O \ (l) \rightarrow 2H_2 + O_2$$

- Anode Reaction:

$$O^{2-} \rightarrow \tfrac{1}{2}O_2 + 2e^-$$

- Cathode Reaction:

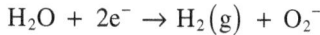

$$H_2O + 2e^- \rightarrow H_2(g) + O_2^-$$

The competitiveness of green hydrogen from electrolysis is associated with scale-up of hydrogen production and the ancillary infrastructure, rather than technology breakthrough. This is because economics of scale will reduce the cost burden of green hydrogen, given the electrolysis industry only represents 5% of the global H_2 production (*The Future of Hydrogen*, 2019). A comparison of the three main electrolyser types is provided in Table 9.4, to highlight their advantages.

Hydrogen generation from coal and other biomass is by gasification, where the carbon source is used to reduce water to hydrogen and carbon monoxide (Higman & van der Burgt, 2008). The process is called gasification, as it transforms a solid material (e.g. coal) into gases (H_2 and CO), which are known as synthetic gas or syngas. Coal gasification was traditional a source of hydrogen and subsequent chemical through syngas and remains widely used in South Africa for synthetic petrol

TABLE 9.4
Comparison of Alkaline, PEM, and Solid Oxide Electrolyser Technologies

	Alkaline	PEM	SOE
Technology readiness level	9	8	4
Electrolyte	Liquid alkaline KOH	Solid proton exchange membrane	Ceramic metal compound
Cathode	Ni, Ni-Mo alloys	Pt, Pt-Pd	Ni-doped ceramic
Anode	Ni, Ni-Co alloys	RuO_2, IrO_2, noble metals	Ni-doped ceramic
Commercial unit size	3.2 MWe	1.5 MWe	Kw range
Hydrogen production	760 Nm³/h	285 Nm³/h	150 Nm³/h
Operational parameters			
Cell temperature (°C)	50–80	20–90	700–1000
Typical pressure (bar)	10–30	20–50	1–15
Current density (A/cm²)	0.25-0.45	1.0–2.0	0.3–1.0
Hydrogen purity (%)	99.9%	99.998%	99.998%
Minimum operation (% capacity)	15	5	15
Cold start-up time	1–2 h	5–10 min	<60 min
Warm start-up time	1–5 min	<10 s	15 min
Performance			
Nominal stack efficiency (LHV)	70%	80%	100%
Specific energy consumption (kwh/Nm³)	4.2–4.8	4.4–5.0	3

production. In recent decades, coal gasification has been replaced by steam reforming of natural gas, which is a comparable process that processes hydrogen and carbon monoxide (e.g. syngas). The production of hydrogen by these methods is carbon intensive, and hence only with carbon capture and storage are they able to demonstrate a low carbon footprint.

Steam reforming of natural gas is the predominate mechanism for producing hydrogen globally (*The Future of Hydrogen*, 2019). The amount of hydrogen produced per hydrocarbon is greater than for coal or biomass gasification. Natural gas reforming technologies are well-known processes for conventional production of hydrogen, undertaken by three main processes: steam methane reforming (SMR), Auto-thermal reforming (ATR) and Partial Oxidation Reforming (POX). Both SMR and ATR require catalyst to achieve hydrogen production, but POX does not as it is a partial combustion process. There is significant difference between these processes in terms of the type of reaction and the conditions of the reactor, which result in different H_2 to CO ratios. A higher H_2/CO ratio of 3 is obtained in SMR reactors compare to the ratio of 2 achievable by ATR and POX. As a result, SMR reactors are the predominant technology for hydrogen production, with a comparison of the three reforming technologies provide in Table 9.5.

The water requirements for large-scale hydrogen production will be significant, with the amount of water requirement per generation process provided in Table 9.6. Water requirements are dependent on the hydrogen generation technology, water purity and the need for additional water for indirect production requirements, such as cooling. For electrolysis, different technologies have different water purity and consumption requirements. For example, alkaline electrolysis purity requirements are lower than PEM and SOEC electrolysis. Coal and biomass gasification water requirements vary widely dependent on the material's moisture content, and the preparation process for the coal (Higman & van der Burgt, 2008). The amount of water for Japan's hydrogen industry has been projected to account for one third of the water currently used in the Australian mining industry (COAG Energy Council Hydrogen Working Group, 2019). So ideally, the water source, especially for electrolysis, would be seawater. However, the desalination requirements add to the cost of production, which have been estimated to be greater than five cents per kilogram of hydrogen.

TABLE 9.5

Comparison of Natural Gas Reforming Processes for Hydrogen Production

	SMR (Natural Gas)	ATR (Natural Gas)	POX
Catalytic	Yes	Yes	No
Reactor	Furnace fired heater	Refractory lined vessel	Refractory lined vessel
Temperature (°C)	700–950	950–1,050	1,200–1,400
Pressure (barg)	20–40	30–50	30–70
Steam/carbon (mol/mol)	1.8–3	0.5–2	0.1–0.5
Oxygen/C	0	0.3–0.5	0.3–0.65
H_2 product (mol%)	68–75	63–65	33–60
H_2/CO product (mol/mol)	3–5	2.5–3.5	1.2–2

TABLE 9.6
Water Requirements to Produce
1 kg of Hydrogen

Electrolysis	9 L
Coal gasification	9 L
Natural gas reforming	4.5 L

9.3.2 METHANOL AND DIMETHYL ETHER

Methanol is currently produced from syngas, at high-pressure conditions (50–100 bar) and high temperatures (493–573 K), with a $Cu/ZnO/Al_2O_3$ catalyst, known as the 'low-pressure method' (Dalena et al., 2017; Tijm, Waller, & Brown, 2001). The syngas is almost exclusively sourced from natural gas (>90%), with the remainder from the gasification of coal or biomass. Methanol is directly synthesised by the following reaction between carbon monoxide and hydrogen:

$$CO + 2H_2 \rightleftarrows CH_3OH$$

The emphasis on CO_2 emissions has made carbon utilisation a potential alternative process for methanol synthesis, known as direct hydrogenation of CO_2, achieved by (Lee et al., 2020):

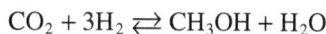

$$CO_2 + 3H_2 \rightleftarrows CH_3OH + H_2O$$

This reaction is catalyst driven and the exothermic equilibrium dictates that high methanol yields are achievable through increased pressure and decreased temperature. However, the relatively inert nature of CO_2 requires an elevated temperature to enable catalyst activation to achieve a reasonable rate of reaction. The methanol economy is dependent on utilising the CO_2 hydrogenation reaction, as this is the only method that ensures the resulting methanol produced has a small carbon footprint.

Dimethyl ether is currently synthesis by two processes, the direct method, and indirect method (requiring two stages) (Catizzone et al., 2018; Semelsberger et al., 2005). The direct method involves the conversion of syngas (CO and H_2) into dimethyl ether directly:

$$2CO + 4H_2 \rightleftarrows CH_3OCH_3 + H_2O$$

In the two-stage indirect method, syngas is reacted to produce methanol, which then undergoes dehydration to dimethyl ether:

$$2CH_3OH \rightleftarrows CH_3OCH_3 + H_2O$$

To produce dimethyl ether with a low carbon footprint, it is necessary to utilise the indirect method and achieve methanol production by CO_2 hydrogenation. Dimethyl ether production from CO_2 is a progression of methanol synthesis from CO_2 and represents a strong component of the methanol economy (Olah, Goeppert, & Prakash, 2009).

There are disadvantages to methanol and dimethyl ether production from CO_2. There is a greater hydrogen consumption rate, because of the additional hydrogen required for the hydrogenation reaction compared to syngas as the feedstock. Methanol yields are not as high from hydrogenation as from syngas, due to the process being less thermodynamically favourable. Hence, methanol and dimethyl ether fuels produced from CO_2 require an additional mechanism to be economically viable, which is widely seen as a cost associated with CO_2 emissions.

9.3.3 AMMONIA

Ammonia can be produced by thermochemical, electrochemical, photochemical and biological methods (Arora, Sharma, Hoadley, Mahajani, & Ganesh, 2018; Banerjee et al., 2015; Bicer, Khalid, Mohamed, Al-Breiki, & Ali, 2020; Giddey, Badwal, & Kulkarni, 2013). The conventional approach for ammonia production is by the Haber-Bosch process, utilising a very high pressure catalysed reaction between hydrogen and nitrogen. The source of the hydrogen is the key to making ammonia the green fuel, with electrolysis sourced hydrogen having a range of advantages (Arora et al., 2018).

Ammonia can also be produced by electrochemical means, which are generally 20% more efficient than the conventional Haber-Bosch process. There are four categories that are defined by operating temperature: liquid electrolytes; molten salt electrolytes, composite electrolytes, and solid electrolytes, which are analogues to electrolyser designs for water splitting. These electrochemical designs are to the same level of technology maturity as the Haber process but have been demonstrated at various scales.

There is also potential for biological ammonia production based on the protein nitrogenases, used by bacteria for nitrogen fixation. The concept is to incorporate nitrogenases with electrodes and fabricate electrocatalysts that can produce 99% pure ammonia. However, this approach has not been expanded beyond the laboratory. Hence, in the immediate future, the Haber-Bosch process represents the only viable method to produce large scale ammonia as a fuel, with electrochemical processes potentially competing in the coming decades.

9.4 TRANSITION TO FUTURE FUELS

Scaling hydrogen projects and infrastructure is the key to developing these future fuels. The most cost-effective route for this infrastructure development is hydrogen hubs. These hubs aggregate users of hydrogen into one area (COAG Energy Council Hydrogen Working Group, 2019). This minimises the cost of providing the infrastructure, such as ports, roads and railway lines, storage tanks and pipelines as well as electrical infrastructure. The hub concept also supports economies of scale by producing hydrogen and hydrogen carriers next to end users. The localised nature of the hub also focuses innovation and ensures a skilled workforce exists nearby, facilitating an industry investment and economic growth. Hydrogen hub locations will be influenced by the following factors:

- Regional demand.
- Land availability.
- Water access and infrastructure.
- Road and rail infrastructure accessibility.
- Electrical and or natural gas grid connectivity.
- Port potential.
- Social factors, such as workforce, safety and community acceptance.
- Environmental factors, such as weather and protections.
- Proximity to hydrogen production regions.
- Potential for hydrogen storage.
- Supportive investment environment.

One key feature of the transition to hydrogen carriers as future fuels is the need to maintain a safe environment for industry and the community. To do this, emergency services need to be properly equipped and trained to deal with hydrogen related incidents, and so minimise the risk to themselves, the community and property. This will require training of firefighters and first responders to handle incidents that involve hydrogen and hydrogen carriers. In the coming few decades, as hydrogen-based transport and industries become more prevalent, it will also be necessary to educate the general public about the dangers of these future futures. Particularly, the low radiative nature of their combustion and resulting lack of visible flame. This will take some adjustment to community understanding of fuel-based fires and the invisible danger such fires represent. The education of first responders and fire fighters will be achieved through new training and assessments, as well as demonstrations of combating hydrogen-based fires. The education of the community will be more difficult, requiring advertising campaigns across multiple media channels as well as likely school-based education workshops, so that the next generation, who will inherit a hydrogen-based future, will instinctively understand the dangers.

Internationally, the trading and use of hydrogen and hydrogen carriers will require transparency and almost certainly a recognised certification scheme, to ensure the future fuel was generated from low carbon processes. The certification scheme will need to focus on carbon emissions produced during the fuels' generation, but given the importance of water, the amount water consumed may also be of relevance. This is especially true for locations where water security is an issue, such as in Australia. Such a scheme will also assist in investment in hydrogen-based fuels. The scheme will need to track and verify the production process and technologies used, scope 1 and 2 carbon emissions as well as any location specific factors. As a result of an international scheme, it is likely that the definition of 'green' and 'low-emission' hydrogen will become more certain and set through international standards.

9.5   CHALLENGES AND OPPORTUNITIES

The biggest challenge facing the future fuels industry is the scale of the transition. The petroleum/gasoline industry has been developing since 1859, when the first oil well erupted in Titusville, Pennsylvania, USA. More than 160 years later, trillions of dollars of investment have constructed a global economy based around petroleum as the predominate fuel. Somehow, hydrogen and associated future fuels need to replicate the success of petroleum in a much shorter time frame and with a low economic burden. This represents a huge engineering challenge, which will require changes in how engineering projects are managed and financed, the business models that underlie energy projects as well as training of a skilled workforce and education of the public.

The scale of the projects that need to be constructed are gigantic. In Australia, a hydrogen hub focused on exporting green hydrogen has been estimated to require a solar photovoltaic farm with the surface area comparable to the major suburban cities of Sydney or Melbourne (Batterham et al., 2022). This represents huge engineering works that have not been attempted at this scale in many countries and is analogous to the construction of the USA interstate highway system and the Netherlands Zuiderzee works for land reclamation.

There are also opportunities, especially in terms of generating a new highly educated workforce. It has been estimated that the transition to net zero emissions will generate between 1 and 1.3 million new workers for Australia (Batterham et al., 2022), which is orders of magnitude greater than those workers currently employed in the petroleum and coal industries. The infrastructure associated with the future fuels will also be benefit as well as the potential to transition many processes to more energy efficient technology based on hydrogen, such as fuel cell designs. A major opportunity is energy independence, as hydrogen can be generated anywhere there is water and hence all countries could become self-sufficient in their energy need. This self-reliance will reduce conflict and geo-politics associated with supplying petroleum for global energy demand. The greatest opportunity will be to the environment and the improvement in the Earth's ecosystem because of less fossil fuel usage and the associated degradation that comes with this, in terms of climate, water quality and land degradation.

9.6   CONCLUSION AND FUTURE PERSPECTIVES

Hydrogen and hydrogen-carriers represent the future of the energy sector, especially as fuels for mobile transportation and storage requirements. The production of these fuels is already established technology, though the green production process is currently more expensive than the conventional carbon methods. Soon, the increased economic burden placed on carbon emissions will result in a complete transition in the fuel industry toward hydrogen and hydrogen carriers. The challenge will be building the infrastructure needed to produce, supply and operate economies based on hydrogen fuels. This is because the existing hydrocarbon fuel industry has been built-up over the past century, and the complete replacement to hydrogen must be done in a shorter period. It is for this reason that blending of hydrogen fuels with fossil fuels will be undertaken as an intermediate step, to ensure the various sectors of the economy have reduced carbon emissions, but without placing to larger financial burden on users.

The transition to hydrogen and carriers is currently underway and represents an outstanding opportunity for various industries and professions. The hydrogen and associated economies will require dedicated engineering expertise in a variety of roles. For example, chemical engineers to produce the fuels, infrastructure engineers for the transportation and storage of these fuels and mechanical engineers to design the devices that will enable efficient use of these fuels. In the long term, society needs to make hydrogen fuels work, to ensure a sustainable and green future for all.

REFERENCES

Arcoumanis, C., Bae, C., Crookes, R., & Kinoshita, E. (2008). The potential of di-methyl ether (DME) as an alternative fuel for compression-ignition engines: A review. *Fuel, 87*, 1014–1030.

Arora, P., Sharma, I., Hoadley, A., Mahajani, S., & Ganesh, A. (2018). Remote, small-scale, 'greener' routes of ammonia production. *J. Clean. Prod., 199*, 177–192.

Banerjee, A., Yuhas, B. D., Margulies, E. A., Zhang, Y., Shim, Y., Wasielewski, M. R., & Kanatzidis, M. G. (2015). Photochemical nitrogen conversion to ammonia in ambient conditions with FeMoS-Chalcogels. *J. Am. Chem. Soc., 137*, 2030–2034.

Batterham, R., Domansky, K., Brear, M., Smart, S., Greig, C., & Bolt, R. (2022). *Interim results*. Retrieved from Melbourne:

Bicer, Y., Khalid, F., Mohamed, A. M. O., Al-Breiki, M., & Ali, M. M. (2020). Electrochemical modelling of ammonia synthesis in molten salt medium for renewable fuel production using wind power. *Int. J. Hydrogen Energ., 45*, 34938–34948.

Catizzone, E., Bonura, G., Migliori, M., Frusteri, F., & Giordano, G. (2018). CO_2 recycling to dimethyl ether: State-of-the-art and perspectives. *Molecules, 23*, 1–28.

Chang, C. D. (1983). Hydrocarbons from methanol. *Catal. Rev., 25*, 1–118.

COAG Energy Council Hydrogen Working Group (2019). *Australia's National Hydrogen Strategy*. Retrieved from https://www.dcceew.gov.au/energy/publications/australias-national-hydrogen-strategy

Dalena, F., Senatore, A., Marino, A., Gordano, A., Basile, M., & Basile, A. (2017). Methanol production and applications: An overview. In A. Basile & F. Dalena (Eds.), *Methanol: Science and Engineering* (pp. 3–28). Amsterdam: Elsevier.

Duijm, N. J., Markert, F., & Paulsen, J. L. (2005). *Safety Assessment of Ammonia as a Transport Fuel*. Roskilde: Forskniningscenter Risoe.

Erisman, J. W., Sutton, M. A., Galloway, J., Klimont, Z., & Winiwarter, W. (2008). How a century of ammonia synthesis changed the world. *Nature Geoscience, 1*, 636–639.

The Future of Hydrogen. (2019). Retrieved from https://webstore.iea.org/the-future-of-hydrogen.

Giddey, S., Badwal, S. P. S., & Kulkarni, A. (2013). Review of electrochemical ammonia production technologies and materials. *Int. J. Hydrogen Energ., 38*, 14576–14594.

Giddey, S., Badwal, S. P. S., Munnings, C., & Dolan, M. (2017). Ammonia as a renewable energy transportation media. *ACS Sustain. Chem. Eng., 5*, 10231–10239.

Graham, P., Hayward, J., Foster, J., & Havas, L. (2022). *Gen Cost 2021–22: Final report*. Retrieved from Australia: https://publications.csiro.au/rpr/pub?pid=csiro:EP2022-2576

Green, L. (1982). An ammonia energy vector for the hydrogen economy. *Int. J. Hydrogen Energ., 7*, 355–359.

Higman, C., & van der Burgt, M. (2008). *Gasification*. Houston, TX: Gulf Professional Publishing.

Isaac, T. (2019). HyDeploy: The UK's first hydrogen blending deployment project. *Clean Energ.*, *3*, 114–125. Retrieved from https://hydeploy.co.uk/.

Lamb, K. E., Dolan, M. D., & Kennedy, D. F. (2019). Ammonia for hydrogen storage: A review of catalytic ammonia decomposition and hydrogen separation and purification. *Int. J. Energ. Res.*, *44*, 3580–3593.

Lee, H. W., Kim, K., An, J., Na, J., Kim, H., Lee, H., & Lee, U. (2020). Toward the practical application of direct CO_2 hydrogenation technology for methanol production. *Int. J. Energ. Res.*, *44*, 8781–8798.

MacFarlane, D. R., Cherepanov, P. V., Choi, J., Suryanto, B. H. R., Hodgetts, R. Y., Bakker, J. M., ... Simonov, A. N. (2020). A roadmap to the ammonia economy. *Joule*, *4*, 1186–1205.

Najjar, Y. S. H. (2013). Hydrogen safety: The road toward green technology. *Int. J. Hydrogen Energ.*, *38*, 10716–10728.

Olah, G. A., Goeppert, A., & Prakash, S. (2009). Chemical recycling of carbon dioxide to methanol and dimethyl ether: From greenhouse gas to renewable, environmentally carbon neutral fuels and synthetic hydrocarbons. *J. Organic Chem.*, *74*, 487–498.

Semelsberger, T. A., Borup, R. L., & Greene, H. L. (2005). Dimethyl ether (DME) as an alternative fuel. *J. Power Sources*, *156*, 497–511.

Stefani, B. V., Hallam, B., & Wright, M. (2022). Solar is the cheapest power, and a literal lightbulb moment showed us we can cut costs and emissions even further. *The Conversation*. https://theconversation.com/solar-is-the-cheapest-power-and-a-literal-light-bulb-moment-showed-us-we-can-cut-costs-and-emissions-even-further-187008

Thambimuthu, K., Soltanieh, M., & Abanades, J. C. (2005). *IPCC Special Report on Carbon Dioxide Capture and Storage.* Cambridge: Cambridge University Press.

Tijm, P., Waller, F., & Brown, D. (2001). Methanol technology developments for the new millennium. *Appl. Cat. A.*, *221*, 275–282.

Yang, F., Wang, T., Deng, X., Dang, J., Huang, Z., Hu, S., ... Ouyang, M. (2021). Review on hydrogen safety isues: Incident statistics, hydrogen diffusion and detonation process. *Int. J. Hydrogen Energ.*, *46*, 31467–31488.

10 Environmental and Social Impacts Assessment and Management

Oswald Eppers
K-UTEC

10.1 SUSTAINABLE GROWTH AND ENVIRONMENT

The annually published Global Risks Report from the World Economic Forum presents the results of the latest Global Risks Perception Survey (GRPS), followed by an analysis of key risks emanating from current economic, societal, environmental, and technological tensions (WEF, 2023). The report published in January 2022 notes that, for the next five years, respondents signaled social and environmental risks as the most concerning. However, over a ten-year horizon, "climate action failure", "extreme weather", and "biodiversity loss" rank as the top three most severe risks (WEF, 2022). People are concerned because the effects of climate change in the form of droughts, fires, floods, resource scarcity, and species loss are already all too evident and increasingly impacting daily life. Hardly a year goes by without new record temperatures being measured around the world or people having to endure new record storms, floods, or draughts. Governments, businesses, and societies are under increasing pressure to halt rapid climate change and prevent the worst consequences.

In the long list of environmental conferences organized by the United Nations, the historical Conference on Environment & Development (UNCED) held at Rio de Janeiro from 3 to 14 June 1992, also known as the "Earth Summit", defined a new blueprint for international action on environmental protection and management. Earth Summits are decennial meetings of world leaders to find ways to take concerted global action to stimulate sustainable development.

The UNCED conference was followed by several international conventions and protocols, which required the signatory countries to implement suitable environmental legislation. Several developed countries also imposed a series of environment-related conditions as non-tariff trade barriers with the objective of obliging developing countries to implement effective environmental regulatory regimes.

In this context, the United Nation's 2030 Agenda for Sustainable Development (UN, 2015) defined 17 Sustainable Development Goals (SDGs) that represent an urgent call to action by all countries – developed and developing – in a global partnership. The basic concept is that fighting poverty, improving health and education, reducing inequality, and boosting economic growth must go hand in hand with combating climate change and the protection of oceans and forests. Table 10.1 shows 17 UN sustainable development goals.

DOI: 10.1201/9781003374879-10

TABLE 10.1

The UN's 17 Sustainable Development Goals (SDGs) to Transform the World

UN's Sustainable Development Goals

Goal 1: No poverty	Goal 7: Affordable and clean energy	Goal 13: Climate action
Goal 2: Zero hunger	Goal 8: Decent work and economic growth	Goal 14: Life below water
Goal 3: Good health and well-being	Goal 9: Industry, innovation, and infrastructure	Goal 15: Life on land
Goal 4: Quality education	Goal 10: Reduced inequality	Goal 16: Peace and justice strong institutions
Goal 5: Gender equality	Goal 11: Sustainable cities and communities	Goal 17: Partnerships to achieve the goal
Goal 6: Clean water and Sanitation	Goal 12: Responsible consumption and production	

Environmental inequality is a growing problem as the world's freshwater resources and access to clean air increasingly become the privilege of the wealthier classes of society. In particular, many developing countries are affected by the results of climate change, pushing these vulnerable states into more poverty and political instability. The SDGs are an urgent call to reduce damage to the environment in order to tackle inequality and poverty.

10.1.1 SUSTAINABILITY IN PRACTICE

The concept of sustainability is based on the four pillars, namely, people, society, economy, and environment, all of which must be taken into account to assure sustainable development. While the strategy of sustainable development was originally developed in response to the extreme commercial exploitation of the world's resources and the continuous degradation of the environment, today the concept is broader. The commonly used definition of sustainability dates back to the so-called Brundtland Report (Brundtland, 1987). Sustainability was defined as "development that meets the needs of the present without compromising the ability of future generations to meet their own needs". While this definition combines two fundamental problems, the problem of environmental degradation and the need for economic growth to alleviate poverty, it has proven too general and imprecise for practical application. A few concepts for the implementation of sustainable development are presented in the following.

10.1.1.1 The Circular Economy Concept

One concept that has proven very useful in practical applications is the idea of the "circular economy", which has been developed to eliminate excessive waste streams, support the economy, and focus on products that are designed to have a positive impact on the environment. The concept of the circular economy was developed in 1990 by British economist David W. Pearce (Pearce and Turner, 1990) as an approach to minimize resource needs and use environmentally friendly technologies.

The natural material cycle was taken as a model to achieve production without waste (zero waste) or emissions (zero emission) through a cascading use of raw materials. A circular economy is based on the elimination of waste and pollution, the circulation of products and materials (at their highest value), and the regeneration of nature. The concept has been included in many environmental policies, including that of the European Union, who, in 2015, established the "Circular Economy Action Plan (CEAP)" as a comprehensive body of legislative and non-legislative actions that aimed to move the European economy from a linear to a circular model. Today, the concept of the circular economy is an integral part of most companies' environmental management strategies.

10.1.1.2   Regenerative Development Concept

Another concept that is derived from sustainability is the "regenerative development" approach. Regenerative development aims to enhance the capacity of all living species to co-evolve so that humanity can express its potential for diversity, complexity, and creativity (Mang and Haggard, 2016; Hes and DuPlessis, 2015). Human activities need to be brought into harmony with the ongoing evolution of life on earth, so that mankind can continue to develop its potential as human beings in a sustainable way at the same time. This puts sustainability in a different perspective. By requiring that resource use be only in harmony with the ecosystem, the ecological component in the concept of sustainable development is weighted much more than economic development. Although this approach is theoretically relevant and extends the concept of environmental sustainability, it has not yet gained widespread policy acceptance due to its rather limited practical applicability.

10.1.1.3   Eco-Sufism

It also is worth mentioning that the sustainability approach in several cultures has a strong religious background. One example is Eco-Sufism, which is based on an ancient *Syair Nasihat* manuscript, written in Arabic and Malay (Wirajaya et al., 2021). The teachings of Eco-Sufism focus on the coexistence of God, man, and nature and require that all human behavior must always be directed toward achieving God's pleasure and ensuring the safety and well-being of humanity and the environment, and the entire universe. Especially in traditional Muslim societies, this concept of sustainability plays an important role in decision-making.

10.1.2   ENVIRONMENTAL PROTECTION AND ECONOMY

The perception that environmental protection only costs an organization money and brings no real benefits is still widespread and is one of the biggest obstacles to implementing an effective environmental management system (EMS). Experience shows that an EMS is only sustainable if management can be convinced that the implementation of costly environmental measures brings economic benefits to the organization, i.e., a win-win situation for the environment and the economy can be achieved. Every environmental manager is therefore faced with the challenge of identifying win-win situations from the measures taken to prevent and mitigate environmental impacts and communicating them to senior management, employees, and the population. Typical examples include reducing freshwater and energy consumption by installing energy-saving technologies or reducing waste disposal costs by replacing hazardous substances in production processes with non-hazardous materials. In particular, implementing the concept of the circular economy offers many opportunities for cost savings by reducing waste streams and optimizing resource use. The steadily growing customer interest in ecologically and sustainably produced products also offers great potential for entering new markets and launching new products, by investing in ecologically and socially sustainable protection.

In this context, digitalization of the economy is sometimes seen as one of the cornerstones for sustainable transformation of the economy, due to the potential cost-saving effects and environmental benefits. The scope and limitations of the digital transformation of the economy for reducing environmental impacts are discussed below.

10.2   DIGITAL ECONOMY AND ENVIRONMENT

Digitalization tends to be associated with the public rather with positive ecological impacts, as there are evident advantages we experience in daily life. Switching from a printed newspaper to a digital one can save enormous amounts of paper and ink, as well as eliminating fuel costs for distribution. Digitalization enables a paperless office, video conferencing across borders and continents without long and expensive travel, and fast and inexpensive information exchange. Online shopping can also reduce fuel consumption and hence CO_2 emissions, as it enables the energy- and time-saving purchase of goods from great distances, without leaving the house.

But this is only part of the truth. On November 21, 2019, the Luxembourg Times reported on Google's plan to build a huge data center in the center of Luxembourg, which was immediately rejected by the population because of the enormous energy and water consumption announced. It was estimated that the data center's operation would require 10 million liters per day, about 10% of this small country's total water consumption. It was also estimated that the center would require up to 12% of the country's energy supply. Noise and air pollution were cited as other potential environmental impacts with unforeseen consequences for the population. This example shows that digital transformation can also have serious impacts on the environment, especially where huge data centers are to be installed.

There are several studies using statistical analysis and attributional process-based life cycle assessments (LCA) whose results support a reduction in emissions and resource consumption from digital products and services, compared to their conventional counterparts (e.g., teleworking vs office work, virtual meetings vs face-to-face, ebooks vs paper books, e-commerce vs traditional shopping, and e-banking vs traditional bank services). These first-order effects are associated with the reduced materiality of digital activities (Pérez-Martínez et al., 2023).

In this respect, digital transformation is an essential condition for transforming the current economic model and at the same time addressing some of the main challenges humanity is facing, including climate change, waste generation, resource depletion and efficiency, economic progress, outdated environmental legislation, and social development (ITU, 2022; Mondejar et al., 2021; O'Sullivan et al., 2021). A large part of this lies in the widespread use of information and communication technologies (ICT), which is seen as one of the most critical aspects of this technological change, as it is seen as an opportunity to achieve an effective decoupling between economic and social development, and environmental degradation (Del Río Castro et al., 2021; Wu et al., 2018, Pérez-Martínez et al., 2023).

But it is also important to consider the whole picture, i.e., the indirect impacts, not only the direct or first-level impacts. For this occasion, a recent study (Don et al., 2022) based on data from 60 countries examined the impact of the development of the digital economy on carbon emissions and the associated transmission mechanisms empirically, using the intermediary effect model. Evidence was found that the development of the digital economy significantly reduces carbon emission intensity but promotes increases in per capita carbon emissions. Economic growth, financial development, and industrial upgrading all play a mediating role between the digital economy and carbon emissions. Thus, digitalization paves the way not only for economic growth but also, as a side effect of economic growth, for increased global carbon emissions.

Brown and Duguid (2000) point out that one of the biggest obstacles to assessing the environmental impact of information technologies is that they catapult us from one to two to a million in terms of benefits. Between users and the imagined benefits stand people and their behavior, which is much harder to predict than the energy consumption of internet servers or pollutants from planes transporting packages around the globe. Excluding human behavior, our technological endeavors often fall victim to the everyday forces that shape daily life.

10.3 ENVIRONMENTAL MANAGEMENT

10.3.1 WHY IS IT IMPORTANT TO IMPLEMENT AN EMS?

So far, only a bird's eye view has been taken of the interrelationships between economic growth and the environment, looking at rather general concepts on a global scale. In practice, however, environmental managers have to ensure that the basis for sustainable economic growth is in place in their own company or organization. This can be achieved by implementing an environmental management system (EMS) as a framework to accomplish the defined environmental goals through consistent review, evaluation, and improvement of environmental performance.

10.3.1.1 Benefits of Implementing an EMS

A company with an implemented, standardized, and, as far as possible, officially certified EMS is recognized as one that takes appropriate action to reduce its environmental footprint. The EMS helps a company build better relationships with its customers and the public and is an important marketing tool. In addition to a good public image, implementing an EMS offers a company many opportunities to save money and resources.

In summary, there are four main benefits:

1. Compliance with legislation – Any fine that comes to public notice can have negative and sometimes even devastating consequences for a company's reputation. A certified EMS ensures that a company can comply better with environmental legislation and is always up to date with legal regulations.
2. Increased reputation by demonstration of leadership – An organization with a certified EMS demonstrates ethical leadership, showing customers, employees, and regulators that it is actively committed to environmental protection. As more and more consumers are taking environmental aspects into account in their purchasing decisions, this increases the company's reputation and can give the required competitive edge. EMS certification can also provide a competitive advantage in tenders, for example.
3. Reduced waste and emissions – The standard helps an organization focus on the right actions to reduce waste and/or effluents, and toxic and greenhouse gas emissions. Toxic substances can be replaced by non-hazardous ones, production processes can be optimized to produce less waste, or waste flows can be recycled. There are many ways to implement the principle of the circular economy. By using more renewable energy sources and energy-saving processes, CO_2 emissions can be reduced while the company becomes more efficient.
4. Reduced costs – An increase in efficiency automatically leads to reduced operating costs and increased productivity. Reducing energy, waste disposal, and wastewater treatment costs; eliminating environmental penalties; and reducing liability insurance costs are examples of potential savings opportunities.

10.3.2 Ten Core Elements of an EMS

An EMS typically consists of ten core elements as shown in Table 10.2. All elements within the system are interconnected like cogs in clockwork, and each is required for effective operation. The EMS should not only consider environmental aspects but also integrate related aspects such as health and safety and socio-economic issues.

This chapter is not intended to present a comprehensive compendium of all the parts required in an EMS, but rather gives a general overview of the EMS structure. The core elements of the EMS formed the basis for developing international EMS standards such as ISO 14001 and EMAS. Sections 10.4 and 10.5 present an introduction to these standards, which are intended to provide basic guidance for establishing policies, plans, programs, and procedures to support EMS implementation.

Each of the core elements represents an essential part of the overall management system.

TABLE 10.2
The Ten Core Elements of an EMS

Element	Title
1	Environmental and social commitment and policy
2	Planning
3	Organization and resources
4	Documents and records
5	Environmental and social risk management
6	Regulatory requirements
7	Monitoring and measurement
8	Emergency preparedness and response
9	Performance evaluation by audits
10	Management review

10.3.2.1 Element 1: Environmental and Social Commitment and Policy

Top management must formally commit to implementing an EMS and to a concerted environmental policy and goals for the organization. The organization's environmental policies, rules, and standards must be defined, including important issues such as who makes the decisions, how they are made and carried out, what sources of information are used to make decisions, and how to involve key stakeholders and the public in decision-making.

Management must make an explicit commitment to comply with or exceed all relevant laws, regulations, and standards designed to protect the environment, i.e., environmental protection and social responsibility should be incorporated into the corporate culture.

10.3.2.1.1 Corporate Culture and Corporate Social Responsibility Goals

Corporate culture encompasses behavioral and procedural norms, which include policies, procedures, ethics, values, goals, and codes of conduct. It defines the "personality" of a company and provides the framework for the working environment. As part of the corporate culture definition, corporate social responsibility goals should be defined if considered relevant for the organization. In defining these goals, the primary focus is often not on shared value. Rather, companies are concerned with issues that are of immediate importance to their business model. A banana plantation owner will want to share information with customers that local farmers are being treated in a socially responsible and fair manner. An oil company, on the contrary, will want to highlight the measures it has taken to mitigate climate change effects. With the right approach to social responsibility, companies can not only make a difference in local communities but also improve their own reputations and ultimately increase revenue.

10.3.2.2 Element 2: Planning

Based on the environmental policy and objectives established, the organization identifies the environmental and social impacts of its activities and assesses compliance with legal and other obligations. Consideration must be given to all aspects that may impact the environment negatively, pose a health risk to employees or the public, or have other negative socio-economic impacts. Examples of such impacts include industrial effluents, hazardous wastes, noise, air pollutants, and soil pollution. Based on an assessment of potential environmental impacts using an appropriate sampling and monitoring program, the organization establishes a prioritized list of environmental and social issues for which a work plan can be developed that sets goals to prevent, reduce, or eliminate environmental impacts. At this stage, it is important to define objectives that are carefully and thoughtfully planned, executable, and traceable and that serve as a roadmap for setting specific targets.

A proven approach to goal setting in this regard is the SMART system, first proposed in 1981 by George T. Doran (1981). Although the system has been applied in practice in many variations, the most common definition of SMART is that goals should be specific, measurable, achievable, realistic, and time-bound.

- *Specific* refers to defining the desired goals as precisely as possible. The more precise and narrow a goal is, the easier it is to identify the steps needed to achieve it.
- *Measurable* refers to the need to monitor progress toward each goal, i.e., the goals must be defined so that they can be tracked to monitor progress.
- *Achievable* refers to the fact that goals must be realistic and achievable within the defined time frame.
- *Relevant* means that the (single) goal itself must be consistent with the values and long-term goals.
- *Time-bound* refers to the need to achieve the goal within a reasonable time frame.

10.3.2.2.1   Environmental Aspects and Impacts

Without an understanding of the environmental implications of an organization's activities, products, and services, planning to avoid, minimize, mitigate, and/or manage environmental impacts or damage will not be possible. A good approach to understanding an organization's environmental implications is by starting to understand the business activities themselves. In the first step, a list of activities, products, and services (aspects) with potential environmental implications is generated. This list is then "sieved" to determine the aspects with major environmental impacts. These are those that require either control or corrective actions to reduce environmental impacts to an acceptable level (see Section 10.6).

10.3.2.2.2   Scope of the EMS

Part of the planning process is also the definition of the EMS's scope, which may consider the whole organization or only parts of it.

The scope of the EMS should include everything that could affect the organization's environmental performance and that the organization can control or influence.

It is important that the scope of the EMS is well defined and does not exclude any aspects that could affect the organization's environmental performance. Turning a blind eye to sensitive issues or trying to hide environmental impacts can be very damaging to an organization's credibility in the view of the public and the authorities.

10.3.2.3   Element 3: Organization and Resources

Without the necessary resources to implement, sustain, and improve the EMS and environmental commitments, the system will not work. Lack of human and financial resources is the main reason why many attempts to implement an EMS fail. In planning, management must ensure that the necessary resources are made available, that the EMS system rules are established sustainably, that they are implemented in such a way that they are known throughout the organization, and that the system is improved continuously over time. The size and characteristics of an organization define the number and type of resources required for both EMS implementation and maintenance. Similarly, principal contractors must demonstrate that they have the necessary resources and organizational structure to meet environmental commitments and project conditions. Responsibilities and accountabilities for environmental management are assigned through management plans, procedures, and position descriptions.

10.3.2.3.1   Process-Based Approach

The process-based or process-oriented management approach is a method of managing an organization based on processes, so that these are all integrated into one system and working toward the same goals.

For a clock to work without breakdowns and delays, all the gears must be perfectly aligned. The same principle can be applied to the management of an organization, which consists of parts and sub-systems that must be optimized to achieve defined business or organizational goals. To ensure optimal performance, everything in the gearbox – employees, divisions, resources, and/or equipment – must be aligned and ideally function synchronously.

It, therefore, makes much sense to integrate the EMS with the organizational or corporate quality management system (QMS). The most applied QMS standard ISO 9001 incorporates the process-based management approach in its strategy, by integrating the plan-do-check-act cycle and risk-based thinking in the management process. This forms the basis on which to align and synchronize the corporate quality and environmental management processes together in one interrelated system.

10.3.2.4   Element 4: Documents and Records

A document control system should be used to ensure that all documents are stored safely in both printed and electronic form, that the most recent and valid versions are always used, and that they are readily retrievable. All environmental records should be legible and identifiable, and traceable to the activity, product, or services provided. Documented programs and procedures to address hazards and risks, regulatory requirements, and operating standards should be maintained, including detailed environmental documentation like plans, procedures, and processes required for the successful implementation of the EMS.

In order to establish a functional EMS, a series of documents needs to be prepared, including, but not limited to:

- Environmental Policy and Commitment,
- Scope of the Environmental Management System,
- Evaluation of Environmental Risks and Opportunities,
- Evaluation of Environmental Aspects and Impacts,
- Environmental Objectives and plans for improvement,
- Operational Control Procedures.

10.3.2.5   Element 5: Environmental and Social Risk Management

Environmental and social risk management (ESRM) can be defined as the conscious and coordinated effort in appraising the potential and/or existing impact of various productive activities on their environment and people.

Site-specific environmental issues such as areas with contaminated soil and groundwater can be dealt with in the ESRM. Environmental, health, and social risks can be characterized, and risk reduction measures defined. Using defined criteria, appropriate risk reduction measures are chosen that reduce the risk to an acceptable level.

In Section 10.8.2, a more detailed description of the "risk" concept and different approaches to assess environmental risks is presented.

10.3.2.6   Element 6: Regulatory Requirements

According to its commitment in the Environmental Policy, the organization needs to meet or exceed all relevant laws, regulations, and standards for the protection of the environment. To this end, it should implement a compliance framework to manage and monitor its regulatory obligations and to help achieve performance expectations.

In order to ensure the achievement of full regulatory compliance, the following actions are recommended:

- Implementation of awareness training for employees and contractors;
- Implementation, use, and maintenance of a regulatory compliance matrix; and
- Regular audits of systems and activities to monitor compliance.

10.3.2.7   Element 7: Monitoring and Measurement

Once the EMS is implemented, it is essential to measure environmental quality parameters and emissions to verify compliance with applicable standards and legislation, determine the root causes of problems, identify areas requiring corrective action, or improve EMS performance and efficiency.

It is therefore important to have a good understanding of the impacts to be monitored and the information needed for the evaluation of environmental performance. The required monitoring, analysis, measurement, and evaluation methods must be defined in advance to ensure that the results are reproducible, reliable, and traceable. Subsequently, definition is required of what, where and at what frequency the monitoring must be performed, and how the results will be interpreted and reported. Support from environmental monitoring specialists might be needed to select the best measurement or monitoring equipment for the task, especially for monitoring and sampling in the field.

Environmental parameter monitoring has several objectives, including:

- Assessing the effectiveness of the EMS;
- Objective evaluation of environmental quality standards and emission threshold values;
- Verification of compliance with legal and in-house standards;
- Assessment of the suitability, adequacy, effectiveness, and efficiency of the EMS;
- Monitoring of environmental planning progress;
- Process performance verification; and
- Identification of corrective actions required.

Operating an EMS without effective monitoring and measurement procedures is like trying to determine a person's blood pressure without a medical device, just by observing the redness of the facial skin. Even if a face is flushed, it cannot be known whether there is really a problem with elevated blood pressure or something else entirely.

Monitoring environmental parameters means using measurable and interpretable parameters to obtain an objective and reproducible indication of the performance of measures taken, for example, to reduce emissions. With modern measurement technology, probes or sensors can be used to measure hazardous chemicals, e.g., NO_x, CO, SO_2, and dioxins, continuously in exhaust gases, or quality parameters such as pH, temperature, conductivity, dissolved oxygen, turbidity, or organic matter in effluents. Where continuous measurement is not possible or necessary, samples must be taken at defined intervals and sampling points and analyzed in an accredited or certified laboratory, to obtain reliable and reproducible results.

Environmental monitoring results can be used to determine program measurement criteria, also known as "performance indicators". These are powerful tools for evaluating the effectiveness and impact of an EMS program. Because an organization's needs may change over time, monitoring objectives may need to be modified from time to time to ensure that the data and results remain valid.

10.3.2.8   Element 8: Emergency Preparedness and Response

An environmental emergency can occur at any time in any organization and can impact the activity or business negatively. Any organization should consider having procedures in place to allow for a planned response to environmental emergencies. For emergency preparedness and response, several measures should be established:

- Identification and characterization of environmental risks, using an environmental risk assessment.
- Implementation of an emergency response plan with defined triggers for response actions.

- Training to ensure that personnel are aware of the risks and know what to do in case of an environmental emergency.
- Provision of necessary equipment, tools, and resources for use in emergencies.
- Assignment of an accountable owner, i.e., a responsible person for each identified risk.
- Assignment of risk mitigation activity owners.
- Provision of information to relevant interested parties of the risks and the emergency response planning.
- Implementation of measures to mitigate environmental risks.

Potential for environmental accidents and emergencies may be identified, using the risk identification and characterization tools discussed in Section 10.8.

10.3.2.9 Element 9: Performance Evaluation by Audit

Once the EMS is implemented, its performance needs to be monitored using systematic and independent internal and inclusive external audits. These, which are performed regularly, have several objectives:

- Evaluation of the performance and effectiveness of the EMS,
- Verification of the organization's compliance with legal requirements, and/ or with other organizational and stakeholder expectations,
- Verification that corrective action has been implemented successfully,
- Assessment of the performance of processes related to the EMS, and
- Identification of weaknesses and opportunities for improvement within the EMS.

A formalized audit schedule should be developed to define audit scope and frequency.

In the first instance, internal audits are carried out, i.e., with personnel designated by the company itself. Those performing the audits must be as impartial and objective as possible. By looking at how processes are organized and operate, and how they should operate according to the EMS, the auditor identifies deficiencies and opportunities for improvement. All audit results should be documented and discussed in regular management meetings.

External audits are performed by qualified auditors and are usually required in the official certification framework of the EMS.

10.3.2.10 Element 10: Management Review

The final element of the EMS implementation and improvement process is the management review, which involves a comprehensive check of compliance with defined rules and goals, taking into account any new environmental issues, changing circumstances in production or business development, and/or changes in legislation. Management reviews should be performed under the supervision of the site director, management representatives, or line managers with specific environmental responsibilities.

The management review should cover as a minimum:

- A review of the results from previous internal audits and reports;
- An assessment of internal and external environmental monitoring reports, to evaluate compliance with defined environmental emission goals and legislation, and definition of follow-up actions;
- The identification of non-conformances and definition of corrective actions required;
- The identification of changes in circumstances due to the implementation of new processes, changes to known environmental issues and/or environmental legislation;
- An assessment of the effectiveness of training provided and definition of new training requirements;
- A review of the organization's environmental policy, objectives and targets, and their consistency with its values; and
- Improvement recommendations.

Management reviews enable managers and key professionals to discuss problems and develop solutions together. They are therefore an important part of the EMS and minutes should be kept of all meetings so that follow-up actions can be planned and monitored.

10.3.3 Seven-Step Strategy for EMS Implementation

Implementation of an effective EMS is the cornerstone for continuous improvement of an organization's environmental performance. It can be achieved in seven steps (USEPA, 2022a):

1. Review of the organization's environmental goals;
2. Analysis of the organization's environmental impacts and legal compliance obligations;
3. Definition of environmental objectives and targets, to reduce environmental impacts and conform with compliance obligations;
4. Establishment of programs to meet these objectives and targets;
5. Monitoring and measuring within a program designed to assess progress in achieving the objectives;
6. Implementation of a training program to ensure employees' environmental awareness and competence; and
7. Review of the progress of the EMS and in achieving improvements.

Frequent review and evaluation of an organization's environmental performance enable the identification of opportunities for the continuous reduction of environmental impacts and improving the management system. The individual level of environmental performance in this context must be determined by each organization according to its individual objectives and is usually based on both legal requirements and in-house standards. An EMS is not a static system that will never change once implemented. In fact, it is essential to follow a continuous improvement strategy to adjust the system to new requirements and challenges.

10.3.4 Continuous Improvement – PDCA Cycle

Products, services, and/or processes in an organization must be improved constantly and continuously, to enhance competitiveness, maintain a dominant market position or, in the case of the EMS, minimize environmental impacts. The measures implemented for this can vary widely in terms of complexity, duration, implementation, and topic. In practice, there will never be a perfect system or product, and company management should always be open to suggestions for improvement. Information and results from internal audits, management reviews, or monitoring and measurement procedures, deliver important inputs for identifying deficits, and developing and initiating the implementation of improvements.

10.3.4.1 The PDCA Cycle

The so-called Plan-Do-Check-Act (PDCA) cycle is a powerful tool for implementing a continuous improvement process. The PDCA cycle, also known as the "Deming Wheel" or "Shewhart Cycle", was designed by the American physicist and management consultant Dr. William Edwards Deming in the 1950s and is an iterative, four-stage approach for achieving continuous business improvement. PDCA maintains systematic testing of possible solutions with a continuous feedback loop, allowing responsible managers to formulate and implement those that have proven successful.

Once the required improvement action has been identified, the cycle starts by planning the action (plan) and continues with its implementation (do). Subsequently, progress is monitored, results are reported (check), and further corrective action is taken if required (act).

In the context of an EMS, PDCA can be interpreted as:

Plan	Do	Check	Act
Environmental policy is defined, and environmental goals and processes established	Implementation of the ten core elements of the EMS	Monitoring, measuring, and evaluating environmental performance	Address nonconformities and continuous improvement

Even though the main objective of an EMS is avoiding and mitigating an organization's environmental impacts, the continuous improvement process also opens

FIGURE 10.1   The PDCA cycle.

many opportunities to decrease operating costs. Examples include energy or fresh-water saving measures, or reducing waste disposal and/or wastewater treatment costs through process improvements. Further positive side effects of an EMS are the opportunity to promote stronger operational control and employee stewardship in the organization.

EMS implementation and maintenance have costs that need to be addressed in the annual budget to ensure the required funding. An EMS has both internal and external costs. Internal costs include the time spent on environmental management and additional staff, if required. External costs include, for instance, the purchase of equipment for environmental monitoring, consulting services, and external training of personnel. Larger investments may be required if processes need to be changed, such as installing new machinery that uses less energy or water, or pro-duces less waste.

To take the importance of management commitment to the functionality of an EMS into account, USEPA (2023) suggests the integration of this topic into the PDCA cycle (Figure 10.2). This modified cycle has been found to give the best long-term cumulative effect of a continuous improvement cycle.

First, the organization establishes an environmental policy, which should also relate to social and cultural impacts, if considered necessary. A work plan is estab-lished, based on this policy, setting targets for continuous environmental perfor-mance improvement. Once it has been implemented, compliance with the work plan is monitored to determine whether the goals and objectives have been achieved, as expected, or if corrective action is required. It is usually the responsibility of top management to conduct an internal audit to verify that the EMS is functioning as

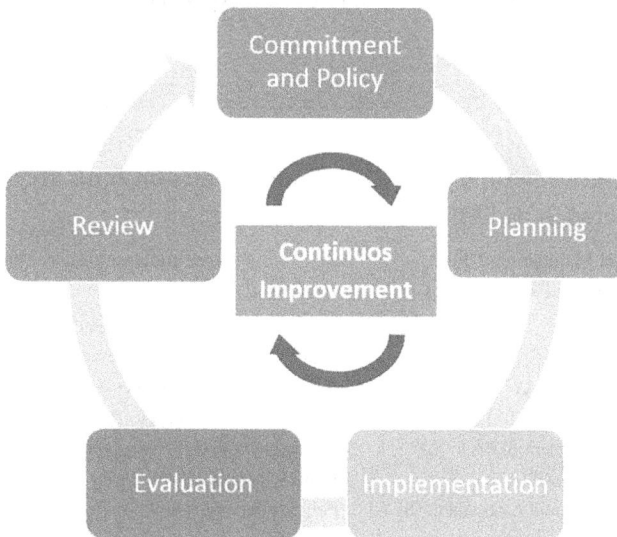

FIGURE 10.2 EMS continuous improvement cycle. (Courtesy of USEPA from Understanding ISO 14001 Environmental Management Systems, 2023. Available online: https://www.eraenvironmental.com/blog/iso-14001-environmental-management-systems.)

expected. Nonconformities are then analyzed by a senior management team, and the work plan is modified accordingly. With this last step, the entire cycle starts from the beginning, ensuring continuous system improvement.

To implement an EMS successfully, achieve the required level of performance, and obtain continuous improvement, effective leadership is needed at all levels, especially involving top management. For this reason, this strategy has proved superior to the basic four-step PDCA cycle for continuous management improvement.

10.3.5 Environmental Management Tools

Several environmental management tools can be incorporated into the EMS to minimize and mitigate environmental risks. Each tool has advantages and limitations in its applicability. The relevance and applicability of the tools vary with the decision-making requirements at different stages of product and/or project development and implementation. According to Petts (1999), there are seven primary environmental management decision tools that may be considered in an EMS (Table 10.3).

Strategic environmental assessment (SEA) involves a systematic assessment of the environmental impacts of a program, plan, or policy. It is important in evaluating and considering the economic and social aspects of a project adequately at the decision-making stage. SEA is particularly useful at the strategic level and allows for effective management of interactions and other cumulative impacts. Examples of SEA applications include the preparation and implementation of programs and plans for industry, energy, water and waste management, transportation, tourism, forestry, agriculture, and the fishery sector (Stinchcombe and Gibson, 2001).

The management tools EIA, RA, and LCA are explained in Sections 10.7 and 10.8.

TABLE 10.3

Seven Primary Environmental Management Decision Tools That Should be Considered in an EMS

Environmental Management Tool	Project Implementation Stage
Environmental Impact Assessment (EIA)	Pre-project level
Risk Assessment (RA)	Part of the environmental impact assessment, at both pre-project and project operation level
Lifecycle Analysis (LCA)	Developmental and project stages
Technology Assessment (TA)	Developmental and project stages
Cost-Benefit Assessment (CBA)	Developmental and project stages
Environmental Audit (EnA)	Project operation level
Strategic Environmental Assessment (SEA)	Policy, planning, and program levels

---

10.3.5.1 Environmental Management Plans (EMP)

The development and implementation of environmental management plans (EMPs) – sometimes also referred to as environmental action plans – are an important part of implementing an EMS and can be seen, formally, as part of the EMS.

An EMP contains important monitoring, prevention, mitigation, and organizational measures to be taken, during the implementation and operation of a project, to eliminate or mitigate adverse environmental and social impacts. For this purpose, the types and magnitudes of anticipated environmental impacts are assessed to determine which management plans are required for a project. EMPs are often submitted at a project planning phase during the environmental impact assessment (EIA) process and may be part of the documentation required by the environmental authority when deciding whether to approve the proposed project. After approval, the organization is obliged to implement the plans according to the specifications defined in the EIA.

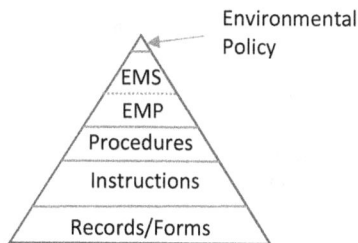

FIGURE 10.3   Typical EMS documentation hierarchy.

Examples of EMPs are:

- Construction Environmental Management Plan,
- Waste Management Plan,
- Air Quality and Noise Abatement Plan,
- Effluent Management Plan,
- Emergency Response Plan,
- Environmental Monitoring Plan,
- Social Management Plan (public and stakeholder engagement).

Environmental objectives and targets should be defined for each EMP, based on:

- The organization's environmental policy;
- Identified environmental and/or social aspects and impacts that need to be addressed;
- Relevant federal, provincial, and municipal standards;
- Measurable objectives; and
- Opportunities for continuous improvement.

10.3.6 Potential Problems with EMS

A certified EMS requires a great deal of bureaucracy and paperwork, which is often met with resistance from employees and managers. Especially in the initial phase, many new documents must be created, and work done to adapt the organizational structure, which overwhelms many employees and managers and makes them doubt the sense of EMS implementation. This perception is amplified by the fact that demonstrable EMS successes only become apparent after a start-up period. Resource allocation is also a common problem, and the organization's limited resources are often not used where they are most needed, but rather to build up administrative overhead without achieving the desired impact.

10.4   THE ISO 14001 STANDARD FOR EMS

The EMS must be adapted to the reality and needs of an organization, and it is recommended that a globally recognized standard that can be certified by an independent institution is used as a guideline. The most frequently used standard, worldwide, is ISO 14001 from the International Standards Organization (ISO). The current version of ISO 14001:2015 was published in September 2015 (ISO, 2015).

ISO 14001 establishes the EMS as part of the general management system and defines it as a "set of interrelated or interacting elements of an organization to establish policies and objectives, and processes to achieve those objectives". It establishes the EMS as "part of the Management System used to manage environmental aspects, fulfill compliance obligations, and address risks and opportunities".

ISO 14001 contains all tools required to manage and improve environmental risks effectively and systematically, resulting in better environmental performance, less waste, and ultimately cost savings. Rather than defining specific requirements for environmental performance, the standard provides a framework for companies and organizations to implement an effective environmental management system.

ISO 14001 provides a flexible framework for environmental management and can be applied in any organization, regardless of size, sector, or geographic location.

With the help of ISO 14001, the environmental manager can implement control instruments systematically to take the necessary measures to prevent and reduce environmental risks and impacts.

Once the policy framework is defined and the necessary organizational procedures have been implemented, the standard requires periodic audits, reviews, and adjustments to improve the system continuously.

ISO 14001 requires organizations to consider all environmentally harmful activities, including resource use and efficiency, water and wastewater issues, air pollution, soil contamination, waste management, and climate change adaptation and mitigation.

10.4.1   THE HISTORY OF ISO 14001

As part of the growing global interest in environmental protection, several international treaties and country-specific "codes of conduct" were adopted in the 1970s to help organizations reduce pollution. The first UN Conference on the Human Environment was held in Stockholm in 1972. At the historic Earth Summit in Rio de Janeiro in 1992, the United Nations agreed to take joint action to reduce the negative impacts of economic activity in order to better protect the environment. Also in 1992, BSI (the British Standards Institution) published the BS 7750 standard for the protection of the environment from the negative effects of economic and human activities. This formed the basis for the development of the first version of ISO 14001 "Environmental Management Systems – specification and guidance for use" in 1996.

ISO 14001 has been updated several times since 1996. ISO standards are reviewed every five years to keep them up to date and relevant. The latest version – ISO 14001:2015 – incorporates a risk-based approach, which enables prioritization of the risks and opportunities relevant to the organization. This helps organizations manage their resources in the best possible way to achieve positive outcomes.

Other important improvements in the 2015 standard include:

- Required appreciation of external and internal elements that may influence the performance of environmental management;
- Assessment of the needs of internal and external parties that may interact with the organization and impact EMS performance;
- Stronger commitment of top management and better integration of environmental management in the organizational structure, to improve alignment with the organization's strategic direction;
- Implementation of a communication plan to improve communication within the organization; and
- A life cycle perspective of the organization.

The success of the standard is demonstrated by a survey conducted by ISO in 2021, which found that nearly 420,000 organizations in 171 countries had achieved ISO 14001 certification (ISO, 2021).

10.5   THE EMAS REGULATION

10.5.1   INTRODUCTION TO EMAS

Although ISO 14001 is the most widely used international standard for EMS implementation, the so-called EMAS regulation was developed in the European Union to address European environmental legislation and to address some of the shortcomings of the ISO standard such as the lack of transparency. EMAS appeared in 1993, specifically in Council Regulation No. 183/93 of June 29, 1993, which allowed voluntary participation by organizations from different industrial sectors in the European Eco-Management and Audit Scheme (European Union, "EMAS"). EMAS requires the various member states of the European Union to establish administrative structures to support the program and to allow organizations to participate on a voluntary basis (European Union, 2023).

The main objective pursued by EMAS is to comply with European Community legislation related to environmental protection and sustainable development, as stipulated in the Treaty of the European Union signed in Maastricht in 1992. The EMAS regulation recognizes that organizations have to take responsibility for managing the environmental impacts of their activities, services, or products, so they need to:

- Implement an active approach within this field;
- Prevent, minimize and, if possible, eliminate pollution;

- Ensure the good management of resources; and
- Use environmentally friendly technologies.

10.5.2   EMAS Versus ISO 14001

EMAS is currently the most credible and robust environmental management tool on the market, adding several important elements to the requirements of ISO 14001:2015 to support organizations in the continuous improvement of their environmental performance.

Any organization, regardless of type or location, can be certified under ISO 14001, but EMAS certification is only available to companies within the member states of the European Union. Initially, EMAS was only applicable to the industrial sector (quarrying, manufacturing, solid and liquid waste, mining, electricity, and water supply), but it has evolved and now service industries and public administrations can be certified as well. EMAS goes beyond ISO 14001, and key differences include the following:

- EMAS has been created specifically for its use in the EU, whereas ISO 14001 is an international standard that can be applied worldwide. For organizations within the European Union, the implementation of the EMAS standard is therefore particularly beneficial, as it has been adapted to the EU environmental regulatory framework.
- EMAS requires a comprehensive assessment of environmental impacts before EMS implementation, whereas ISO 14001 only recommends performing an initial review to identify significant environmental aspects and impacts before implementation, without prescribing what should be covered. EMAS also generally has more stringent requirements for measuring and evaluating environmental performance parameters.
- ISO 14001 can be applied to all sectors of an organization, whereas EMAS is applied specifically to an organization's "operation sites", making it more specific and targeted.
- While the ISO 14001 certification process does not require a full compliance audit, the EMAS Regulation demands that organizations demonstrate compliance with environmental legislation, including permits. Since compliance is verified both by the environmental verifier on site and by the competent body during the registration process, a close relationship with the competent authorities quickly develops. This facilitates the work of the environmental authorities such as environmental monitoring and the processing of pending environmental permit procedures and reduces time and costs.
- EMAS decrees that the audit of the EMS and environmental performance must be carried out at least every three years, while ISO 14001 only requires their performance at planned intervals.
- EMAS requires active involvement by employees and their representatives, an issue not considered specifically in ISO 14001.
- EMAS defines requirements for transparent external communication with the community, which is not considered in ISO 14001. EMAS requires an open dialogue with external stakeholders, with external reporting based on a regularly published environmental statement.

- EMAS states that environmental policy must include a commitment to continual improvement of environmental performance with the aim of reducing impacts to acceptable levels through economically justifiable application of the best available technology, generally interpreted as the technology with the best environmental, health, and safety performance. The ISO 14001 standard in comparison is less stringent, requiring only that the EMS "promotes" the use of the most advanced technology, provided it is appropriate and economically viable for the organization.
- Choosing the best available technology in EMAS involves several steps:
  - Investigation of environmental aspects and impacts of operations – Before the best available technology for environmental mitigation can be chosen, the activities that have important impacts on the environment and the specific environmental issues that need to be addressed must be identified.
  - Research into available technologies – Once the environmental issues to be addressed have been identified, a study of available technologies is needed, using information from technology suppliers, trade publications, industry associations, government agencies, conferences, etc.
  - Evaluation of technologies – Once potential technologies have been identified, they need to be evaluated against the organization's specific needs and requirements. Evaluation needs to take factors such as cost, effectiveness, ease of implementation, and compatibility with existing processes into account.
  - Technology piloting – Before implementing a new technology at a large scale, it is good practice to pilot it in a small-scale test. This will help identify issues or challenges that may arise and refine the technology before implementing it in production.
  - Monitoring and evaluation of results – Once the new technology has been implemented, its performance must be monitored and evaluated to ensure that it is achieving the desired results – e.g., collecting data on energy consumption, emissions, waste generation, and other environmental factors.

10.6   ASSESSING ENVIRONMENTAL ASPECTS

Any activity, product, or service that interacts with the environment is referred to as an "aspect", with a potential positive or negative effect. Virtually, any activity generates some kind of impact, of which some may be considered significant and related to the realized or planned activities, and others not.

ISO 14001 describes an environmental aspect as those elements of an organization's activities, products, or physical resources that impact the environment and include utilities, products, processes, and activities such as storage and disposal. The identification of environmental aspects is also included in the so-called life cycle assessment (LCA), a methodology for assessing potential environmental impacts throughout a whole product life cycle ("cradle-to-grave"). The assessment starts with raw material extraction and continues through product manufacture and distribution, application, or use, to final product disposal or recycling. Any of these can cause impacts due to air emissions, effluent discharges, waste generation and disposal, use of renewable or non-renewable energy resources, water and/or soil pollution, etc. More details of LCA are given in Section 10.7.

The potential magnitude of impact depends very much on the type of activity, and the reference value used to identify whether or not the impact is measurable. The EMS must address all environmental aspects and impacts associated with the organization's activities, setting targets and objectives to manage them.

In general, a four-step procedure can be applied to identify environmental aspects:

1. Review/assess activities (at different project stages, including conceptual design, construction, operation, maintenance, abandonment, dismantling, or restoration);
2. Identify all environmental aspects;
3. Identify all possible negative and positive environmental impacts, associated with the activity, product, or service; and
4. Select a suitable measurement or detection system with environmental reference values to distinguish significant aspects from insignificant ones.

ISO 14001 defines the following factors to be considered while assessing aspects:

- Ecological effects
- Resource depletion
- Influences on human health
- Catastrophic effects
- Severity and duration of impacts
- Probability of occurrence
- Possible legal, financial, business risks
- Other business concerns associated with the aspects.

Under the EMS, an organization can control aspects to minimize impacts, such as reducing air polluting emissions, improving wastewater pretreatment, implementing waste recycling and waste reduction programs, and/or using environmentally friendly energy resources.

Not all aspects are under the company's full control, however. In some cases, decades of consistent environmental data are required to establish a comprehensive baseline and identify an activity's real impacts. Examples are assessments of the marine ecosystem or mining impacts on groundwater.

Some aspects can only be influenced in coordination with other actors in the supply chain, such as raw material and equipment suppliers, contractors, distributors, and, last but not least, the product customer or end user. Environmentally, it may be desirable to reduce the ecological fingerprint by extending product life, using biodegradable packaging, or replacing some non-biodegradable components with more environmentally friendly materials. This is only possible if the customer and other actors in the supply chain work together to try to sell a more environmentally sustainable product, as such changes may be accompanied by possible price premiums or changes in product characteristics and/or quality.

ISO 14001 requires the use of a systematic approach to determine aspects and impacts by implementing a documented procedure which:

- Identifies the environmental aspects of products, services, and activities, taking into account ongoing and planned activities, and in which all aspects that can be controlled or influenced need to be considered;
- Identifies the environmental impacts of each aspect; and
- Evaluates the significance of these aspects and impacts.

10.7 ASSESSING ENVIRONMENTAL IMPACTS

According to ISO 14001, an environmental impact is "any change to the environment, whether adverse or beneficial, wholly or partially resulting from an organization's activities, products or services". Once the environmental aspects and their causes have been identified, identification of the potential negative impacts on human health and the ecosystem in general can be initiated. Most of the environmental impacts are likely to be negative but not necessarily all of them.

Two basic questions need to be answered to identify the best procedure for evaluating the environmental aspects and impacts of an organization's activities, services, and products:

1. Whether or not a particular aspect has significant environmental impacts?
2. Whether or not it is possible to control or influence those impacts in the organization, and how this could be done?

To conduct the assessment, all key personnel should be involved in identifying potential environmental issues, in addition to the environmental manager. Activities, services, and products where significant impacts have been identified are then reassessed to set specific targets to control or mitigate the impacts. It is recommended that the process is conducted at least annually to identify changes in the aspects identified, especially if a new activity, service, or product has been implemented. Examples of typical environmental impacts are shown in Table 10.4.

To avoid social problems in implementing a planned intervention (programs, plans, projects), a related environmental impact assessment should always consider the potential social consequences, both positive and negative. For greenfield projects, the inclusion of social and cultural impacts, as well as mitigation measures is mandatory. In many countries, a comprehensive assessment of the archaeological heritage in the affected areas is also required, including defining measures to protect it.

Social impact may involve any of:

- Future fears and hopes – Changes in society lead to both fears for the future and hopeful expectations for a better life for children.
- Cultural impact – An activity can have a positive or negative impact on the way of life, including customs, values, language, archaeological heritage, and/or shared beliefs.

TABLE 10.4
Examples of Environmental Impacts

Aspect	Description	Potential Impacts
Industrial waste	Industrial waste is non-hazardous – e.g., construction debris, cardboard, pallets, and scrap.	Storage management and disposal costs
Asbestos-containing construction material	Several warehouse walls were constructed using attic and wall insulation containing white asbestos (chrysotile).	Chrysotile is carcinogenic to humans (mesothelioma), and can cause diseases like asbestosis and laryngeal effects; staff exposure to asbestos fibers must be avoided; high remediation costs likely.
Wastewater discharge	Operations discharge wastewater into the sewer, with only basic treatment by sedimentation tanks and oil separators. The storm and industrial wastewater sewer and not completely separate.	Risk of fines; risk of social problems with community; contamination of surface- and ground-water resources.
Contaminated land	Historical photographs show potentially waste-containing hazardous materials were buried in the area now covered by a car park.	Human health impact; groundwater and soil contamination; site investigation and potential remediation costs.
Dust emissions	A busy, unpaved road on the premises is a source of dust emissions, which sometimes reach adjacent houses and provoke complaints.	Air pollution, human health impact, social problems with community, and fines.
Noise and vibration	Noisy machinery with deficient sound-absorbing material	Human health impacts, social problems due to vibrations affecting neighborhood, fines, costs of noise, and vibration mitigation
Freshwater consumption	High freshwater consumption for daily process operations	Resource depletion, high wastewater treatment costs
Energy consumption	High energy demand for industrial process; energy source is largely non-renewable (coal and gas).	Greenhouse gas emissions, air pollution by emission of combustion gases like NO_x, SO_2, and CO_2

- Changes in social structure – The social cohesion or character of the evolved structure of a community, with its facilities and services, may change.
- Changes in the political system – The activity may have an impact on the local and regional political system, by changing the established bases for participation, decision-making, and freedom of choice.
- Impact on health and health care – A change in lifestyle can also have an impact on health and the health care system. Health depends to a large extent on the state of physical, mental, social, and spiritual well-being of the population.
- Impacts on property and personal rights – Especially for projects that require large amounts of land, there may be restrictions on property rights and

therefore personal rights. The impact can be particularly negative if people experience financial disadvantage, or their civil liberties are restricted. It can also be positive if the changes can be made in a socially acceptable way.

10.7.1 Six-Step Approach to Identify Environmental Impacts

Environmental impacts in an organization can be identified following a six-step approach:

1. A cross-functional team led by the environmental manager is formed to conduct the assessment. The broader the composition of the team chosen, the better, as each team member can bring their own perspective to the table.
2. The evaluation should consider normal operating conditions, as well as emergencies or crises.
3. The team uses brainstorming to identify possible environmental aspects in the organization.
4. Using defined criteria (see below) to select potential and actual environmental aspects, the key ones are determined.
5. The same criteria are used to assess the degree and severity of impacts (positive or negative).
6. If a threshold in the application of the criteria is exceeded, an environmental impact is classified as "significant" (see below).

Four criteria can be used to evaluate environmental interactions:

1. Legal impact – The organization must know and understand the legal framework and requirements it must comply with, as well as the potential consequences of non-compliance.
2. Impact on the environment – Understanding of the potential negative or positive impacts of the activity on human health, living organisms, or the ecosystem in general. All relevant environmental factors must be considered: air quality, air emissions, noise and vibration, radiation, surface-, ground- and waste-water, soil, solid waste (including contagious material and asbestos), and contaminated land.
3. Economic impact on the organization – Assessment of the potential economic impacts for the organization arising from production disruption, remediation costs, or fines.
4. Social impact – Assessment of whether and to what extent an activity has potential impacts on the opinions of the local community or authorities, and whether cultural issues are involved.

Once the environmental impacts have been identified and assessed, the next step is to quantify them. Because it is difficult to quantify the impact of non-compliance with laws and regulations (and other obligations), potential negative impacts are rated "yes" and "no" if no negative impact is expected.

TABLE 10.5

Example Impact Values for Environmental Impacts

Impact Level	Impact Value	Description
Severe	(5)	Very strong or disastrous impact, or very high probability of occurrence; irreversible over time and/or large areas affected
High	(4)	Strong impact or high probability of occurrence; partially or completely irreversible over time, and/or manageable areas affected
Medium	(3)	Moderate impact or probability of occurrence; reversible over time and/or smaller areas affected
Low	(2)	Low impact or probability of occurrence; reversible over time and/or small area affected
Insignificant	(1)	Negligible impact or probability of occurrence

It is common practice to use five categories for impact intensity (insignificant, low, moderate, high, and severe) on the environment (and human health), the organization, and the community, and assign impact values between 1 and 5 to them. These values can be positive for positive impacts and negative for negative ones (Table 10.5).

Whereas only negative impact effects are described in Table 10.6, positive impacts can occur as well, including wetland restoration, natural reserve installation, reforestation, the discharge of clean/treated water into a polluted river or irrigation with such wastewater.

The severity of the economic impacts very much depends on the size and economic situation of the organization. Whereas for some small companies a payment of US$ 100,000 could be disastrous, this amount might be insignificant for a large corporation.

Examples for positive social impacts of projects include the creation of jobs, support of education programs with scholarships, construction of schools and hospitals, increasing access to energy and freshwater in rural areas, and financial support of other welfare programs.

TABLE 10.6
Impact Classification Examples

Impact		Impact Level			
	Insignificant	Low	Medium	High	Severe
Environ-mental	Negligible impact or negligible probability of occurrence.	Weak impact on human health, ecology, or welfare; reversible over time and/or very small areas affected.	Moderate impact on human health, ecology, or welfare; reversible over time and/or medium-size areas affected.	Strong impact on human health, ecology, or welfare; irreversible over time and/or manageable areas affected; high remediation costs involved.	Very strong or disastrous impact on human health, ecology, or welfare; irreversible over time and/or several large areas affected; remediation impossible or extremely expensive.
Financial	Negligible probability of an interruption of normal operations and no significant economic losses for the organization.	Low probability of an interruption of normal operations and/or cost of damage, fine, or remediation is below $100k.	Moderate probability of an interruption of normal operations and/or cost of damage, fine, or remediation is between $100k and $1 million; recoverable impact without significant repercussions.	High probability of an interruption of normal operations and/or cost of damage, fine, or remediation is between $1 and $5 million.	Very high probability of an interruption of normal operations and/or cost of damage, fine, or remediation is more than $5 million.
Social	negligible or non-existent public interest	Only occasional isolated negative public critics; temporary.	Moderate negative public response (e.g., back-page press coverage); temporary disapproval.	Significant negative public response (e.g., lawsuit, front-page press coverage); likely persistent and long-lasting disapproval.	Very negative to disastrous public response (e.g., lawsuit, social unrest); persistent and long-lasting disapproval.
Legal	"yes" if there is a negative expected impact; "no" if no negative impact is expected				

TABLE 10.7
Example of a Team Aspect Identification and Impact Valuation Form

Activity/ Aspect Product/ Service	Impact Value					Objective	Target/Corrective Action
	Environmental	Economic	Social	Legal			
				Yes	No		
1.							
2.							

10.7.1.1 Aspects, Impacts, Objectives, and Target Identification Form

The use of a form as shown in Table 10.7 is recommended to enable the systematic identification of aspects, impacts, objectives, and targets. The form or analysis worksheet should be filled out by the team and is an important tool for setting priorities and allocating resources accordingly. Each aspect identified should be evaluated according to its legal, environmental, financial, organizational, and community impacts. The form should be reviewed and updated annually.

10.7.1.2 Outcome Evaluation

The total impact of each interaction will be the sum of the scores for the individual criteria. The potential impact of an activity is considered significant if one or more of the following conditions are met:

- There are legal obligations to comply with when conducting the activity,
- The sum of the calculated scores is 10 or higher, or
- Neither of the above conditions is met, but the activity must be considered according to the EMS.

The relevant scores of all significant activities are used as the basis for defining a priority list. Scores above 25 might be selected for "severe", and between 10 and 25 for "significant" impacts. The higher an activity is rated, the more resources must be devoted to managing it, in order to mitigate environmental impacts.

10.7.2 DEVELOPMENT PROJECT EIAS

National, regional, or local environmental regulation may require an environmental clearance or permit for all new public and private development projects (greenfield) or the expansion of existing projects (brownfield). The decision is usually based on an Environmental Impact Assessment (EIA), wherein which the likely adverse and beneficial environmental, human health, socio-economic and cultural impacts of a proposed project or development are analyzed. The assessment's detail and depth depend on the expected impacts and, in most legislations, are defined according to the activity sector and type. A shopping center will have much less potential impact than a large open pit copper mine and therefore require a less rigorous EIA process.

The aim of an EIA is to identify the environmental, human health, social and economic impacts of a project prior to decision-making, at an early stage in project planning and design. Only at this stage can the project planning be adjusted easily to suit local conditions, to reduce negative impacts, and to present the different options to legal authorities.

The EIA process for development projects is more comprehensive than that described in chapter 10.7.1 for a company or organization's ongoing activities. In general, it involves ten steps:

1. Screening – Screening determines whether an EIA is necessary, and, if yes, which type is required for a particular project. According to the applicable environmental legislation, this may be an environmental declaration, for low-impact projects, or a semi-detailed or detailed EIA for larger projects with higher impact probability. Two approaches are applied for screening:
   - Prescriptive or standardized approaches – These are based on environmental legislation, where types of development proposals are linked to different types of EIA. For instance, according to EU legislation, 89 types of projects require an EIA under certain conditions (mainly defined by size and project extension), including waste treatment plants, thermal power plants, roads and railroad lines, raw material extraction, hydroelectric power plants, intensive animal husbandry and industrial facilities (including chemical plants, iron and steel mills, pulp and paper mills, foundries, cement plants, refineries, breweries, etc.).
   - Customized approaches – The type of EIA required is selected case-by-case, individually, applying indicative guidance with categories.
2. Scoping – Once the type of EIA has been identified, the project's potential impacts, influence area, mitigation measures, and requirements for monitoring are defined.

   In this stage, potential environmental impacts on environmental parameters such as water, air, and soil, as well as socio-economic impacts in the different project phases (e.g., design, construction, operation, and demolition) are assessed, based on professional judgment and experience from similar projects. A wealth of information is available on the internet for this from relevant environmental authorities like USEPA and the European Environmental Agency (see Section 10.9).
3. Baseline data collection – Before starting with the activities, it is important to determine the project area's environmental and socio-economic baseline. This information is required so that the monitoring results can be compared with the situation before project implementation, to identify positive or negative deviations related to project activities.
4. Impact prediction – In this phase, the project's impacts are predicted, based on information from similar projects and the evaluator's experience. Therefore, a good technical and logistical knowledge of the planned project is required. Impacts can be positive or negative, reversible or irreversible, and/or temporary or permanent.
5. Mitigation measures – Based on the identification of potential impacts, preventative and mitigation measures must be defined. Part of this phase is also the definition of compensation for probable environmental damage or loss.

6. Public participation – Public participation is typically conducted through public hearings, a form of community involvement that brings stakeholders together to learn about project planning, anticipated impacts, and planned mitigation measures, and to listen to questions and concerns. This is an opportunity for the community to influence the decision-making process by expressing opinions and submitting proposals for project modifications.

   Public participation is an essential part of project planning, and, especially in developing countries, it is important to inform the local population as early as possible about the project and talk openly and seriously about their concerns and fears. All too often, companies fail to do so and face strong opposition from local communities, which can derail the entire project – especially in the case of large mining and infrastructure projects.

7. Decision-making – During the decision-making process, the environmental authority reviews the EIA to determine whether all relevant impacts have been considered, whether appropriate mitigation measures have been defined, and whether the potential adverse impacts of the project are in the public interest and remain within legal limits with the proposed mitigation measures. Not only direct environmental factors need to be considered, but also impacts on landscape, land-use, and cultural and archeological heritage need to be considered.

8. Monitoring – To verify compliance with environmental legislation and compare real impacts with those predicted in the EIA, monitoring of environmental and social-economic parameters is required during the construction, operation, and closing/restoration phases.

9. Assessment of alternatives and delineation of mitigation measures – For every project, possible alternatives with potentially lower impacts need to be identified. Project alternatives include, among other things, location, technology, and project size.

10. Risk assessment – Assessment of environmental and human health risks must be included. Risk assessment is an essential part of emergency response plan development.

More detailed information on the legal requirements and the methodology and tools of the EIA can be found in Section 10.9.

10.7.3   Impact Assessment with the Leopold-Matrix

10.7.3.1   Introduction to the Leopold-Matrix

The Leopold Matrix (Leopold et al., 1971) is still the most widely used method for identifying and weighting the potential environmental impacts of development or planned projects.

The original matrix consists of a grid of 100 rows with project activities in different project phases on the horizontal axis, and 88 columns with the potential environmental and socio-economic impacts of these activities on the vertical axis. Leopold et al. (1971) recommend eight steps for assessing environmental impacts:

1. Definition of the proposed project's main objective,
2. Evaluation of technological options to achieve the objective,
3. Discussion of one or more project options or measures that may cause environmental impacts,
4. Analysis of baseline conditions prior to initiating action,
5. Development of technical proposals for actions, including a cost-benefit analysis,
6. Analysis of the proposed action's environmental impacts,
7. Numerical assessment of the proposed action's environmental impacts, and
8. Summary and recommendations.

The EIA must define both the "magnitude" of the impact and the "importance" or significance of the various environmental factors. Whereas the magnitude can be evaluated on the basis of objective characteristics like comparison of analytical data with environmental quality specifications, the importance of impacts is more subjective, depending on the evaluator team's experience and the availability of information from similar projects. Hence, numerical assessment of each action's environmental impacts depends on quantitative values of magnitude and rather qualitative information about impact importance.

In its original format the Leopold matrix contains about 8,800 possible interactions, but, in practical applications, usually only 20 to 50 interactions merit comprehensive consideration because of their explicit magnitude and importance. For each expected effect, the corresponding matrix cell is divided diagonally from upper right to lower left. In the upper (left) half, the magnitude is indicated on a scale from −5 to +5 and in the lower half the significance from 1 to 5. In addition to assigning numerical values, plus (+) or minus (−) signs can be used to indicate whether an effect is beneficial or detrimental.

Separate impact factors are assessed for each relevant environmental component identified. The impact magnitude ratings are: 0 – no observable effect; 1 – low impact; 2 – acceptable impact; 3 – medium impact; 4 – significant impact; 5 – very high effect (devastation).

The impact significance scale ratings are: 1 – limited impact on location; 2 – impact of importance for municipality; 3 – impact of regional character; 4 – impact of national character; 5 – impact of cross-border character.

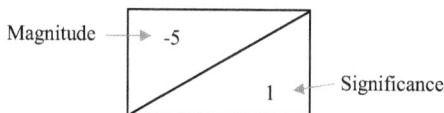

To simplify the matrix, impacts should only be selected if they have a relatively large magnitude and significance and are identified as interacting. Each impact identified must then be justified sufficiently for the reviewer to understand and systematically track the evaluator's viewpoints, and on that basis identify areas of agreement and disagreement.

10.7.3.2 Application of the Leopold-Matrix – Wind Farm

10.7.3.2.1 General Potential Impacts of Wind Farms

Wind energy is one of the core technologies for reducing carbon emissions because no carbon dioxide is released directly when electricity is generated from wind energy. It is a comparatively clean energy source that releases no pollutants during operation and produces no waste. However, quite significant environmental impacts and carbon dioxide emissions can occur during manufacture, delivery, installation, operational maintenance, and end-of-life dismantling and waste disposal.

Several local environmental impacts during operation are related to flora. Some birds and bats are at risk from wind turbines due to collisions, disturbance, or habitat destruction. Turbines must therefore be sited and designed to minimize impacts on bird populations. Wind farms must also not be installed near important migration routes and/or habitats. It also means ensuring a minimum distance from woodlands, hedgerows, and trees.

Strategies to avoid collisions with bats include avoiding particularly hazardous locations for these animals and shutting down turbines at certain times of the year and during night when wind speeds are low and bat activity is high. Bats can also be impacted even without direct contact with rotor blades, as they can suffer often deadly "barotrauma" caused by pressure differences, especially at the rotor blade ends.

To make the most of wind resources, turbines must be set high, and heights of 130–230 m are now standard. Even though turbines are installed at minimum distances from towns and villages, their height causes visual impact, whether in the open countryside or a populated area. Impacts can be minimized by screening turbines with landscape features such as trees and/or hills, and not siting them in sensitive landscapes.

Wind turbine noise is considered a nuisance, especially in rural areas with little background noise. However, modern turbines emit very little noise, and, on a windy day, the wind will mask the turbine's noise emissions.

Another disturbing effect is the so-called shadow flicker, a flicker effect caused by the sun shining through the turbine blades and casting a moving shadow. Modern turbines are designed to shut themselves down when the potential for shadow flicker exists.

Other potential negative impacts on communication links, such as television signals or cellular networks, as well as radar and aircraft navigation systems, must also be considered when determining location.

A wind farm can also have positive socio-economic impacts, of course, especially if the operator allows residents to invest preferentially in the turbine and thereby secure a share of the profits. For this purpose, developers of large-scale projects often offer a community fund in which individuals or entire communities can participate. Owners of land on which such plants are installed usually receive generous lease terms. Farmers on whose land wind turbines are installed can also continue to farm the land around the turbine base, giving them a double source of income.

Another positive effect is that the installation of a wind farm provides employment and business opportunities for the local community.

10.7.3.2.2   Leopold Impact Matrix

Considering these and other potential impacts, ten project phases with relevant environmental impact factors can be identified (adapted from Josimović et al., 2014):

1. Location selection
2. Column foundations
3. Construction materials
4. Substation construction
5. Transmission line construction
6. Access route construction
7. Construction equipment operation
8. Waste management
9. Project exploitation
10. Dismantling/restoration

In principle, it would be possible to reduce these components and define aggregate or average ratings. It is also important in the analysis to consider that some components may be synergistic and reinforce each other's effects. A Leopold-Matrix is presented below in which the implementation of mitigation measures is ignored (see Table 10.8).

Noise emission, mainly during construction, is the impact with the highest score, followed by landscape impacts. Soil and fauna impacts are also considered significant. Mitigation measures need to be implemented in the corresponding project phases for all impacts with relatively higher scores.

10.7.4   Life Cycle Assessment, LCA

10.7.4.1   Basics of LCA

A powerful tool for assessing environmental and human health impacts is life cycle assessment (LCA), a methodology standardized in ISO 14040 (ISO, 2006).

ISO 14040 defines LCA "as the compilation and evaluation of the inputs, outputs and the potential environmental impacts of a product system throughout its life cycle". This method is sometimes referred to, therefore, as "cradle-to-grave" assessment.

According to the European Union (European Commission, 2019a), it is a tool to:

- Enable quantification of potential environmental impacts of a product, process, or system throughout its life cycle, from the extraction of raw materials to the end of life.
- Quantify all relevant flows of raw materials consumed and pollutants emitted throughout the supply chain.
- Assess the potential impacts on the environment and human health of the entire supply chain of a product comprehensively and identify environmental impact hotspots across the supply chain.
- Identify trade-offs between life-cycle stages, impact categories or regions that can lead to a shifting of environmental burdens.
- Provide insight into upstream and downstream trade-offs related to environmental impacts, human health, and resource consumption with a single tool. Social and economic assessments can be added to complement these results.

TABLE 10.8

Leopold Matrix for hypothetical Wind Park Project

	Impact Factors	Location selection	Column foundation	Construction materials	Substation construction	Transmission line construction	Access route construction	Construction equipment operation	Waste management	Project Exploitation	Dismantling/Restoration Works	SUM
Physical Components	Land/Soil	-1 /1	-2 /1	-1 /1	-1 /1	-1 /2	-1 /1	-2 /1	-2 /1	-1 /1	-1 /1	-13
	Air	0	0	0	-1 /1	0	-2 /1	-3 /1	-1 /1	0	-2	-9
	Noise	-1 /1	-1 /1	-1 /1	-2 /1	-1 /1	-2 /1	-3 /1	0	-2 /1	-2 /1	-16
	Water	0	0	0	0	0	0	0	0	0	0	0
	Erosion	0	0	0	0	0	0	0	0	0	0	0
	Micro-climate	0	0	0	0	0	0	0	0	0	0	0
Biological Components	Diversity of Flora	0	-1 /1	0	0	-1 /1	0	-1 /1	-1 /1	0	-1	-4
	Diversity of Fauna	-2 /1	-1 /1	-1 /1	-1 /1	-1 /1	0	-2 /1	-2 /1	-2 /2	0	-12
	Birds	-2 /1	-1 /1	-1 /1	-1 /1	-1 /1	0	-2 /1	-2 /1	-2 /2	0	-12
	Bats	-2 /1	-1 /1	-1 /1	-1 /1	-1 /1	0	-2 /1	-2 /1	-2 /2	0	-12
	Barriers/Corridors	-2 /1	-1 /1	-1 /1	-1 /1	-1 /1	0	-1 /1	-1 /1	-2 /2	-1	-11
	Economy	0	0	0	0	0	0	0	0	+1 /2	0	+1
	Land use	-1 /1	-2 /1	-1 /1	-1 /1	-1 /1	-1 /1	-1 /1	-2 /1	-1 /1	-1 /1	-12
	Landscape	-2 /1	-2 /1	-1 /1	-2 /1	-1 /1	-1 /1	-1 /1	-3 /1	-2 /1	0 /1	-15
	Cultural heritage	0	-2 /2	0	0	0	0	0	0	-2 /2	0	-4
	Accidents	-2 /1	0	0	-1 /1	0	0	0	-2 /1	-2 /1	-2 /1	-9
	TOTAL	-15	-14	-8	-12	-9	-7	-18	-18	-17	-10	

Life cycle assessment has established itself as an important tool in the industry for:

- Reducing environmental impacts throughout the life cycle of goods and services,
- Improving product competitiveness,
- Improving communication with government agencies,
- Decision-making on improving product design (e.g., use of sustainable materials, consideration of recycling), and
- Decision-making on procurement and technology investments, etc.

10.7.4.2   LCA Methodology Phases

LCA implementation is divided into four phases: (1) goal and scope definition; (2) life cycle inventory (LCI); (3) life cycle impact assessment (LCIA); and (4) result interpretation. Figure 10.4 shows that the phases are inter-dependent, not isolated. With the first three phases complete, the results are refined using an iterative approach. A LCA can become very complex depending on its scope, and it is important to interpret the results continuously as the process is worked through, to optimize the outcome.

10.7.4.2.1   Goal and Scope Definition

In the first phase, the analytical goals and scopes are defined, i.e., what exactly needs to be analyzed, and what results are expected at what level of detail. Assuming a new fruit-juice product needs to be assessed, first the quantity produced and marketed (in functional units, i.e. volume, weight, number of packages, etc.) must be defined. The ecological footprint and environmental impacts differ if the juice is bottled in 1 L glass or plastic bottles, or in a plasticized cardboard carton, and whether the

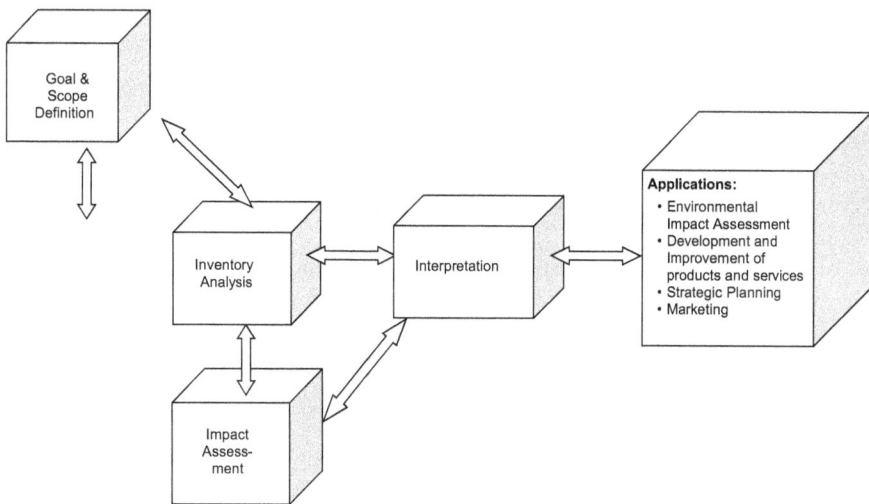

FIGURE 10.4   Four phases of an LCA.

container is disposable or returnable, and where the packaging is recycled. The definition of a Functional Unit (FU) as a quantified description and reference base of a product's function is important for building and modeling a product system in LCA. Common examples of food-related FUs are 1 L, 1 kg, 1000 calories of product or 1 ha of land, to assess the area needed for food production.

Defining the goal and scope requires answers to three questions:

10.7.4.2.1.1   What Will be Assessed?

In order to evaluate the impact of a product, the amount of product to be considered must be defined. To do this, it is necessary to determine the product's FU and define the assessment depth. Considering the environmental, social, and economic impacts in the raw materials' countries of origin, can become very complex, so it is important to determine where to draw the analysis boundaries. It can be difficult and, in some cases, even impossible to assess all the environmental and social impacts of goods produced, for instance, in Africa, Asia, or South America thoroughly. Sometimes the only option is to require suppliers to meet sustainability criteria when purchasing raw materials and rely on the information they provide.

10.7.4.2.1.2   What Impact Category Will be Focused On?

Impact-causing emissions have different shapes and formats. In the LCIA phase of a life cycle assessment, different types of emissions are converted into numerical values to quantify impacts. Emissions from a fruit-juice factory are very different from those from a chemical plant, and impact categories need to be defined accordingly. European Standard EN 15804+A2 (European Commission, 2019a) defines criteria for LCAs in the construction sector which can be adapted easily to other sectors. EN 15804+A2 establishes 15 impact categories as shown in Table 10.9.

If different emissions are reflected in the same impact, the impacts should be related to a single unit so that they can be compared (Table 10.10). A typical example is the "global warming potential" (GWP) of different greenhouse gases, defined by the Intergovernmental Panel on Climate Change, IPCC (2019). The GWP expresses the warming effect of a certain amount of greenhouse gas over a fixed period of time (usually 100 years) compared to that of CO_2. For example, methane has a 28-times greater climate impact than CO_2 but remains in the atmosphere for less time. Table 10.11 shows the GWPs with atmospheric retention times for selected greenhouse gases like nitrous oxide, hydrochlorofluorocarbons (HCFCs), and hydrofluorocarbons (HFCs).

In Table 10.10 more detailed descriptions of the units are presented. More information about characterization factors for LCI can be found in the literature (Leiden University, 2016; EC, 2016).

Table 10.11 shows the lifetimes and direct (except for CH_4) 100-year GWPs relative to CO_2 for selected GHG and ozone-depleting substances and their replacements (Smith et al., 2021).

TABLE 10.9
Environmental Impact Categories in the Construction Sector

Impact Category/ Indicator	Unit	Description
Climate change – total, fossil, biogenic and land use	kg CO_2-eq	Potential global warming due to emissions of greenhouse gases from fossil or bio-based resources, or land use changes.
Ozone depletion	kg CFC-11-eq	Stratospheric ozone depletion emissions to air
Photochemical ozone formation	kg NMVOC-eq	Formation of ozone at ground level in the troposphere is caused by photochemical oxidation of volatile organic compounds (VOCs) and carbon monoxide (CO), in the presence of nitrogen oxides (NO_x) and sunlight. Ozone in the troposphere attacks organic compounds in animals and plants and is responsible for respiratory problems during photochemical smog ("summer smog") phases in cities.
Acidification	kg mol H^+	Potential acidification of soil and water is caused by air pollutants such as nitrogen oxides and sulfur dioxide. Acidification is partially responsible for the decline of coniferous forests, coral depletion, and an increase in fish mortality. Combustion processes in power and heat generation, as well as transport, produce acidic gases such as NO_x and SO_2, which are also chemically modified in the atmosphere and finally washed out of the air into the soil and surface water.
Eutrophication – freshwater	kg PO_4-eq	Pollution of freshwater resources with nutritional elements – e.g., nitrogen- or phosphate-containing compounds – leads to excessive algae growth with resulting oxygen depletion and causing fish mortality
Eutrophication – marine	Kg N-eq	Impacts caused by eutrophication of the marine ecosystem, with nutritional elements from effluents or residues.
Eutrophication – terrestrial	mol N-eq	Impacts caused by excessive eutrophication of the terrestrial ecosystem with nutritional elements from fertilizers, residues, sewage, and/or wastewater.
Depletion of abiotic resources – minerals and metals	kg Sb-eq	Depletion of natural, non-fossil resources. Extracting large quantities of non-fossil resources will force future generations to extract lower-concentration or lower-value resources.
Depletion of abiotic resources – fossil fuels	MJ, net calorific value	Depletion of natural fossil fuel resources.

(Continued)

TABLE 10.9 (*Continued*)
Environmental Impact Categories in the Construction Sector

Impact Category/ Indicator	Unit	Description
Human toxicity – cancer, non-cancer	CTU_h CTU_h = Comparative toxic unit for human	Impact on humans of toxic substances emitted to the environment.
Freshwater eco-toxicity	CTU_e Comparative toxic unit for eco-toxicity	Impact on freshwater ecosystems of toxic substances emitted to the environment.
Water use	m^3 world eq. deprived	Relative amount of fresh or industrial water used, in relation to regionalized water scarcity factors.
Land use	Dimensionless	Changes in soil quality (biotic production, erosion resistance, mechanical filtration).
Ionizing radiation, human health	kBq U-235	Radionuclide emissions can damage human health and ecosystems, when critical target doses are exceeded.
Particulate matter emissions	Disease incidence	Particulate matter emissions exceeding threshold levels can cause health and ecosystem damage.

TABLE 10.10
Description of Units of Impact Categories

Unit	Description
kg CO_2-eq	The impact category 'climate change' is expressed in kg CO_2 equivalents, i.e., the impact of greenhouse gases is related to the impact of CO_2 (for examples, see Table 10.11).
kg CFC-11-eq	Ozone depletion potential (ODP) is expressed relative to the impact of trichlorofluoromethane, also called freon-11, CFC-11, or R-11. Examples (USEPA, 2023b): CFC-11 ODP=1 1,1,1,2-tetrafluoroethane (CF_3-CH_2F): 0.000015 ODP bromochlorodifluoromethane (BCF, Halon-1211): 6.9 ODP Carbon tetrachloride (CCl_4): 0.82 ODP Methyl bromide (CH_3Br): 0.57 ODP N_2O: 0.017 ODP Alcanes, ammonia, CO_2, nitrogen: 0.0 ODP
kg NMVOC-eq	Photochemical ozone formation is expressed relative to the impact of non-methane volatile organic compounds, NMVOC (e.g., alcohols and aromatics). Examples, unspecified emissions to air, unspecified location (Van Zelm et al., 2008) 1 kg nitrogen oxides = 1 kg NMVOC eq. 1 kg carbon monoxide (fossil) = 0.0456 kg NMVOC eq. 1 kg acetic acid = 0.164 kg NMVOC eq.

(Continued)

TABLE 10.10 (*Continued*)
Description of Units of Impact Categories

Unit	Description
kg mol H⁺ eq.	To express the potential impact of substances to acidification, the contribution to acidification of a substance is converted to the equivalent of moles of protons, i.e., the cationic form of atomic hydrogen, mol H⁺ per kg of substance.
	Examples are:
	1 kg ammonia = 3.02 mol H⁺ eq.
	1 kg nitrogen oxides = 0.74 mol H⁺ eq.
	1 kg sulfur oxides = 1.31 mol H⁺ eq.
kg PO₄-eq	The potential impact of substances contributing to eutrophication is converted to the equivalent of kilograms of phosphorus (kg P eq).
	For example, 1 kg ammonia has a eutrophication potential of 0.35 kg PO₄-equivalents, i.e., the eutrophication potential of 1 kg ammonia and 0.35 kg phosphate are identical.
kg N-eq	The potential impact of substances contributing to marine eutrophication is usually converted to the equivalent of kilograms of nitrogen (kg N eq).
	Examples, unspecified emissions to air, unspecified location:
	1 kg nitrogen oxides = 0.389 kg N eq.
	1 kg ammonia = 0.092 kg N eq
kg Sb-eq	The amount of materials contributing to resource depletion are converted into equivalents of kilograms of antimony (kg Sb eq).
	Examples:
	1 kg antimony = 1 kg Sb eq.
	1 kg aluminum = 1.09 * 10⁻⁹ Sb eq.
	1 kg silver = 1.18 kg Sb eq.
MJ, net calorific value	The amounts of materials contributing to resource depletion are converted into MJ.
CTUₕ	This indicator refers to potential impacts on human health caused by exposure to substances through the air, water, and soil. The unit of measurement is comparative toxic unit for humans (CTUₕ), based on a model called USEt$_{ox}$ (https://usetox.org/model). ESEtox contains a toxicity database of hundreds of chemical compounds.
CTUe	The unit of measurement is Comparative Toxic Unit for ecosystems (CTU$_e$), based on the model USE$_{tox}$.
m³ world eq. Deprived	A m³-world eq. represents the "importance" of a m³ of freshwater consumed on average in different places in the world. As freshwater is scarce in many places, a "characterization factor" (CF) is defined as expressing the scarcity footprint of a specific area. This factor is multiplied by the inventory of m³ consumed in order to obtain the m³-world equivalent. For more detail, refer to the specialized literature (e.g., Anne-Marie Boulay et al., 2018).
kBq U-235	The potential impact on human health of different ionizing radiations is converted to the equivalent of kilobecquerels of Uranium 235 (kg U235 eq).

TABLE 10.11
100-year GWP and Lifetime of Greenhouse and Ozone-Depleting Substances

Gas	kg CO_2-Equivalent (100-year GWP)	Lifetime (years)
Carbon dioxide (CO_2)	1	300–1,000 (NASA, 2023)
Methane (CH_4)	28	12
Nitrous oxide (N_2O)	273	109
Carbon tetrachloride (CCl_4)	2,200	32
Trichlorofluoromethane, CFC-11 (CCl_3F)	6,230'	52
Methyl bromide (CH_3Br)	2,43	0.8
Methyl chloroform (CH_3CCl_3)	161	5
Sulfur hexafluoride (SF_6)	24,300	1,000
Nitrogen trifluoride (NF_3)	17,400	569
Methylene chloride (CH_2Cl_2)	11.2	0.493
Hydrofluorocarbon-23 (CHF_3)	14,600	14,600
Hydrofluorocarbon-32 (CH_2F_2)	771	5.4
Perfluoromethane (CF_4)	7,380	50,000
Perfluoroethane (C_2F_6)	12,400	10,000
Perfluoropropane (C_3F_8)	9,290	2,600
Perfluorobutane (C_4F_{10})	10,000	2,600
Perfluorocyclobutane (c-C_4F_8)	13,900	3,000
Perfluoropentane (C_5F_{12})	9,220	4,100
Perfluorohexane (C_6F_{14})	8,620	3,100

10.7.4.2.1.3   What Is the Assessment's Scope?
Because the value chain can be very deep and the analysis very complex, it is very important to define a clear framework for the assessment. Some raw materials required will have a complex value chain that might not be known in detail. Examples include microchips or specialty chemicals that have gone through hundreds of manufacturing steps before arriving at the warehouse as finished product. To facilitate such an analysis, it is best not to go into too many details, but rather apply generic information about the environmental impacts of such product groups. A thorough assessment of the social impact of the organization's activities can also be complex, especially if all activities related to services and raw material production must be included. Therefore, it is very important to define what is truly relevant to the expected product life cycle outcomes. In this respect, legal obligations might play the dominant role.

10.7.4.2.2   Life Cycle Inventory
LCI analysis is essentially the data collection phase of LCA, involving the compilation and quantification of inputs and outputs for a product or service throughout its life cycle.

LCI is the core element of LCA, as data collection and validation, and the calculation procedures are integrated in this phase. It is important to proceed systematically, either working with data collection sheets or using appropriate LCA software. Frequently used software packages include for instance Simapro, GaBi, CMLCA, Umberto LCA, openLCA, or One Click LCA. All data collected should be reviewed critically and validated by a process expert before proceeding. Once validated, the data are converted into units based on their FUs. In the case of reuse or recycling, or when analyzing the impact of multiple products, the mass or energy balance over a sufficiently large period of time must be taken into account to obtain a representative result.

The inventory requires the compilation of various environmental inputs and outputs involved in the product life cycle, including typically as a minimum:

- Energy requirements,
- Raw material needs,
- atmospheric emissions,
- Waterborne emissions,
- Emissions to land,
- Solid wastes.

According to ISO 14040, an LCI analysis should follow the four steps below:

1. Development of a flow diagram of processes within the defined system boundary,
2. Development of a data collection methodology,
3. Collection of the relevant data, and
4. Evaluation and reporting of results.

A flow diagram showing interrelationships between individual unit processes, like that in Figure 10.5, is useful in mapping the inputs and outputs to a process or system.

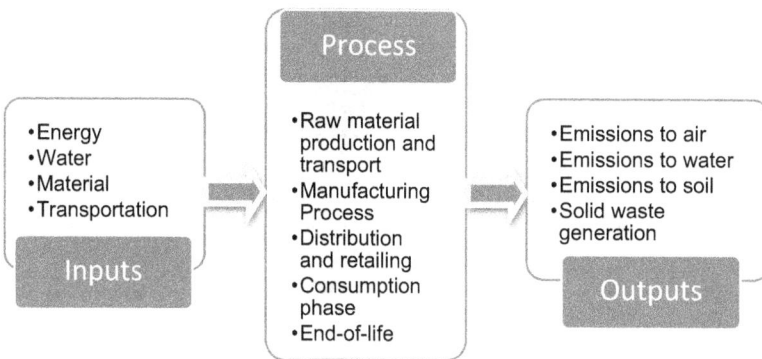

FIGURE 10.5   Typical process flow diagram with generalized unit processes.

Manufacturing processes and supply chains can be complex, making assessment difficult and time-consuming.

10.7.4.2.3   Life Cycle Impact Assessment Phase

LCIA is the third phase of LCA and deals with the evaluation of environmental impacts of products and services – e.g., pollution, toxicity, and/or climate change – over the whole life cycle. Here, the significance of identified impacts is assessed and, where appropriate, environmental impact categories and appropriate "SMART" indicators are assigned. Emissions are first divided into impact categories and then converted to common units, to enable comparison.

10.7.4.2.4   Life Cycle Interpretation Phase

According to ISO 14044:2006, LCA interpretation should include:

- Identifying significant issues based on the LCI and LCIA assessments;
- Evaluating the study itself, i.e., how complete it is, whether it has been done sensitively and consistently; and
- Conclusions, limitations, and recommendations.

Thus, in the interpretation phase, LCI and LCIA results need to be interpreted in accordance with the stated goals and scopes. It is also necessary to include completeness, sensitivity and consistency checks, and an analysis of the uncertainty and accuracy of the results. Conclusions can only be drawn from the study if it can be shown that the results are fit for purpose. The answers to the following questions can be used to guide the process:

- Are there any emissions of concern in the process?
- How do the emissions compare to those of other products in this or competitor plants?
- What is the best way to reduce the environmental impact?
- Is it possible to make production more efficient and environmentally friendly?

10.8   ENVIRONMENTAL RISK ASSESSMENT

10.8.1   Environmental Risks Versus Impacts

At first glance, environmental impact and risk assessments appear to serve similar purposes. In fact, each addresses a different critical aspect of environmental management. Risk assessments analyze potential threats and their likelihood of occurrence, while environmental and social impact assessments explain the impact of specific accidents and their severity.

In practice, environmental impact assessment and risk assessment should be considered separate processes, although, they are by no means independent. An environmental impact assessment is essentially an extension of a risk assessment. While the aim of a risk assessment is to identify risk factors, that of the environmental impact assessment is to predict how the risks identified will actually affect the company and/or the environment, if an event occurs.

Environmental risk can be defined as the probability that an undesired event will cause an adverse effect. Probability is a mathematical measure of risk. Therefore, information on the historical frequency of adverse events in the organization or the industry in general is required, or other measures to estimate the probability of failure based on information from, for example, equipment or machinery suppliers. Finally, risk assessment requires characterization and evaluation of the hazard posed by an adverse event.

Different methods of assessing the risk of environmental events are available, depending on the scope. Qualitative risk assessments are sufficient for the day-to-day operations of a business, but remediation of contaminated sites, for example, requires a much more complex and costly quantitative risk assessment as the basis for calculating site-specific remediation targets.

Risk assessment matured as an engineering discipline assisted by seminal reports by the United States National Research Council and the Royal Society in Britain. Environmental agencies, most notably the USEPA, have embraced it as an objective tool to enable them to set standards and priorities and provide assistance in decision-making. It has long been applied in this way to evaluate the risks to human health arising from radionuclides and chemicals in the environment. Hazard assessment has been used to study natural hazards and assist in preparing for them. The more recent application of risk assessment techniques to flora and fauna is termed ecological risk assessment (USEPA, 2022b).

Different definitions may also be used of the terms around the risk evaluation arising from environmental hazards, depending on the country, and national or regional legislation. USEPA uses risk assessment to "characterize the nature and magnitude of health risks to humans and ecological receptors, from chemical contaminants and other stressors that may be present in the environment" (USEPA, 2022c). This interpretation is broadly accepted internationally. Sometimes the term risk analysis is used as a synonym for risk assessment. Generally, however, risk analysis is considered only part of the risk assessment procedure, which itself is part of the overall risk management system.

The US has a major program using risk assessment to determine environmental priorities at different levels. In this form, it is called comparative risk assessment. It involves ranking issues on the basis of their likelihood of occurrence, and the magnitude of the actual or perceived consequences.

An environmental manager must be able to identify the risks present in the organization, understand the options by which they can be managed, when and how the public must be informed of health and environmental risks, and the consequences if a risk assessment is either not carried out or carried out incorrectly. All these aspects must be considered an integral part of a quality-assured EMS. Due to the differences in terminology and legal requirements, it is essential to know and understand the legislation applicable to the organization in order to evaluate potential legal risks.

Because risk is a positive or negative deviation from an expected value, managing it sometimes presents new opportunities, such as developing new products, technologies, or processes, or entering new market segments. On this background, the foundation stone for implementing risk-based thinking in the EMS can be found in the international quality management standard ISO 9001 (ISO, 2015).

The risk-based thinking approach from ISO 9001 has been applied widely in ISO 14001 as a blueprint for dealing with environmental problems, before, during and after their occurrence. Other standards involve the same philosophy, including international standard AS9100, a widely adopted and standardized quality management system for the aerospace industry. AS9100 requires identification, assessment, and communication of risks related to the aerospace industry, throughout product realization, implementation, and action management.

ISO 31000 was developed specifically for organizational risk management and contains important concepts such as:

- Avoiding activities associated with a particular risk,
- When to accept or not accept a risk when an important opportunity arises, and
- Determining the best way to eliminate sources of risk.

ISO 9004 also addresses risk impacts on strategy and innovation, as well as other aspects of risk management. The focus of this chapter is rather on the qualitative assessment of environmental risks, and quantitative risk assessment to the environment and human health is covered in the relevant literature (e.g. USEPA, 2022c; Robinson, 2018).

10.8.2 Concept of Risk

Environmental risk deals with the probability of an event causing a potentially undesirable effect. Quantitative risk assessment thus deals with statistics, because probability is the mathematical measure of risk. The other part of the story is the presence of a hazard that determines the nature of the undesirable effect.

Environmental practitioners need to know the means by which risk can be managed and communicate that to the public, as well as the consequences of not conducting an adequate risk assessment or conducting one incorrectly. These issues are an integral part of a quality-assured EMS.

10.8.2.1 The Distinction between Risk and Uncertainty

Imagine in a traffic accident, the hood, fender, and bumper of a car are damaged. The vehicle is already in the repair shop and the owner is waiting for a price quote. An initial unofficial estimate was given by a mechanic friend, who estimated the damage at between $3000 and $4000. The owner would then be in a state of uncertainty. However, it could not be said that he really is exposed to risk. So, the concept of risk implies both uncertainty and some kind of loss or damage that someone might suffer (Kaplan and Garrick, 1981).

Symbolically, this can be expressed as:

$$Risk = Uncertainty + Harm$$

10.8.2.2   The Distinction between Risk and Hazard

To understand, for instance, the newly flared-up controversial discussion about electricity generation with nuclear power, it is helpful to clarify the terms risk, hazard, and danger. In linguistic usage, hazard is the sense of a "source of hazard". Risk, on the contrary, is the "possibility of loss or damage" and the "degree of probability of such loss or damage". Thus, while hazard simply exists as a source, risk encompasses the likelihood of transformation of that source into actual loss, form of harm, or injury. Applied to nuclear power, a nuclear power plant can be considered an enormous source of danger, for example in the event of a reactor rupture. However, if the plant's design and technology are state of the art, the likelihood of such an accident with a core meltdown, and the resulting horrendous damage to the population and the environment is very, if not extremely, low. This example also shows that public opinion about risks depends on the observer's point of view. While the risk from nuclear energy is considered too high to be acceptable by some of the population, others consider it a safe and acceptable alternative to mitigate climate change effects, and an important part of the energy mix to solve the energy and climate crisis. In other words, risk perception is subjective and depends on who is looking at it.

This idea can be expressed symbolically as:

$$Risk = Hazard/Safeguards$$

This equation also makes it clear that risk can be minimized by improving safeguards, i.e. minimizing the probability of failure, but it can never be reduced to zero, although it can be very small.

Risk is, thus, best expressed in terms of a combination of the consequences of an event (including changes in circumstances) and the associated "probability" of its occurrence (as defined in ISO Guide 73: 2009, 3.6.1.1).

Mathematically, risk can be expressed as the product of severity and probability:

$$Risk = Severity \times Probability$$

Some events have extremely high severity, like a crack in a nuclear reactor with the massive release of radioactive material, but as such an event is combined with a very low statistical probability, the overall risk is considered very small.

10.8.3   Environmental Risk Assessment (ERA)

10.8.3.1   Risk Identification

ERA is used to evaluate the likelihood of an impact on living organisms, natural habitats, and ecosystems, by exposure to one or more environmental stressors, such as chemicals, radiation, noise, climate change, pathogens, or invasive species. Sources of such impacts might be inappropriate disposal of chemical or biological waste, contaminated areas with soil and groundwater pollution, asbestos-containing material, or emission of air pollutants.

Environmental emergency situations might include:

- Accidental emissions to the atmosphere,
- Accidental discharges to water and/or land,
- Exposure to human health and safety hazards, and
- Fire and explosion.

Precautions can be taken, based on risk analysis, to mitigate these risks and measures can be taken to comply better with environmental policies and regulations.

An environmental risk assessment comprises six steps:

1. Identification of areas where there might be risk issues,
2. Characterization of the hazards,
3. Characterization of the target organisms or ecosystem, and exposure routes,
4. Characterization of the risk,
5. Documentation of the assessment and implementation of precaution measures, and
6. Performance of periodic monitoring.

Environmental risks (and opportunities) must be identified and means of addressing them determined.

Several general questions are useful when starting a risk assessment:

- How can potential threats be identified?
- How can threats be prevented or reduced?
- What actions should be taken to ensure that the intended outcomes can be achieved?
- Who will be responsible for addressing the risk management process?
- When and how are the risk management actions triggered?
- What priorities have the threats and what costs are expected in case of an event?
- What are the identified threats' root causes?
- Which actors are needed to identify and manage the risks?
- How should the risk management system be evaluated, tested, and updated, to ensure its operability in case of emergency?

Tools and strategies that can be useful in identifying risks and opportunities include:

- Brainstorming meetings,
- Review of accident statistics for the organization and related industries,
- Review of technical documentation and audit reports,
- Interviews with key personnel and external experts,
- Applying standard methodologies such as failure modes, effects and criticality analysis (FMECA), cause trees, etc.,
- Using standard methodological approaches – e.g., fault tree analysis (FTA) or failure mode effects analysis (FMEA), and
- Using pre-established checklists or questionnaires covering the different areas of the project (risk breakdown structure or RBS).

		IMPACT				
		Negligible	Low	Medium	High	Severe
PROBABILITY	Very likely	Medium	High	Very high	Extreme	Extreme
	Likely	Medium	Medium	High	Very high	Extreme
	Moderate	Low	Medium	Medium	High	Very high
	Unlikely	Very Low	Low	Medium	Medium	High
	Rare	Very Low	Very Low	Low	Medium	Medium

FIGURE 10.6   Risk matrix.

An important part of risk management is the creation of a risk register, which must be constantly updated as part of the continuous improvement process. The risk register helps in risk identification, analysis, prioritization, and mitigation, even before they develop into real problems.

10.8.3.2   Risk Assessment with the Risk Matrix (5 × 5 Impact Matrix)

A risk matrix, also known as a "5×5 impact matrix" or "probability matrix", is a two-dimensional tool for assessing and visualizing the likelihood that an adverse event or accident will occur, and the associated severity of adverse consequences on the environment, human health, economy, and community.

In the case of risks related to exposure to hazardous substances, severity can be interpreted as "toxicity" using as evaluation criteria toxicological data, like lethal doses, maximum permissible daily doses, or safe environmental pollutant concentrations, defined as environmental quality standards in national environmental legislation. The probability of a severe incident in this case is the likelihood that a sensitive receptor will be exposed to the hazardous substance well above the established reference dose or concentration.

Once a risk has been identified, the matrix helps to characterize the level of risk and to take appropriate preventive and mitigating actions. The matrix is represented visually as a table with five probability categories on the x-axis and five for severity on the y-axis. Each box in the matrix represents the risk rating calculated on the basis of its particular levels of probability and severity. In most cases, numerical values are used in the matrix to represent risk ratings better.

For risk management, the following actions may be considered:

Risk Rating:	Action:
Extreme:	Stop activities, and inform management and authorities of the risk
Very high:	Stop activities and make immediate improvements
High:	Inform staff and look to improve within a defined timescale
Medium:	Look to improve at the next review
Low:	No further action, but ensure controls are maintained
Very low:	No further action

The risk matrix is a comprehensive tool to assess environmental risks in an operation qualitatively as part of the EIA, at the pre-project, operation, and post-project level. It also serves as a supplementary tool in evaluating the possible damage or disruption arising from risks.

The method is also commonly used in quality management, such as project planning, operations management, or workplace hazard analysis, where a risk matrix can be used to determine the likelihood and degree of impact of environmental damage and a worker's exposure to risk associated with workplace hazards.

The standard risk matrix uses five assessment levels for each component but can be tailored to meet an organization's own criteria. The use of the 5×5 risk matrix has two main advantages:

1. Allowing the simplified and visual representation of the different risk levels.
2. Often replacing or complementing time-consuming and costly quantitative risk analysis.

10.8.3.2.1 Risk Probability (Likelihood)

The risk ratings under this component are classified between rare and almost certain. Ideally, a statistical, semi-quantitative, event assessment is used to rate event probability (see Table 10.12).

Adequate use of statistics is complex, as statistics from organizations in the same field or business line depend on many factors, including organization size, production capacity, location, etc. Hence, generally, only a qualitative or reduced semi-quantitative assessment can be performed due to the lack of reliable statistical data.

10.8.3.2.2 Risk Severity (Impact)

Severity (y-axis) determines the level of environmental impacts that an undesired event or accident may cause. Table 10.13 shows an example of how severity can be characterized, using five levels: negligible, low, medium, high, and severe (Figure 10.7).

An environmental impact's severity depends on many factors, including toxicity, contaminant concentrations, exposure (the dose a potential receptor receives), types of receptors, and frequency and duration of exposure.

For a more objective characterization of severity in the case of an accidental release of toxic substances into the air, water. wastewater or soil, as well as the emission of noise and vibrations, several quantifiable parameters are available, including:

TABLE 10.12

Example of Definition of Risk Probabilities

Probability	Qualitative Description	Average Events Per Year in Own or Related Organizations
Rare	Never happened within the past 50 years	Close to zero
Unlikely	Unlikely to happen; i.e. one event was reported within the past 50 years in the industry	<0.02
Moderate	Could/can happen; i.e. one event was reported within the past 15 years in this organization	0.02–0.1
Likely	Likely to happen; i.e., may happen annually on average	0.1–1
Very likely	Almost certain to happen; happens several times a year in this organization	>1

TABLE 10.13

Example of the Characterization of Severity in Risk Assessment

Severity	Criteria
Negligible	The risk is so small that it can be ignored; it will not cause any environmental impact.
Low	Can cause minor impacts; the risk is insignificant and can be managed with routine procedures; remediation is not required.
Medium	Can cause impacts that may cause environmental or human health damage; injuries or illnesses may require medical attention but limited treatment; the risk is significant but manageable, with additional controls or mitigations; limited areas affected that can be remediated at moderate cost.
High	Can cause major impacts, with reversible environmental impacts and injuries that require constant medical attention; the risk is severe and requires immediate corrective action; limited affected areas can be remediated with much effort and at high cost.
Severe	Can cause severe impacts, with the risk of human fatalities and irreversible ecological damage; large affected areas that cannot be remediated; catastrophic risk.

   i. Permissible emission and discharge concentrations specified in the applicable regulations,
  ii. Environmental baseline concentrations,
 iii. Legally defined environmental quality standards (EQSs), and
 iv. Toxicological reference data (dose-response assessments, LD50, LC50, etc.).

A powerful tool to predict the special impact of contaminants in case of an accidental release are quantitative mathematical fate and transport models. Such models include:

  a. Models for predicting the dispersion of air pollutants near the ground at various distances from a point source, such as a smokestack, in the form of isopleths.
  b. Noise prediction models that simulate sound wave propagation.
  c. Hydrogeological models that predict and express the plume extension of contaminants like heavy metals or organic compounds in groundwater.

Legally established EQSs are available for most of the relevant pollutants in water, soil, and air. A common interpretation of these threshold numbers is that concentrations below an EQS will not cause any significant harm to the environment or human health. To assess the negligible and low-risk levels, the EQS levels can, therefore, be applied safely. It is not that straightforward, however, to quantify trigger levels to define high and severe risks to human health or the ecosystem. Once environmental concentrations exceed legally defined EQSs or receptors receive a dose of a hazardous substance exceeding a critical threshold, a quantitative assessment of potential impacts requires the performance of a comprehensive human health and ecological risk assessment (see, for instance, enHealth or USEPA guidelines).

Assessment Title:

Location: Address:
Date: Prepared by:
ENVIRONMENTAL RISK ASSESSMENT
Problem formulation:
Identification of potential receptors (who or what needs to be protected?):

HAZARD CHARACTERIZATION		EXPOSURE CHARACTERIZATION		
Hazard description	Specify the type of risk associated	Probability of exposure		
		Low	Medium	High
		☐	☐	☐
		☐	☐	☐
		☐	☐	☐

RISK CHARACTERIZATION

SEVERITY

Hazard #	Negligible	Low	Medium	High	Severe
1	☐	☐	☐	☐	☐
2	☐	☐	☐	☐	☐
3	☐	☐	☐	☐	☐

LIKELIHOOD

Hazard #	Rare	Unlikely	Moderate	Likely	Very likely
1	☐	☐	☐	☐	☐
2	☐	☐	☐	☐	☐
3	☐	☐	☐	☐	☐

RISK RATING

Hazard #	Risk rating			Control measures
	Low	Medium	High	
1				
2				
3				

NAME AND SIGNATURE OF INSPECTOR:

FIGURE 10.7 Sample Template for ERA

However, for most applications in an environmental manager's practice, a qualitative ERA is sufficient to identify and prioritize risks and take the necessary actions to minimize them.

10.8.4 RISK MANAGEMENT

Risk management requires an organization to implement an end-to-end risk management process, which must be executed consistently, carefully, and extensively. There are many different situations that cause risks, so the risk management process must be sufficiently flexible to cover all these potential events. This was the reason that ISO developed a comprehensive reference set for this area, including:

- ANSI/ASSE Z690.1–2011 Vocabulary for Risk Management (U.S. Adoption of ISO Guide 73:2009),
- ANSI/ASSE Z690.2–2011 Risk Management Principles and Guidelines (U.S. Adoption of IEC/ISO 31000:2009),
- ANSI/ASSE Z690.3–2011 Risk Assessment Techniques (U.S. Adoption of IEC/ISO 31010:2009,

Once environmental and social risks (and opportunities) have been identified, a risk treatment plan should be developed. Its objective is to implement prevention, mitigation, and corrective actions. Most important is to prevent environmental accidents well in advance, to avoid environmental, economic, and social damage. For the case in which an incident has occurred already, an emergency response plan must be developed.

Risk prevention and mitigation measures can usually reduce the probability of failure and thus the overall risk. A confirmed risk needs to be managed and becomes an issue in the risk management plan according to the defined priority.

The approach to risk management depends on both the characteristics and scope of the activities performed and the nature of the risks to be managed. Some businesses such as offices, convenience stores, or cafes pose low environmental risks, whereas businesses such as mining or chemical industry pose much higher risks.

In particular, businesses with medium to high-risk activities are required to implement a risk management process as part of their EMS. Activities that may be considered to eliminate or mitigate risks include:

- Industrial waste storage and disposal;
- Transport, storage, and disposal of hazardous liquids (including waste oil, fuel, organic solvents, acids, bases);
- Storage and disposal of solid hazardous material;
- Substitution of hazardous with non-hazardous materials;
- Process optimization for waste and risk reduction;
- Adaptation of infrastructure measures to reduce fire hazards and improve spill management (secondary containment, sewer, drainage, UST removal, etc.).
- Staff training in risk management.

10.8.4.1   The Environmental Emergency Response Plan (EERP)

If, despite the precautions taken to prevent and mitigate environmental accidents, such an uncontrolled and unplanned event occurs, it must be possible to respond quickly and effectively with the appropriate measures. This is what an EERP is designed for. The EERP contains procedures for prevention, preparedness, response, recovery, and reporting events like spills, or accidental releases of hazardous liquids or gases (for instance caused by fire). A typical EERP provides measures covering topics like:

1. Prevention of accidents;
2. Measures ensuring an effective response to an incident;
3. Response measures to accidents;
4. Recovery, remediation, and restoration measures; and
5. Reporting protocol.

Failure to plan and prepare for environmental disasters can have unpredictable consequences for the health and safety of workers, the public, and the environment. Pollution of the air, soil, surface water, or groundwater, destruction of wildlife habitat, and possible damage to property are violations of the law that can result in heavy fines, serious public relations problems, and even existential threats to the entire operation.

REFERENCES AND RESOURCES

Boulay, A.-M., Bare, J., Benini, L., Berger, M., Lathuillière, M. J., Manzardo, A., Margni, M., Motoshita, M., Núñez, M., Pastor, A. V., Ridoutt, B., Oki, T., Worbe, S., Pfister, S. (2018). The WULCA consensus characterization model for water scarcity footprints: Assessing impacts of water consumption based on available water remaining (AWARE). *The International Journal of Life Cycle Assessment*, 23(2), 1–11. https://doi.org/10.1007/s11367-017-1333-8

Brown, J. S., Duguid, P. (2000). *The Social Life of Information*. Harvard Business School Press, Cambridge, 336.

Brundtland, G. H. (1987). *Our Common Future*. Oxford University Press, for the World Commission on Environment and Development, Oxford, 43.

Del Río Castro, G., González Fernández, M. C., Uruburu Colsa, Á. (2021). Unleashing the convergence amid digitalization and sustainability towards pursuing the Sustainable Development Goals (SDGs): A holistic review. *Journal of Cleaner Production*, 280, 122204. Available online: https://doi.org/10.1016/j.jclepro.2020.122204

Doran, G. T. (1981). There's a S.M.A.R.T. way to write management's goals and objectives. *Management Review*, 70, 35–36.

European Union (2023). EMAS, eco-management and audit scheme. Available online: www.ec.europa.eu/environment/emas/index_en.htm

European Union (2016). European platform on life cycle assessment, EN 15804 reference package. Available Online: https://eplca.jrc.ec.europa.eu/LCDN/EN15804.xhtml

European Union (2019a). European platform on life cycle assessment, life cycle assessment. Available online: https://ec.europa.eu/environment/ipp/lca.htm

European Union (2019b). CSN EN 15804+A2, sustainability of construction works – Environmental product declarations – Core rules for the product category of construction products. https://www.en-standard.eu/

Dong, F., Hu, M., Gaoa, Y., Liu, Y., Zhu, J., Pan, Y. (2022). How does digital economy affect carbon emissions? Evidence from global 60 countries. *Science of the Total Environment*, 852, 15 December, 158401.

Hes, D., DuPlessis, C. (2015). *Designing for Hope: Pathways to Regenerative Sustainability*. Routledge, Abingdon, Oxon.

International Standards Organization, ISO (2015). ISO 9001 and related standards, quality management. Available online: https://www.iso.org/iso-9001-quality-management.html

International Standards Organization, ISO (2021). The ISO survey 2021. Available online: https://www.iso.org/the-iso-survey.html

International Standards Organization, ISO 14001:2015. Environmental management systems – Requirements with guidance for use. Available online: www.iso.org/standard/60857. html

International Standards Organization, ISO 14040:2006. Environmental management – Life cycle assessment – Principles and framework. Available online: www.iso.org/standard/37456.html

ITU (2022). Digital technologies to achieve the UN SDGs. The UN International Telecommunication Union (ITU). Available online: shorturl.at/sGOYZ

Josimović, B., Petric, J., Milijic, S. (2014). The use of the Leopold matrix in carrying out the EIA for wind farms in Serbia. *Energy and Environment Research*, 4(1), 43–54.

Kaplan, S., Garrick, B. J. (1981). *On The Quantitative Definition of Risk, Risk Analysis*, Vol. I, No. I, Wiley Online Library. Available online: shorturl.at/bfm28

Leiden University (2016). Department of Industrial Ecology, Belgium, CML-IA Characterisation Factors, 05 September. Available online: https://www.universiteitleiden.nl/en/research/research-output/science/cml-ia-characterisation-factors

Leopold, L. B., Clarke, F. E., Hanshaw, B. B., Balsley, J. R. (1971). *A Procedure for Evaluating Environmental Impact*. U.S. Geological Survey, Washington, DC.

Luxembourg Times (21 November 2019). Village worries about environment in first Google face-off.

Mang, P., Haggard, B. (2016). *Regenerative Development and Design: A Framework for Evolving Sustainability*. John Wiley & Sons, Hoboken, NJ.

Mondejar, M. E., Avtar, R., Diaz, H. L. B., Dubey, R. K., Esteban, J., Gómez-Morales, A., Hallam, B., Mbungu, N. T., Okolo, C. C., Prasad, K. A., She, Q., Garcia-Segura, S. (2021). Digitalization to achieve sustainable development goals: Steps towards a Smart Green Planet. *Science of The Total Environment*. Abstract: https://www.sciencedirect.com/science/article/pii/S0048969721036111

NASA (2023) by Alan Buis. The atmosphere: Getting a handle on carbon dioxide, NASA's Jet Propulsion Laboratory. Available online: https://climate.nasa.gov/news/2915/the-atmosphere-getting-a-handle-on-carbon-dioxide/

O'Sullivan, K., Clark, S., Marshall, K., MacLachlan, M. (2021). A just digital framework to ensure equitable achievement of the Sustainable Development Goals. *Nature Communications*, 12, 1–4. Available online: https://www.nature.com/articles/s41467-021-26217-8

Pearce, D. W., Turner, R. K. (1990). *Economics of Natural Resources and the Environment*. Harvester, Hemel Hempsted (Kapitel 2: The Circular Economy), ISBN 978-0-7450-0225-5

Pérez-Martínez, J., Hernandez-Gil, F., Miguel, G. S., Ruiz, D., Arredondo, M. T. (2023). Analysing associations between digitalization and the accomplishment of the Sustainable Development Goals. *Science of The Total Environment*, 857(Part 3), 20 January. Abstract: https://www.sciencedirect.com/science/article/abs/pii/S0048969722068000

Petts, J. (1999). *Handbook of Environmental Impact Assessment*. Blackwell Science, London.

Rathi, A. K. A. (2021). *Handbook of Environmental Impact Assessment – Concepts and Practice, The Basics of Environmental Impact Assessment*. Cambridge Scholars Publishing, Cambridge, ISBN (10): 1-5275-6664-1

Robinson, L. (2018). *Public Health Risk Assessment for Human Exposure to Chemicals (Environmental Pollution)*. Springer Dordrecht, ISBN 1402009216

Smith, C., Nicholls, Z. R. J., Armour, K., Collins, W., Forster, P., Meinshausen, M., Palmer, M. D., Watanabe, M. (2021). The Earth's energy budget, climate feedbacks, and climate sensitivity supplementary material. In *Climate Change 2021: The Physical Science Basis. Contribution of Working Group I to the Sixth Assessment Report of the Intergovernmental Panel on Climate Change (IPPC)*. Available from https://www.ipcc.ch/ (shorturl.at/hmuS7)

Stinchcombe, K., Gibson, R. B. (2011). Strategic environmental assessment as a means of pursuing sustainability: Ten advantages and ten challenges. *Journal of Environmental Assessment Policy and Management*, 3(3), 343–372. https://doi.org/10.1142/S1464333201000741

United Nations (UN) (2015). Department of Economic and Social Affairs, Sustainable Development, the 17 goals. Available online: https://sdgs.un.org/goals

USEPA (2022a). Learn about environmental management systems. Available online: https://www.epa.gov/ems

USEPA (2022b). Ecological risk assessment. Available online: https://www.epa.gov/risk/ecological-risk-assessment

USEPA (2022c). Conducting a human health risk assessment. Available online: https://www.epa.gov/risk/conducting-human-health-risk-assessment USEPA (2023). Understanding ISO 14001 environmental management systems. Available online: https://www.era-environmental.com/blog/iso-14001-environmental-management-systems

USEPA (2023b). Ozone depleting substances. Available online: https://www.epa.gov/ozone-layer-protection/ozone-depleting-substances

Van Zelm, R., Huijbregts, M. A. J., Den Hollander, H. A., Van Jaarsveld, H. A., Sauter, F. J., Struijs, J., Van Wijnen, H. J., Van de Meent, D. (2008). European characterization factors for human health damage due to PM10 and ozone in life cycle impact assessment. European characterization factors for human health damage of PM10 and ozone in life cycle impact assessment. *Atmospheric Environment*, 42, 441–453. Available online: https://www.sciencedirect.com/science/article/abs/pii/S1352231007008667

Wirajaya, A Y, Sudardi, B., Istadiyantha, Warto (2021). Eco-Sufism concept in Syair Nasihat as an alternative to Sustainable Development Goals (SDGs) policy in the environmental sector. *IOP Conference Series: Earth and Environmental Science*, 905, 012081. Available online: https://iopscience.iop.org/article/10.1088/1755-1315/905/1/012081/meta

World Economic Forum (WEF) (2022). The Global Risks Report 2022, 17th Edition. Available online: https://www3.weforum.org/docs/WEF_The_Global_Risks_Report_2022.pdf

World Economic Forum (WEF) (2023). Global Risk Report 2023, Available online: https://www.weforum.org/reports/global-risks-report-2023/

Wu, J., Guo, S., Huang, H., Liu, W., Xiang, Y. (2018). Information and communications technologies for Sustainable Development Goals: State-of-the-art, needs and perspectives. *IEEE Communications Surveys & Tutorials*, 20, 2389–2406. https://ieeexplore.ieee.org/document/8306870

USEFUL RESOURCES FOR ENVIRONMENTAL IMPACT ASSESSMENT

Donelly, A., Dalal-Clayton, B., Hughes, R., International Institute for Environment and Development, a Directory of Impact Assessment Guidelines. https://europa.eu/capacity4dev/file/25411/download?token=mVmSdPgR

European Commission Environmental Impact Assessment Guidelines, https://environment.ec.europa.eu/law-and-governance/environmental-assessments/environmental-impact-assessment_en

USEPA, A Software Guidance System for Choosing Analytical Subsurface Fate and Transport Models Including a Library of Computer Solutions for the Analytical Models, https://cfpub.epa.gov/ncer_abstracts/index.cfm/fuseaction/display.abstractDetail/abstract_id/1153

USEPA, Technical Review Guidelines for Environmental Impact Assessments in the Tourism, Energy and Mining Sectors, https://www.epa.gov/international-cooperation/technical-review-guidelines-environmental-impact-assessments-tourism enHealth Guidance – Health Impact Assessment Guidelines, https://www.health.gov.au/sites/default/files/documents/2022/07/enhealth-guidance-health-impact-assessment-guidelines.docx

IISD, EIA-7 Steps, https://www.iisd.org/learning/eia/eia-7-steps/

TOOLS FOR ENVIRONMENTAL MANAGEMENT

Esdat.net, Environmental Management Database Software, Assessment Levels for Soil, Sediment and Water, https://support.esdat.net/Environmental%20Standards/australia/wa/assessment%20levels%20-%202010.pdf

USEPA Sustainable Materials Management Tools. Available online: https://www.epa.gov/smm/sustainable-materials-management-tools

USEFUL GENERAL ENVIRONMENTAL DATA RESOURCES

UN Environmental Programme GRID-Geneva: https://unepgrid.ch/

The UNEP Programme GRID-Geneva contains numerous high quality geospatial data sets at various scales (global, continental, national and subnational) on a variety of environment related themes, including: Ramsar sites information service, Global Risk Data Platform of natural hazards, GridCore Datasets (basic fundamental geospatial data) and live environmental monitoring data.

GEOSS Portal https://www.geoportal.org/

The Global Earth Observation System of Systems (GEOSS) Portal offers a single Internet access point to Earth observation data, information and knowledge from all over the world for users with different backgrounds and from different disciplines. It has been developed by the European Space Agency (ESA) and coordinated by the Group on Earth Observations (GEO).

World Resources Institute (WRI) Data platform: https://www.wri.org/data/data-platforms

The World Resources Institute (WRI) proposes a suite of data platforms that enable users to monitor forests with satellites, track the drivers of climate change, understand indigenous communities land rights, plan for water scarcity, examine the cross-section of global environmental issues and more.

Climate TRACE Independent satellite greenhouse gas emissions tracking. https://climatetrace.org/

USDA Forest Service, Forest Service Geodata Clearinghouse. Includes data and maps. https://data.fs.usda.gov/geodata/edw/datasets.php

Data Portals, https://datacatalogs.org/, contains 597 open data portals from around the world, including geospatial and environmental data from universities and public institutions.

ECHA – European Chemicals Agency, https://echa.europa.eu/information-on-chemicals; This is a unique source of information on chemicals manufactured in or imported into Europe. It covers their hazardous properties, their classification and labeling, and information on their safe use.

USEPA, National Ambient Air Quality Standards (NAAQS), NAAQS Table | US EPA

USEPA, Water Quality Criteria, https://www.epa.gov/wqc

USEPA, Superfund Soil Screening Guidance, https://www.epa.gov/superfund/superfund-soil-screening-guidance

European Commission, Environmental Noise Guidelines for the European Region, https://www.euro.who.int/__data/assets/pdf_file/0008/383921/noise-guidelines-eng.pdf

European Commission, Air quality standards - European Environment Agency, https://www.eea.europa.eu/themes/air/air-quality-concentrations/air-quality-standards

European Commission, Environmental quality standards applicable to surface water, https://eur-lex.europa.eu/EN/legal-content/summary/environmental-quality-standards-applicable-to-surface-water.html

European Commission, European Standards on Soil Quality, https://www.en-standard.eu/csn-standards/83-environment-protection-working-and-personal-protection-safety-of-machine-equipment-and-ergonomics/8361-soil-quality/

ISRIC, "International" Soil Standards, comprehensive catalogue of 'international' standards for soil sampling, analysis, data interpretation, etc. https://www.isric.org/international-soil-standards#OGC_standards

USEPA, Environmental Topics, most popular pages in your topic of interest, https://www.epa.gov/environmental-topics

USEPA Data Catalogs, https://www.epa.gov/data/data-catalogs,

EPA provides access to environmental data through Environmental Dataset Gateway (EDG), Geospatial Applications, Registry of EPA Applications, Models, and Data Warehouses (READ), and the System of Registries (SoR). Specific data sets can be queried by subject area, environmental issue, or file type.

Esdat.net, ESdat is intuitive and user-friendly software that helps scientists and engineers manage environmental and earth-science data from laboratories, field programs, data loggers, sensors, historical sources, and regulatory standards. All in one place. www.esdat.net

LCA SOFTWARE

Simapro	https://simapro.com/
GaBi	https://sphera.com/life-cycle-assessment-lca-software/
CMLCA	http://cmlca.eu/
Umberto LCA	https://www.ifu.com/umberto/lca-software/
OpenLCA	https://www.openlca.org/ (open source software)
One Click LCA	https://www.oneclicklca.com/

Index

For Product Safety Concerns and Information please contact our EU
representative GPSR@taylorandfrancis.com
Taylor & Francis Verlag GmbH, Kaufingerstraße 24, 80331 München, Germany

www.ingramcontent.com/pod-product-compliance
Lightning Source LLC
Chambersburg PA
CBHW060812220326
41598CB00022B/2594